Innovative Approaches in Computational Systems and Smart Applications

Mosiur Rahaman
Global Research Center for Technology and Science, India & International Center for AI and Cyber Security Research and Innovations, Asia University, Taiwan

Faruk Azam Shaik
Centre National de la Recherche Scientifique, France

IGI Global
Scientific Publishing
Publishing Tomorrow's Research Today

Vice President of Editorial	Melissa Wagner
Director of Acquisitions	Mikaela Felty
Director of Book Development	Jocelynn Hessler
Production Manager	Mike Brehm
Cover Design	Jose Rosado

Published in the United States of America by
IGI Global Scientific Publishing
701 East Chocolate Avenue
Hershey, PA, 17033, USA
Tel: 717-533-8845
Fax: 717-533-7115
Website: https://www.igi-global.com E-mail: cust@igi-global.com

Library of Congress Cataloging-in-Publication Data

LCCN: 2024055144 (CIP Data Pending)
ISBN13: 979-8-3693-9846-3
Isbn13Softcover: 979-8-3693-9847-0
EISBN13: 979-8-3693-9848-7
British Cataloguing in Publication Data
A Cataloguing in Publication record for this book is available from the British Library.

Table of Contents

Detailed Table of Contents

Over five years, significant advancements in image-processing techniques have emerged for assessing biotic and abiotic stresses in rice plants. This review compiles and analyzes recent research developments, highlighting innovative methodologies in stress detection and management. It covers a range of techniques, including hyperspectral imaging, thermal imaging, and machine learning approaches, which have been crucial for accurately identifying and quantifying stress factors. The integration of these methods with modern computational tools allows for precise monitoring and early detection of stress, facilitating timely interventions. Additionally, the review addresses the challenges and limitations of current methodologies and suggests potential research directions. By providing an extensive overview of state-of-the-art techniques, this study serves as a valuable resource for researchers and practitioners in agricultural science and technology.

Rooted in the idea that machines could mimic cognitive functions associated with the human mind. Today, AI encompasses a variety of technologies and methodologies that empower process information, recognize patterns, and make decisions autonomously or semi-autonomously. language translation, or game playing. These systems

Due to insufficient cancer literacy and lack of understanding of breast cancer risk, Indian women face an higher incidence of breast cancer, regardless of socioeconomic status or education (Gupta et al., 2015). Breast cancer is a common and serious threat to women's physical and emotional well-being. Despite technological advancements in India, many cancer patients cannot access or afford modern treatment (Consensus document for management of breast cancer, Indian Council of Medical Research). Current computer vision (CV) models focus on image quality, while natural language processing (NLP) models fail to extract key information from mammography reports. This paper proposes an Explainable AI (XAI) risk assessment system for early breast cancer diagnosis using factors like patient history, socioeconomic status, symptoms, and lifestyle. XAI will help highlight significant features, aiding medical decisions. The model will classify a person's breast cancer risk as low or high.

The architecture of IoT can be categorized into several layers: data collection through sensors that detect environmental conditions, user inputs, or device states. improving urban living conditions and resource efficiency. In agriculture, IoT solutions enhance crop monitoring and irrigation management, enabling farmers to maximize yield while minimizing water usage. Transportation systems benefit from IoT through connected vehicles that provide real-time traffic updates and enhance safety features, ultimately leading to more efficient logistics and reduced congestion. proliferation of particularly regarding security With billions of connected devices. This shift is particularly beneficial for applications requiring immediate feedback, In conclusion,, enabling automation, and fostering innovation. As we continue to explore and harness the capabilities of IoT, it is essential to prioritize responsible implementation practices that address these challenges while maximizing the benefits for society as a whole.

The use of IoT in drip irrigation has been widely applied in previous studies in agriculture to overcome water management problems well in agriculture by utilizing certain sensors in IoT. The method used in this study to classify IoT-based drip irrigation data is the Support Vector Machine (SVM) method. The SVM method is used to find the best classification and distinguish be-tween the two classes in the training data. From the results of the data using the SVM method, the original data before diff has an accuracy of 0.8334 and after diff is done on the data it results in an increase in accuracy to 0.9127. It can be concluded that the data after diff has greater accuracy. Cloudy weather results in faster watering and longer drying. With the rbf graph, it can be concluded that the higher the temperature, the more watering is required and will result in high soil moisture, while if the temperature de-creases, the soil will dry out.

Intelligence is the ability to think, experience learning, problem-solving, and adaptation to new situations. Thus, intelligence is an essential foundation in the learning process. Logical-mathematical intelligence (KLM) is a strong indicator for assessing individual intelligence levels and learning achievements. KLM significantly correlates to the ability to solve story problems (KSC). However, exploring these two variables to group students to help educators prepare good learning strategies has not yet been done. Therefore, this study applied the K-means algorithm to reveal it all by utilizing KLM and KSC data. Silhouette index (SI) and Davies-Bouldin (DBI) were used to validate the results of the K-means cluster. The results show that K-means can classify students based on their KLM and KSC data into three groups: high, medium, and low. The SI and DBI values are 0.46 and 0.75, respectively. It indicates the formed clusters are quite good and there is still an opportunity for improvement.

Machine learning methods are now recognized as mighty instruments for bolstering network security in our progressively interconnected world. This article presents an extensive overview of machine learning's utilization in network security, encompassing critical areas like intrusion detection, malware detection, phishing detection, anomaly detection, and beyond. The paper starts with a summary of machine learning fundamentals, it is included supervised learning, unsupervised learning, reinforcement learning, and evaluation metrics. It explores how machine learning is applied across different domains of network security, highlighting the character of machine learning techniques in bolstering consecutive security measures and addressing emerging threats. The paper identifies challenges and limitations in applying machine learning to network security, encompassing concerns about data quality and quantity, feature selection, model interpretability, adversarial attacks, scalability and real time processing constraints.

Currently, the development of Internet of things technology has become so rapid, it can be seen from every aspect of life, the Internet of things is used to help hu-man life. With rapid development, Security Internet of things is necessary to en-sure that there is no interference in the use of this technology. This research con-ducted a research gap analysis of 2069 journal articles indexed by Scopus from 2017 to 2023. Research analysis using VOSviewer and Biblioshiny, the analysis produced a visual depiction of the relationship between journal articles, global re-search trends, the most cited journal articles, resulting in 3 information The jour-nal with the largest H index is the "IEEE Internet of Things" journal at 94, the "Sensor" Journal at 57,

Preface

INTRODUCTION

It is with great enthusiasm that we present *Innovative Approaches in Computational Systems and Smart Applications*, a comprehensive volume that reflects the dynamic evolution and interdisciplinary nature of today's computational and intelligent systems. This book is a curated collection of high-quality research papers originally presented at the Mandalika International Multi-conference on Science and Engineering (MIMSE) 2023, and it embodies the forward-thinking spirit and collaborative efforts of the global scientific community.

The rapid expansion of artificial intelligence (AI), machine learning, the Internet of Things (IoT), and related computational technologies is reshaping how we approach challenges across a broad spectrum of industries. From optimizing healthcare delivery and enhancing precision agriculture to streamlining network management and transportation systems, the transformative potential of intelligent and data-driven systems continues to unfold in compelling ways. This volume aims to capture these innovations and explore the methodologies, frameworks, and real-world applications driving progress in these fields.

Our primary objective with this publication is to bridge the gap between cutting-edge theoretical research and its practical deployment. By showcasing case studies, experimental results, and innovative models, we seek to offer valuable insights into how intelligent systems are being implemented across domains. Each chapter is selected not only for its technical merit but also for its relevance to solving contemporary issues with scalable and sustainable solutions.

This book is designed to serve a wide readership. For academic researchers and graduate students in computer science, data science, and engineering disciplines, it provides a robust foundation of current trends and future directions. For industry professionals and developers engaged in smart applications, IoT, and AI systems, it offers practical strategies and design considerations. Policymakers and business leaders will also find in these pages a useful guide to understanding the broader implications and potential of digital innovations in transforming society.

We are deeply grateful to all the authors who contributed their work and to the peer reviewers who provided insightful feedback to ensure the quality and impact of this volume. We also extend our appreciation to the organizers of MIMSE 2023 for providing a platform that fosters the exchange of innovative ideas and collaborative research.

We hope that this book will inspire further exploration, spark meaningful collaborations, and contribute to the advancement of intelligent systems that serve humanity in impactful and sustainable ways.

ORGANIZATION OF THE BOOK

Chapter 1: Advancements in Image Processing Techniques for Assessing Biotic and Abiotic Stress in Rice Plants: A Comprehensive Review

This chapter presents a rigorous review of cutting-edge image processing technologies developed over the past five years to assess biotic and abiotic stress in rice cultivation. By examining hyperspectral imaging, thermal imaging, and machine learning applications, the authors provide a comprehensive account of how these tools have transformed stress detection, monitoring, and management in agriculture. The discussion extends to computational integrations, current limitations, and future research pathways, making this chapter an essential reference for scholars and practitioners seeking precision agriculture solutions.

Chapter 2: Applications of AI In Healthcare, Agriculture, Finance, and Other Industries

Exploring the evolution and widespread influence of Artificial Intelligence, this chapter delves into the transformative applications of AI across diverse sectors such as healthcare, agriculture, finance, and more. It highlights the rise of deep learning, natural language processing, and computer vision, with a particular focus on advanced models like GPT and BERT. Through real-world examples, the chapter illustrates how narrow AI excels in domain-specific tasks, offering insights into both the potential and constraints of AI integration in industry and society.

Chapter 3: Blockchain Applications for Transparent Fund Management, Donor Accountability, and Intellectual Property Rights in Digital Libraries

In this chapter, the authors investigate the growing role of blockchain technology in promoting transparency and trust within digital libraries. Through compelling case studies and practical examples, they explore how blockchain enhances fund management, donor accountability, and the protection of intellectual property. The discussion also addresses challenges, interoperability issues, and legal frameworks, making this a forward-looking examination of blockchain's transformative power in knowledge institutions.

Chapter 4: Breast Cancer Risk Estimation Using Explainable Artificial Intelligence Based Machine Learning Techniques

Focusing on the healthcare domain, this chapter introduces an Explainable AI-based machine learning model for assessing breast cancer risk in Indian women. By incorporating diverse factors such as medical history, lifestyle, and socioeconomic indicators, the model offers high interpretability to support early diagnosis and clinical decision-making. The authors also critically examine limitations in existing CV and NLP approaches, proposing a more holistic and accessible solution for cancer risk estimation.

Chapter 5: IoT Applications in Agriculture, Healthcare, and Industry: IoT Security and Privacy Challenges

This chapter investigates the expansive role of the Internet of Things (IoT) across critical sectors, including agriculture, healthcare, and industrial systems. It outlines the layered architecture of IoT, emphasizing real-time data collection and responsive automation. While showcasing its benefits, the authors also raise important concerns regarding IoT security and privacy, advocating for responsible implementation strategies that safeguard both technology and user trust.

Chapter 6: IoT-Based Drip Irrigation: Sequential Data Classification Using SVM

Addressing smart agriculture, this chapter evaluates the effectiveness of using Support Vector Machines (SVM) to classify data collected from IoT-based drip irrigation systems. The results demonstrate improved classification accuracy when preprocessing is applied, offering a robust method for optimizing irrigation schedules based on environmental parameters like temperature and soil moisture. The insights presented support the integration of intelligent analytics in water resource management.

Chapter 7: K-Means Clustering for Profiling Logical-Mathematical Intelligence and Problem-Solving Abilities

This chapter applies K-means clustering to analyze students' logical-mathematical intelligence and problem-solving capabilities, aiming to inform more personalized educational strategies. Validation metrics like the Silhouette Index and Davies-Bouldin Index confirm the efficacy of the clustering. The findings categorize students into high, medium, and low ability groups, offering a data-driven approach to enhancing learning outcomes and instructional design.

Chapter 8: Machine Learning Techniques for Network Security: A Comprehensive Review

In a comprehensive survey of machine learning applications in cybersecurity, this chapter reviews models and methods for detecting and mitigating threats such as intrusions, malware, phishing, and anomalies. It explains fundamental machine learning paradigms and evaluates their strengths and limitations in real-time security applications. Challenges such as adversarial attacks, scalability, and data quality are explored, providing readers with a grounded understanding of the field's complexities and potential.

Chapter 9: Security Analysis in Internet of Things Research Perspective Using Bibliometric Analysis

This chapter presents a bibliometric analysis of global research trends in IoT security using data from over 2,000 Scopus-indexed journal articles. Employing tools like VOSviewer and Biblioshiny, the authors visualize research gaps, influential publications, and emerging themes. The analysis identifies top-performing journals and provides strategic direction for future studies on IoT security, making it a valuable resource for scholars and technologists alike.

Chapter 10: Sentiment Analysis in Transportation Apps: Machine Learning for User Feedback Classification

This chapter explores the application of machine learning techniques to analyze user sentiment in Indonesian transportation apps such as Gojek, Grab, and Maxim. Using natural language processing tools and classification algorithms like SVM, thc study identifies dominant sentiment trends and categorizes reviews with high accuracy. The insights gained provide valuable feedback for service providers aiming to improve customer satisfaction and user experience through data-driven decision-making.

Chapter 11: Vesting Smart Irrigation-Farming Infusing Blockchain and Internet of Agricultural Things (IoAT): Focusing Ground Water Scarcity Ensuring Space-Age Sustainable Farming

This chapter presents a transformative vision for sustainable agriculture by integrating smart irrigation systems with blockchain technology and the Internet of Agricultural Things (IoAT). Amidst global groundwater depletion and climate-induced water crises, the authors propose a future-forward solution where real-time IoAT sensor data informs precision irrigation, significantly improving water use efficiency. Blockchain's secure and decentralized architecture enforces fair water distribution through smart contracts, ensuring transparency and accountability. The chapter highlights how this fusion of technologies empowers farmers to optimize water use, mitigate over-extraction of aquifers, and enhance crop yields sustainably. Ultimately, it offers a scalable model for water-secure, tech-enabled farming in an environmentally fragile era.

Chapter 12: YOLO-based CCTV Surveillance Advanced Techniques for Cheating Detection

Focusing on the growing need for academic integrity in digital education environments, this chapter explores the application of advanced computer vision techniques to detect cheating during examinations. The authors employ the YOLO (You Only Look Once) object detection algorithm—across its v5, v6, and v7 iterations—to monitor and classify suspicious student behaviors from CCTV footage. With cheating methods evolving beyond the capabilities of human invigilators, the study demonstrates how AI-powered surveillance can enhance fairness and transparency in assessments. While challenges such as class imbalance affect model performance, the research shows that strategic data augmentation can lead to meaningful accuracy improvements, with YOLOv7 achieving a notable 60% detection rate. This chapter underscores the promise and limitations of AI surveillance in educational quality assurance.

CONCLUSION

As editors of this volume, we have witnessed an inspiring convergence of interdisciplinary ideas, technological ingenuity, and societal relevance throughout the chapters presented. Each contribution within this reference book reflects the ever-evolving relationship between digital innovation and real-world challenges—from enhancing agricultural sustainability through blockchain and IoAT to bolstering academic integrity with AI-driven surveillance systems.

This curated collection showcases not only the breadth of emerging digital applications but also the depth of commitment from researchers and practitioners striving to make these technologies both functional and ethically grounded. Whether addressing environmental crises, educational vulnerabilities, or the broader pursuit of sustainable development, the chapters collectively demonstrate that innovation must serve a higher purpose—one rooted in equity, transparency, and long-term resilience.

We hope that this volume serves as a catalyst for further exploration, collaboration, and implementation of the ideas it contains. As the digital era continues to reshape our global landscape, we invite readers—academics, policymakers, technologists, and students alike—to engage with these insights and carry forward the mission of responsible, impactful innovation.

Chapter 1
Advancements in Image Processing Techniques for Assessing Biotic and Abiotic Stress in Rice Plants:
A Comprehensive Review

Prabira Kumar Sethy
https://orcid.org/0000-0003-3477-6715
Guru Ghasidas Vishwavidyalaya, Bilaspur, India

Jagamohan Padhi
https://orcid.org/0000-0001-8080-6597
Sambalpur University, India

Santi Kumari Behera
https://orcid.org/0000-0003-4857-7821
VSSUT Burla, India

Baishnu Devi
https://orcid.org/0009-0003-7740-7685
VSSUT Burla, India

ABSTRACT

Over five years, significant advancements in image-processing techniques have emerged for assessing biotic and abiotic stresses in rice plants. This review compiles and analyzes recent research developments, highlighting innovative methodologies

DOI: 10.4018/979-8-3693-9846-3.ch001

in stress detection and management. It covers a range of techniques, including hyperspectral imaging, thermal imaging, and machine learning approaches, which have been crucial for accurately identifying and quantifying stress factors. The integration of these methods with modern computational tools allows for precise monitoring and early detection of stress, facilitating timely interventions. Additionally, the review addresses the challenges and limitations of current methodologies and suggests potential research directions. By providing an extensive overview of state-of-the-art techniques, this study serves as a valuable resource for researchers and practitioners in agricultural science and technology.

1. INTRODUCTION

Rice (Oryza sativa) is a staple food for more than half of the world's population, with more than 100 nations engaged in its cultivation. Asia accounts for 90% of the global rice output. Despite the existence of over 110,000 cultivated rice varieties, white and brown rice are the two primary categories post-harvest, (Fukagawa & Ziska, 2019). While white rice is preferred for its longer shelf life, better taste, and ease of cooking, brown rice is richer in bioactive compounds but less popular. White rice's higher glycaemic load can affect glucose levels, yet it remains a staple in balanced diets. Promoting brown rice production and preserving nutrients lost during milling are crucial for a nutritious and sustainable rice supply. Collaboration among farmers, healthcare practitioners, food producers, and nutritionists is essential to ensure this goal.

Rice is not only a significant source of calories but also provides essential nutrients such as magnesium, phosphorus, manganese, selenium, iron, folic acid, thiamine, and niacin. However, it is low in fiber and fat. The growing global population underscores the importance of maintaining a stable and nutritious rice supply.

Rice plants constantly face numerous environmental stressors that severely limit their agricultural production. These stressors are broadly classified into abiotic and biotic stresses (Figure 1).

Figure 1. Classification tree of stressors of rice plant

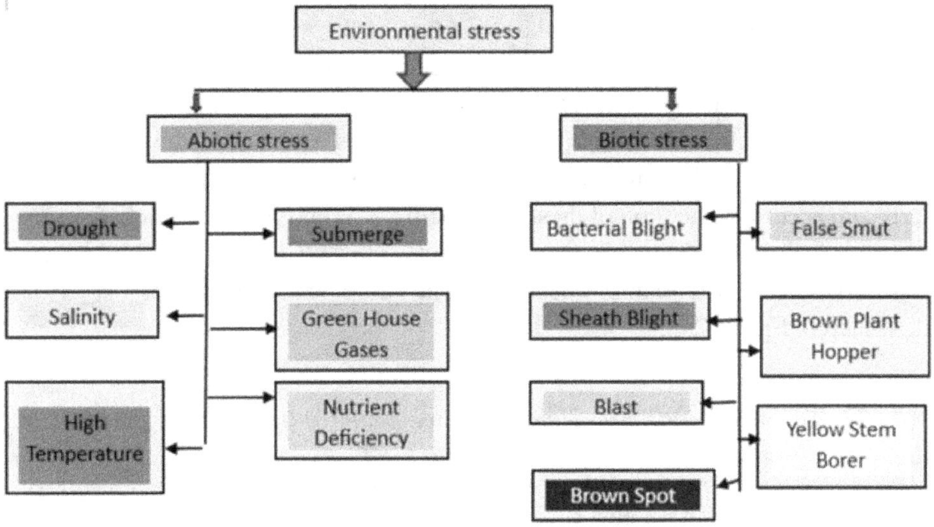

Abiotic stress includes high temperature, drought, salinity, submergence, greenhouse gases, and nutrient deficiencies. The growing desertification of landmasses, increasing salinization of soil and water, depletion of water supplies, and rising environmental pollution are predicted to worsen these stresses.

Biotic stress results from pathogen attacks by bacteria, fungi, oomycetes, nematodes, and herbivores. Major biotic stresses such as BLB, SHB, blast, BS, FS, BPH, YSB, and GM significantly decrease the productivity and quality of rice grains.

Plants, being immobile, have developed various defence systems to withstand these pressures. They absorb the external stress environment, become activated, and trigger appropriate cellular responses. These responses function by transmitting stimuli from sensors on the cell surface or within the cytoplasm to transcriptional machinery in the nucleus. This transmission is facilitated by various signal transduction pathways, leading to differential transcriptional changes that enhance plant stress tolerance, (Verma, Nizam, & Verma, 2013).

The field of Oryza sativa stress detection has undergone remarkable transformations with the advent of DL and image processing techniques. The integration of sophisticated algorithms, particularly DNNs, has shown immense potential for automating the detection of stress levels and their types. This comprehensive review aims to thoroughly analyse the research papers contributing to the understanding and advancement of automated detection of Oryza sativa stress conditions, (Zhu, 2002) (Caiming *et al.,* 2024).

2. BIOTIC STRESS DETECTION IN RICE PLANTS

Rice is a staple food crop for a large portion of the world's population, making its cultivation and protection crucial for global food security. However, rice plants are susceptible to various biotic stresses, primarily caused by living organisms such as fungi, bacteria, viruses, and insects. These stresses can lead to significant losses in crop yield and quality, impacting both farmers' livelihoods and food availability. Early and accurate detection of these biotic stresses is essential for timely intervention and effective management of rice crops, (Dubey & Choubey, 2024) (Kulkarni & Shastri, 2024) (Rahman *et al.,* 2024). The sample images of biotic stress are illustrated in Figure 2.

Figure 2. Sample of biotic stress images of rice plant (a) Bacterial blight (b)sheath blight (c) Blast (d) Brown Spot (f) False smut (g) Tungro

2.1. Image Processing and Feature Extraction Techniques

Researchers have explored various image processing techniques for rice disease detection. Pothen *et al.* (2020) proposed a method using Otsu's segmentation, followed by feature extraction using LBP and HOG, and classification using SVM. Manohar *et al.* (2020) presented an image processing system for automatic identification and classification of bacterial leaf blight, brown spot, and leaf smut using pre-processing, segmentation, feature extraction, and k-nearest neighbors classification. Saputra *et al.* (2020) employed the GLCM method for feature extraction and the KNN algorithm for classification, achieving moderate accuracy. Nigam *et al.* (2020) proposed a system using image processing techniques to identify and classify paddy leaf diseases, achieving high accuracy and low Cross Entropy Loss.

2.2. Machine Learning Approaches

Machine learning techniques have been widely applied to rice disease detection and classification. Ahmad *et al.* (2021) proposed an automated approach using pre-processing, segmentation, feature calculation, feature selection, and classification using SVM. Patidar *et al.* (2020) used RNN for classification, finding it more efficient and effective than CNN and SVM. Ganatra *et al.* (2020) proposed a prediction model for plant leaf disease detection and classification using computer vision and ML methods, with the RF classifier performing best. Setiawan *et al.* (2023) investigated the application of ML for classifying rice leaf diseases using the Nu-SVM algorithm.

2.3. Deep Learning Approaches

DL techniques, particularly CNNs, have shown promising results in rice disease detection. Ghosal *et al.* (2020) developed a DL model based on VGG-16 using TL, achieving an accuracy of 92.46%. Andrianto *et al.* (2020) developed a DL-based rice disease detection system using the VGG16 architecture, which achieved a train accuracy of 100% and test accuracy of 60%. Bhattacharya *et al.* (2020) developed a CNN framework for automatic classification of rice leaf diseases, achieving 94% accuracy in distinguishing healthy and diseased leaves and 78.44% accuracy in categorizing three types of diseases. Bhuyan *et al.* (2023) proposed a novel DL approach using a SPnet with a SE architecture, achieving the highest accuracy, sensitivity, specificity, precision, recall, and F1-score.

2.4. Transfer Learning and Hybrid Approaches

Researchers have explored transfer learning and hybrid approaches to improve the performance of rice disease detection systems. Sharma *et al.* (2022) conducted a comparative analysis, demonstrating the superiority of TL, with InceptionResNetV2 and XceptionNet achieving the best results. Jiang *et al.* (2021) proposed an improved VGG16 model using multi-task learning, achieving high accuracies for both rice and wheat leaf diseases. Ramesh *et al.* (2020) proposed a method for recognizing and classifying paddy leaf diseases using an optimized DNN with the Jaya algorithm, achieving high accuracy rates for various diseases. Daniya *et al.* (2023) proposed a hybrid optimization algorithm for categorizing plant diseases, achieving high performance metrics.

2.5. Recent Advancements (2020-2024)

Dubey *et al.* (2023) presented a ML-based method for early plant disease detection, involving pre-processing, feature extraction, and classification using an optimized ANN. Kulkarni *et al.* (2024) proposed a novel CNN model for classifying prevalent rice leaf diseases, achieving 95% accuracy. Rahman et al. (2024) developed an automated image acquisition system and algorithm to estimate the severity of disease in paddy leaves, demonstrating the potential of image processing. Ritharson *et al.* (2024) proposed a DL approach using TL and custom CNN models, achieving an exceptional accuracy of 99.94%.

Simhadri *et al.* (2024) reviewed various DL approaches for rice leaf disease detection, including TL, EL, and Hybrid approaches, and provided future research directions. Stephen *et al.* (2024) proposed a novel approach using 3D 2D CNN and an improved backtracking search algorithm-optimized DGAN, achieving 98.7% accuracy. Lu *et al.* (2024) proposed an ACNN-TL model combining Atrous convolution and TL, achieving high accuracy rates. Dubey *et al.* (2024) developed an automatic leaf detection approach using ML, achieving a maximum accuracy of 97.94%.

Tyagi *et al.* (2024) proposed an advanced approach for accurate detection of rice leaf diseases, achieving 99% accuracy with their Light Weighted CNN model. Mehta *et al.* (2024) proposed a lightweight deep CNN method that outperforms state-of-the-art models, achieving high accuracy, precision, recall, and F1-score. Tripathy *et al.* (2024) proposed a novel classification scheme for rice leaf diseases using Federated Learning, achieving improved performance metrics. Mahadevan *et al.* (2024) proposed a novel approach using DSGAN2 with Improved Artificial Plant Optimization, achieving a lower false rate compared to existing systems.

Pahlawanto *et al.* (2024) employed CNN models, specifically VGG16, to improve the accuracy of rice plant disease detection, demonstrating enhanced performance compared to previous studies. Jesie *et al.* (2024) proposed an automated disease classification model for paddy crops using image processing and DL techniques, outperforming previous approaches. Firnando *et al.* (2024) conducted a comparative study on DL architectures, finding InceptionV3 as the best model for classifying rice diseases. Chaudhari *et al.* (2023) proposed a hybrid CNN (Inception-ResNet)-SVM model for detecting and treating damaged rice leaves, achieving high performance.

Dogra *et al.* (2023) proposed a CNN-VGG19 model leveraging TL for rice leaf disease classification, demonstrating improved performance compared to existing baseline models. Aggarwal *et al.* (2023) proposed a DL-based system for predicting rice leaf diseases, achieving 94% accuracy on segmented data. Rajpoot *et al.* (2023) proposed a novel approach using VGG-16 TL and Faster R-CNN, achieving an average predicting accuracy of 97.3%. Haruna *et al.* (2023) proposed a novel data augmentation approach using SG2-ADA and the Laplacian filter, significantly improving the performance of Faster-RCNN and SSD models.

Singh *et al.* (2023) proposed a custom CNN architecture for detecting and classifying common rice plant diseases, achieving high accuracy using optimization techniques. Peng *et al.* (2023) proposed a deep residual network model, RiceDRA-Net, for accurate identification of rice leaf diseases with complex backgrounds, achieving high recognition accuracy. Ahad *et al.* (2023) compared the performance of DL architectures, finding the ensemble framework achieved the best accuracy of 98%. Zoang *et al.* (2023) proposed a high-quality image augmentation (HQIA) method using a dual GAN, improving recognition accuracy compared to standard data augmentation.

Khan *et al.* (2023) proposed a hybrid model combining CNN, LSTM, and Self-Attention modules to diagnose rice leaf diseases, achieving high accuracy, precision, recall, F1-score, and well-fitted performance. Sankareshwaran *et al.* (2023) proposed a novel approach using the crossover boosted artificial hummingbird algorithm-based AX-RetinaNet, achieving an accuracy rate of 98.1% and outperforming existing methods. Gogoi *et al.* (2023) proposed a 3-Stage CNN architecture with TL and fine-tuning, achieving 94% accuracy on a small dataset. Mohapatra *et al.* (2023) proposed a TL-based AlexNet model for automatic detection and diagnosis of rice plant diseases, outperforming other approaches.

Das *et al.* (2023) proposed a DL-based model for large-scale monitoring of leaf blast disease in rice using remote sensing parameters, achieving high training and validation accuracy. Kumar *et al.* (2023) proposed a multi-scale feature fusion-based RDTNet for rice leaf disease identification, achieving high average precision, F1-score, and accuracy. Kukreja *et al.* (2023) introduced a novel hybrid CNN-LSTM model for multi-classification of rice leaf blight disease, achieving 95.25% accuracy. Yogeshwari *et al.* (2023) developed a novel DCNN-based framework for detecting plant leaf diseases, demonstrating high classification accuracy.

Petchiammal *et al.* (2023) introduced "Paddy Doctor," a visual image dataset for identifying paddy diseases, and benchmarked it using CNN and TL models, with ResNet34 achieving the highest F1-score. Verma *et al.* (2023) developed a lightweight CNN model for disease identification in corn, rice, and wheat, outperforming several benchmark models. Pan *et al.* (2023) developed a two-stage method, RiceNet, to identify rice diseases in complex scenes, achieving 99.03% accuracy. Bhimavarapu *et al.* (2023) proposed an Improved Activation and Optimizer Function (IAOF) in a CNN model for rice disease detection, outperforming existing methods.

Balaji *et al.* (2023) reviewed various methodologies and strategies used to detect rice diseases, proposing a model using an enhanced CNN. Rachman *et al.* (2024) addressed the limitations of CNNs and proposed the use of Vision Transformer (ViT) for rice disease recognition, outperforming traditional CNN models. Soydan *et al.* (2022) proposed an image processing approach to identify and classify rice leaf blight disease, achieving high accuracy in identification and classification. Pallathadka *et al.* (2022) presented a ML-based framework for leaf disease classification and detection, with potential to increase crop yield.

Kundu *et al.* (2022) focused on developing techniques for detecting diseases in plant leaves using image processing, contributing to the agriculture sector's strength and sustainability. Kiratiratanapruk *et al.* (2022) proposed a computer vision technique using CNNs and image tiling to detect rice diseases, achieving high performance in leaf width prediction and disease prediction. Tejaswini *et al.* (2022) investigated various ML and DL approaches for detecting rice leaf diseases, finding that DL models outperformed ML methods. Kumar *et al.* (2022) developed a prototype for detecting rice plant diseases using ML and image processing, achieving 98.8% accuracy.

Poornappriya *et al.* (2022) reviewed the use of image processing techniques and DL models for early and accurate identification of rice diseases, highlighting the potential of CNNs, TL, and data augmentation. Upadhyay *et al.* (2022) proposed a CNN approach for efficient rice plant disease detection, achieving an accuracy of 99.7%. Bhartiya *et al.* (2022) proposed a ML approach for automated detection and diagnosis of rice leaf diseases, achieving an accuracy of 81.8%. Shrivastava *et al.* (2022) proposed a decision support system for Indian rice farmers to identify diseases, with ANN showing superior accuracy and time consumption compared to SVM.

Ökten *et al.* (2022) developed a computer-aided classification system to detect rice plant diseases from leaf images, achieving an accuracy rate of 76.8%. Tusher *et al.* (2022) proposed a novel DL-based technique for diagnosing and classifying Bacterial Blight and leaf brown spot diseases, achieving an accuracy rate of 95.26%. Archana *et al.* (2022) proposed a method using modified K-means segmentation and a novel SVM-based probabilistic neural network, achieving high accuracies for various rice diseases.

Sudhesh *et al.* (2023) addressed the challenge of rice leaf disease identification using transfer-learned DL models, proposing a DMD-based attention-driven preprocessing method and achieving a test accuracy of 100%. Ibrahim *et al.* (2022) proposed the use of image processing and data mining algorithms to analyze and detect diseases in rice leaves, comparing the performance of various ML algorithms.

Iqbal *et al.* (2023) developed an exploration system to identify and classify disease symptoms on rice leaves, with VGG19 achieving a higher accuracy of 97.94%.

Azim *et al.* (2021) proposed a model for detecting three rice leaf diseases, removing image backgrounds and segmenting disease-affected areas, and achieving 86.58% accuracy using an extreme gradient boosting decision tree ensemble. Moorthy *et al.* (2021) proposed the use of InceptionResNetV2 with TL for automatic identification of rice leaf diseases, achieving 95.67% accuracy. Mekha *et al.* (2021) employed various image processing and classification algorithms, finding that the RF algorithm achieved the best performance with 69.44% accuracy.

Bari *et al.* (2021) proposed a machine-driven disease diagnosis system using Faster R-CNN, achieving high accuracies ranging from 98.09% to 99.25% for various rice diseases. Kartikeyan *et al.* (2021) provided an overview of established methods for plant disease detection and classification systems, discussing various techniques. Nalini *et al.* (2021) developed a novel DNN classification model for paddy leaf disease identification, demonstrating superior performance compared to SVMs. Shrivastava *et al.* (2021) developed an image-based automated system for classifying rice plant diseases using color features, achieving a highest accuracy of 94.65% with SVM.

Jiang *et al.* (2021) proposed an improved VGG16 model using multi-task learning, achieving high accuracies for both rice and wheat leaf diseases. Sharma *et al.* (2022) conducted a comparative analysis, demonstrating the superiority of TL, with InceptionResNetV2 and XceptionNet achieving the best results. Vishnoi *et al.* (2021) provided a comprehensive review of existing research on automated detection of plant diseases through leaf image analysis. Ahmad *et al.* (2021) proposed an automated approach using pre-processing, segmentation, feature calculation, and classification using SVM, achieving high accuracy.

Pothen *et al.* (2020) proposed a method for rice leaf disease classification, utilizing Otsu's segmentation and feature extraction using LBP and HOG, achieving 94.6% accuracy with SVM. Patidar *et al.* (2020) used Residual Neural Network (RNN) for classification of rice plant diseases, achieving 95.83% accuracy and finding RNN more efficient than CNN and SVM. Ramesh *et al.* (2020) proposed a method for recognizing and classifying paddy leaf diseases using an optimized DNN with the Jaya algorithm, achieving high accuracy rates.

Ghosal *et al.* (2020) addressed the challenge of manual disease identification by farmers, utilizing TL to develop a DL model based on VGG-16, achieving 92.46% accuracy. Andrianto *et al.* (2020) developed a DL-based rice disease detection system using the VGG16 architecture, achieving a train accuracy of 100% and test accuracy of 60%. Padilla *et al.* (2020) investigated the detection of maize leaf diseases using CNN and OpenMP, achieving high accuracy rates. Manohar *et al.* (2020) proposed an image processing system for automatic identification and classification

Manohar *et al.* (2020) proposed an image processing system for automatic identification and classification of bacterial leaf blight, brown spot, and leaf smut, utilizing texture and shape features for leaf disease recognition. Sulistyaningrum *et al.* (2020) proposed an automatic detection system using digital image processing and SVM classification, achieving an accuracy of 86.10%. Saputra *et al.* (2020) employed data mining classification methods to detect and classify rice leaf diseases, achieving an accuracy of 65.83% and a kappa value of 0.485.

Jiang *et al.* (2020) proposed a method combining CNNs for feature extraction and SVM for classification and prediction of rice leaf diseases, achieving an average correct recognition rate of 96.8%. Wadhawan *et al.* (2020) reviewed various methods for detecting crop diseases in rice plants using traditional image processing techniques and neural networks, exploring the use of semantic segmentation. Nigam *et al.* (2020) proposed a system using image processing techniques to identify and classify paddy leaf diseases, achieving high accuracy and low Cross Entropy Loss.

Ganatra *et al.* (2020) proposed a prediction model for plant leaf disease detection and classification using computer vision and ML methods, finding the RF classifier performed best. Das *et al.* (2020) developed a DL-based automated feature engineering approach for rice leaf disease prediction using CNNs, demonstrating effective performance in disease classification. Ajra *et al.* (2020) proposed a plant leaf disease detection and prevention technique using image processing and CNN models, achieving high accuracy rates.

Matin *et al.* (2020) applied the AlexNet technique to detect three prevalent rice leaf diseases, achieving a remarkable accuracy rate of over 99%. Sakhamuri *et al.* (2020) discussed the limitations of manual plant disease identification and highlighted the potential of image processing techniques in automating multi-disease models and pest prediction. Sazzad *et al.* (2020) proposed a framework for automated detection and classification of rice blast disease using RGB color images, achieving an accuracy rate of 97.43%.

Sharma *et al.* (2020) proposed a novel approach for detecting and categorizing rice crop diseases, finding that Bayes' Classifiers and MDC achieved overall accuracy rates of 69.358% and 81.06%, respectively. Syamila *et al.* (2020) developed a system to detect paddy leaf diseases, achieving an average accuracy of 88.57%, precision of 82.97%, and misclassification rate of 11.43%. Singh *et al.* (2020) highlighted various image processing techniques for paddy disease identification and classification, emphasizing the need for advanced techniques.

Arora *et al.* (2020) proposed an image processing approach to detect and classify three key rice diseases, employing various advanced classification models. Sagarika *et al.* (2020) proposed an innovative approach using image processing and CNN to classify paddy plants into disease classes, achieving 94.12% accuracy. Parven *et al.*

(2020) proposed an early detection approach using simple image processing and ML techniques to recognize four types of paddy leaf diseases, achieving 94% accuracy.

Patil *et al.* (2020) proposed an automatic technique for rice disease detection and identification using multiclass SVM, utilizing image processing techniques for disease classification. Rahman *et al.* (2020) proposed a DL-based approach for detecting diseases and pests in rice plants using CNNs, achieving 93.3% accuracy with a significantly reduced model size. Mallick *et al.* (2020) explored the integration of AI, image processing, and data mining in agricultural practices, achieving 80.6% accuracy in predicting diseases on out-of-sample data.

This comprehensive summary covers a wide range of approaches and methodologies used in rice disease detection, from traditional image processing techniques to advanced DL models. The research in this field has shown significant progress over the years, with recent studies achieving high accuracy rates and developing more efficient and practical solutions for real-world applications. However, challenges remain in terms of dataset limitations, real-world applicability, early detection capabilities, and model interpretability. Addressing these challenges will be crucial for the widespread adoption of automated disease detection systems in rice cultivation. The summary is detailed in Table 1.

Table 1. ***Summary of studies of biotic stress on plant using ML and DL techniques***

Study	*Techniques Used*	*Performance Metrics*	*Remarks*
Dubey et al. (2023)	*Image pre-processing, Feature Extraction, ANN, ASFO*	*Accuracy: 97.94%*	*Effective in early plant disease diagnosis*
Kulkarni et al. (2024)	*Convolutional Neural Network (CNN)*	*Accuracy: 95%*	*Automated system for identifying rice plant problems*
Rahman et al. (2024)	*Automated image acquisition, Pixel counting*	*Range: 15.53% - 41.23% infected*	*Outperforms traditional leaf area meter method*
Ritharson et al. (2024)	*Data Augmentation, VGG16, TL*	*Accuracy: 99.94%*	*High accuracy with interpretable AI*
Simhadri et al. (2024)	*TL, EL, Hybrid Approaches*	*Varies per model*	*Review of existing techniques and future directions*
Stephen et al. (2024)	*3D 2D CNN, Improved Backtracking Search (IBS) algorithm*	*Accuracy: 98.7%*	*Overcomes limitations of traditional methods*
Lu et al. (2024)	*Atrous Convolution, TL, ResNet-34, VGG-16*	*Accuracy: 98.4% (ResNet-34)*	*Effective for small-sample datasets*

continued on following page

Table 1. Continued

Study	Techniques Used	Performance Metrics	Remarks
Tyagi et al. (2024)	CLAHE, HSV, K-means, Light Weighted CNN	Accuracy: 99%	Includes an Android application for practical use
Mehta et al. (2024)	Deep CNN, Manual Image Annotation	Accuracy: 99.81%	Efficient with fewer parameters, offers crop monitoring API
Tripathy et al. (2024)	Kalman Filter, LeNet, SHAO	Precision: 91.3%, Recall: 92.2%	Utilizes Federated Learning framework
Mahadevan et al. (2024)	ITNN, SMNS, S2AFS, DSGAN2	Lower false rate	Advanced feature selection and disease detection
Pahlawanto et al. (2024)	VGG16	Higher accuracy than previous studies	Enhances detection accuracy for rice plant diseases
Jesie et al. (2024)	CLAHE, GLCM, Hybrid CNN	Not specified	Early detection model for improving crop production
Firnando et al. (2024)	InceptionV3, InceptionResNetV2, EfficientNetB3, etc.	Accuracy: 99.53% (InceptionV3)	Best performance with InceptionV3 model
Chaudhari et al. (2023)	Grab-Cut Segmentation, Inception-ResNet, SVM	Accuracy: 0.97	Reliable and efficient for disease detection
Dogra et al. (2023)	CNN-VGG19 with TL	Accuracy: 93.0% Sensitivity: 89.9% Specificity: 94.7% Precision: 92.4% F1-Score: 90.5%	Improved performance compared to baseline models.
Aggarwal et al. (2023)	EfficientNetV2B3 with ET and HGB Classifiers	Accuracy: 93-94% on segmented data Identification Accuracy: 90-91%	Demonstrates the effectiveness of a DL-based system for identifying rice leaf diseases.
Rajpoot et al. (2023)	VGG-16 with Faster R-CNN and RF	Average Predicting Accuracy: 97.3%	Effectiveness in detecting rice diseases like bacterial leaf blight, brown spot, and leaf smut.
Daniya et al. (2023)	Hybrid Optimization with Deep Neuro-Fuzzy Network and DRN	Accuracy: 93.04% Sensitivity: 94.59% Specificity: 83.83% F1-Score: 91.42%	Effective for automatic classification of rice diseases.
Haridasan et al. (2023)	SVM Classifier and CNN	Validation Accuracy: 91.45%	System helps in predicting remedies for rice crop diseases.

continued on following page

Table 1. Continued

Study	*Techniques Used*	*Performance Metrics*	*Remarks*
Tholkapiyan et al. (2023)	*Hybrid DL Model*	*Not specified in study*	*Focuses on automated diagnosis using image processing and DL.*
Mohammed et al. (2023)	*Various Techniques including K-means, Naive Bayes, CNN*	*Not specified in study*	*Critical review emphasizing the need for technological advancements in disease detection.*
Setiawan et al. (2023)	*Nu-SVM Algorithm with 5-fold Cross-Validation*	*Accuracy: 52.12%-53.81%*	*Moderate accuracy indicating a need for advanced techniques.*
Kumar et al. (2023)	*Multi-scale YOLO v5 with DenseNet-201*	*Average Precision: 82.8% Accuracy: 94.87%*	*Addresses early-stage diagnosis and environmental impact reduction.*
Bharanidharan et al. (2023)	*Modified Lemurs Optimization Algorithm*	*Balanced Accuracy: 90%*	*Uses thermal images for early disease detection in paddy.*
Haruna et al. (2023)	*SG2-ADA with Faster-RCNN and SSD*	*Mean Average Precision (mAP): 0.93 (Faster-RCNN), 0.91 (SSD)*	*Augmentation technique for handling small and imbalanced datasets.*
Singh et al. (2023)	*Custom CNN Architecture*	*Accuracy: 99.83% (Adam), 99.66% (SGDM)*	*Demonstrates the potential of CNNs for high accuracy in rice disease detection.*
Peng et al. (2023)	*RiceDRA-Net with Res-Attention Module*	*Accuracy: 99.71% (CBG-Dataset), 97.86% (SBG-Dataset)*	*High recognition accuracy with complex backgrounds.*
Ahad et al. (2023)	*Ensemble Framework with TL*	*Accuracy: 98%*	*Enhances performance with ensemble techniques.*
Zoang et al. (2023)	*High-Quality Image Augmentation using Dual GAN*	*Accuracy Increase: 4.57% (ResNet18), 4.1% (VGG11)*	*Focuses on generating high-quality training samples.*
Khan et al. (2023)	*Hybrid Model (CNN + LSTM + Self-Attention)*	*Accuracy: 100% (Training), 97.51% (Validation) Precision, Recall, F1-Score: 100%*	*Combines multiple DL modules for high accuracy.*
Bhuyan et al. (2023)	*SE_SPnet with Squeeze-and-Excitation*	*Accuracy: 99.2% Sensitivity: 98.2% Specificity: 98.5%*	*Outperforms existing CNN models in disease identification.*
Gogoi et al. (2023)	*3-Stage CNN with TL*	*Accuracy: 94%*	*Early detection with reduced need for large labeled datasets.*

continued on following page

Table 1. Continued

Study	*Techniques Used*	*Performance Metrics*	*Remarks*
Mohapatra et al. (2023)	*AlexNet with TL*	*High accuracy (not specified)*	*Outperforms other ML approaches.*
Das et al. (2023)	*DL, Spectral indices (NDVI, EVI, NDMI, SAVI)*	*Training accuracy: 90.02%, Validation accuracy: 85.33%*	*Focuses on large-scale monitoring of leaf blast disease using remote sensing*
Kumar et al. (2023)	*RDTNet (LBP, gray, HOG images, Transformer, Convolution blocks)*	*Average precision: 99.55%, F1-score: 99.54%, Accuracy: 99.53%*	*Multi-scale feature fusion approach for real-time rice disease diagnosis*
Kukreja et al. (2023)	*CNN, LSTM*	*Accuracy: 95.25%*	*Focuses on multi-classification of rice leaf blight disease*
Yogeshwari et al. (2023)	*DCNN, GLCM, PCA*	*High classification accuracy (exact values not specified)*	*Framework for detecting plant leaf diseases using deep CNN*
Petchiammal et al. (2023)	*CNN, TL (VGG16, MobileNet, Xception, ResNet34)*	*Highest F1-score: 97.50% (ResNet34)*	*Introduces a new dataset and benchmarks various models for paddy disease identification*
Verma et al. (2023)	*Lightweight CNN*	*Accuracy: 84.4% (Overall), 99.74% (Corn), 82.67% (Rice), 97.5% (Wheat)*	*Focuses on three major crops: Corn, Rice, and Wheat for disease identification*
Pan et al. (2023)	*RiceNet (YoloX, Siamese Network)*	*Accuracy: 99.03%*	*Effective in rapidly identifying rice diseases in complex scenes*
Bhimavarapu et al. (2023)	*CNN, IAOF*	*Outperforms existing methods (exact values not specified)*	*Focuses on early detection of rice leaf infections using IAOF-CNN*
Balaji et al. (2023)	*Enhanced CNN*	*Highest accuracy: 78.2% (5-layer convolution model)*	*Survey of methodologies for rice disease detection*
Rachman et al. (2024)	*Vision Transformer (ViT)*	*High recall, precision, specificity, f1-score, accuracy*	*Proposes a ViT-based model for improved rice disease recognition*
Soydan et al. (2022)	*Clustering, ANNs, Levenberg-Marquardt*	*Identification: R2: 0.9999, Error: 0.019%*	*Proposes image processing approach for rice leaf blight disease classification*

continued on following page

Table 1. Continued

Study	*Techniques Used*	*Performance Metrics*	*Remarks*
Pallathadka et al. (2022)	*SVM, Naïve Bayes, CNN*	*Performance metrics not specified*	*Automated leaf disease diagnosis system using ML*
Kundu et al. (2022)	*Image processing, Pesticide-free crop production techniques*	*Performance metrics not specified*	*Focuses on reducing chemical usage in agriculture*
Kiratiratanapruk et al. (2022)	*CNN (18-layer ResNet), Image Tiling, YOLOv4*	*MAPE: 11.18% (leaf width), mAP: 91.14% (disease prediction)*	*Computer vision technique for rice disease detection using image tiling*
Tejaswini et al. (2022)	*ML, DL (VGG16)*	*Highest accuracy: 78.2% (5-layer convolution model)*	*Compares various ML and DL approaches for rice leaf disease detection*
Kumar K et al. (2022)	*Median filtering, DWT, AdaBoostSVM*	*Accuracy: 98.8%*	*Prototype for detecting rice diseases using ML and image processing*
Poornappriya et al. (2022)	*CNN, TL, Data Augmentation*	*Performance metrics not specified*	*Early detection of rice diseases using image processing techniques*
Upadhyay et al. (2022)	*CNN, Otsu's thresholding*	*Accuracy: 99.7%*	*Efficient method for identifying rice plant diseases*
Bhartiya et al. (2022)	*ML, Quadratic SVM*	*Accuracy: 81.8%*	*Automated detection of rice leaf diseases using shape features*
Shrivastava et al. (2022)	*SVM, ANN, Canny edge detection, LBP*	*ANN shows superior performance (exact values not specified)*	*Decision support system for identifying rice diseases using ML*
Ökten et al. (2022)	*Median filtering, OTSU, GLCM, Probabilistic Neural Network (PNN)*	*Accuracy: 76.8%*	*Computer-aided system for rice disease detection using image processing*
Tusher et al. (2022)	*CNN*	*Accuracy: 95.26%*	*Focuses on early detection of leaf diseases in multiple crops*
Archana et al. (2022)	*Modified K-means, NIBCFE, GLCM, NSVMBPNN*	*Accuracy: 95.20%-99.20%*	*Enhanced feature extraction and classification for rice disease identification*
Study	*Methodology*	*Performance*	*Remarks*

continued on following page

Table 1. Continued

Study	Techniques Used	Performance Metrics	Remarks
Sudhesh et al. (2023)	*DMD-based attention-driven preprocessing, Deep CNN models (DenseNet121, XceptionNet), SVM, RF*	*DenseNet121: 93.87% accuracy; XceptionNet with DMD: 100% in some cases*	*Significant improvement in performance metrics; effective preprocessing and models.*
Ibrahim et al. (2022)	*Image processing techniques, RF, J48, SVM classifiers*	*Evaluation of classifiers; specific accuracy not provided*	*Emphasizes need for early detection.*
Iqbal et al. (2023)	*Feature extraction with image processing, Classifiers (Inception v3, VGG19)*	*VGG19: 97.94% accuracy*	*Effective for classifying healthy vs. diseased leaves.*
Azim et al. (2021)	*Image background removal, segmentation, feature extraction (color, shape, texture), Extreme Gradient Boosting Decision Tree*	*86.58% accuracy*	*Effective for bacterial leaf blight, brown spot, leaf smut detection.*
Moorthy N et al. (2021)	*InceptionResNetV2 with TL*	*95.67% accuracy after parameter optimization*	*Effective automatic identification with parameter tuning.*
Mekha et al. (2021)	*Image processing, RF, Decision Tree, Gradient Boosting, Naïve Bayes*	*RF: 69.44% accuracy*	*Various algorithms compared; highlights algorithm performance.*
Bari et al. (2021)	*Faster R-CNN with RPN architecture*	*Up to 99.25% accuracy for healthy leaves; high accuracy for various diseases*	*Effective for detecting rice blast, brown spot, and hispa.*
Kartikeyan et al. (2021)	*Various image processing techniques, classifiers (SVM, ANN, KNN)*	*Overview of methods; specific performance not detailed*	*Need for early-stage detection and automation emphasized.*
Nalini et al. (2021)	*DNN-CSA architecture, feature extraction, classification*	*High classification accuracy; reduced computational workload*	*Effective feature-based classification with reduced computational demands.*
Shrivastava et al. (2021)	*Color features, 14 color spaces, 7 classifiers*	*94.65% accuracy with SVM*	*Effective use of color features for disease classification.*
Jiang et al. (2021)	*Improved VGG16 model with multi-task learning*	*VGG16: 97.22% accuracy for rice; 98.75% for wheat*	*Superior performance over other models for both rice and wheat diseases.*
Sharma et al. (2022)	*TL vs. conventional ML methods, InceptionResNetV2, XceptionNet*	*TL showed superiority; best results with InceptionResNetV2 and XceptionNet*	*TL demonstrated best performance.*

continued on following page

Table 1. Continued

Study	*Techniques Used*	*Performance Metrics*	*Remarks*
Vishnoi et al. (2021)	*Review of plant disease detection techniques*	*Highlights modern feature extraction techniques for effective detection*	*Comprehensive review; emphasizes advanced techniques.*
Ahmad et al. (2021)	*Pre-processing, segmentation, feature extraction, SVM classification*	*98.79% accuracy*	*Effective feature-based identification method.*
Pothen et al. (2020)	*Otsu's method for segmentation, LBP, HOG feature extraction, SVM*	*94.6% accuracy with HOG features*	*Contributes to disease control procedures with effective feature extraction.*
Patidar et al. (2020)	*RNN classification*	*95.83% accuracy; RNN more efficient than CNN and SVM for larger datasets*	*RNN showed better performance for larger datasets.*
Ramesh et al. (2020)	*Optimized DNN with Jaya algorithm*	*High accuracy for various diseases; outperforms other neural network models*	*Effective model optimization for paddy leaf disease recognition.*
Ghosal et al. (2020)	*VGG-16 with TL*	*92.46% accuracy for healthy vs. diseased leaves; 78.44% for disease classification*	*Effective for identifying rice leaf diseases with limited datasets.*
Andrianto et al. (2020)	*VGG16 architecture, cloud server, smartphone application*	*100% train accuracy; test accuracy of 60%*	*Potential for improvement with better datasets; effective in specific scenarios.*
Padilla et al. (2020)	*CNN and OpenMP*	*High accuracy for different maize leaf diseases*	*Effective for maize leaf disease classification.*
Manohar N et al. (2020)	*Image processing, Otsu thresholding, feature extraction, k-NN*	*Rapid and accurate disease classification*	*Effective use of texture and shape features for disease detection.*
Sulistyaningrum et al. (2020)	*Multi-SVM strategy (One Vs. All)*	*86.10% accuracy; linear kernel outperformed other kernels*	*Effective SVM approach for disease detection.*
Saputra et al. (2020)	*GLCM feature extraction, KNN classifier*	*65.83% accuracy*	*Explores data mining methods for classification.*
Jiang et al. (2020)	*CNN feature extraction, SVM classification*	*96.8% accuracy*	*Effective combination of CNN and SVM for disease diagnosis.*
Wadhawan et al. (2020)	*Semantic segmentation, ngrok and Twilio servers*	*Efficient detection and estimation; facilitates real-time monitoring*	*Effective for real-time disease monitoring.*

continued on following page

Table 1. Continued

Study	*Techniques Used*	*Performance Metrics*	*Remarks*
Nigam et al. (2020)	*Image processing, HSV model, PCA, BFO-DNN*	*98% accuracy*	*Effective automated system for disease detection.*
Bhattacharya et al. (2020)	*CNN framework*	*94% accuracy for healthy vs. diseased leaves; 78.44% for disease classification*	*Effective for disease classification; specific accuracies noted.*
Ganatra et al. (2020)	*Pre-processing, segmentation, feature extraction (shape, color, texture), RF classifier*	*RF achieved best performance*	*Highlights potential of computer vision and ML.*
Das et al. (2020)	*DL (CNN), 4 convolutional layers, 2 fully connected layers*	*Effective performance with 10,500 infected leaves*	*Contributes to automated feature engineering for disease prediction.*
Ajra et al. (2020)	*CNN models (AlexNet, ResNet-50), tested on Kaggle datasets*	*High accuracy rates for both AlexNet and ResNet-50*	*Effective for detecting healthy vs. unhealthy leaves.*
Matin et al. (2020)	*AlexNet technique, image augmentation*	*Over 99% accuracy*	*Highly effective technique for rice leaf disease detection.*
Sakhamuri et al. (2020)	*Various image processing techniques, focus on automation*	*Improved automation; various techniques discussed*	*Effective for automating disease detection.*
Kumar et al. (2020)	*CNNs, TL, feature extraction techniques*	*High accuracy rates achieved*	*Emphasizes effectiveness of DL for leaf disease classification.*

Table 1 provides a comprehensive overview of various studies on plant disease detection using ML and DL techniques. The table highlights a range of methodologies and their performance metrics, reflecting the evolving landscape of plant disease detection.

CHALLENGES AND FUTURE DIRECTIONS

Despite the significant progress in rice disease detection using computational methods, several challenges remain:

1. Dataset limitations: Many studies are constrained by small or imbalanced datasets, which can affect the generalizability of the models. Future research should focus on creating larger, more diverse datasets that represent a wide range of disease types, severity levels, and environmental conditions.

2. Real-world applicability: While many models show high accuracy in controlled environments, their performance in real-world field conditions may vary. Future studies should emphasize testing and validating models in actual farming scenarios.
3. Early detection: Improving the ability to detect diseases at very early stages remains a challenge. Advanced sensor technologies and multi-spectral imaging could potentially enhance early detection capabilities.
4. Interpretability: As models become more complex, ensuring their interpretability becomes crucial for gaining farmers' trust and adoption. Future research should focus on developing explainable AI models for disease detection.
5. Integration with other technologies: Combining disease detection systems with other precision agriculture technologies, such as drones and IoT devices, could provide more comprehensive crop management solutions.
6. Addressing multiple stresses: Most current studies focus on individual diseases. Future research should aim to develop models that can simultaneously detect and differentiate between multiple biotic and abiotic stresses.
7. Computational efficiency: As models become more sophisticated, ensuring they can run efficiently on mobile devices or with limited computational resources is crucial for widespread adoption.

3. ABIOTIC STRESSES IN RICE PLANTS

Abiotic stresses significantly impact rice plants, leading to reduced yields and compromised crop quality. These stresses, which include factors such as drought, salinity, temperature extremes, and nutrient deficiencies, can hinder plant growth and development. Identifying and managing abiotic stresses is crucial for maintaining rice productivity and ensuring food security. Advanced diagnostic techniques, including ML and DL models, offer promising solutions for the early detection and assessment of these stresses, (Gao *et al.*, 2020) (Gou *et al.*, 2024) (Walsh, Mangina, & Negrão, 2024). The samples of abiotic stress in rice plants are illustrated in Figure 3.

Figure 3. Samples of abiotic stress in rice plant (a) Drought (b) Salinity (c) High temperature (d) Nitrogen deficient (e) Magnesium deficient (f) Potassium deficient

3.1. Machine Learning and Deep Learning Techniques for Abiotic Stress Detection

Researchers have explored the use of DL techniques for the detection and classification of abiotic stresses in plants. Gao *et al.* (2020) reviewed recent DL techniques for image analysis in crop stress diagnosis, compiling current sensor tools and DL principles in plant stress phenotyping. Goyal *et al.* (2024) developed a DL-based custom CNN (CNN) framework for in situ identification and classification of drought stress in maize crops, achieving high accuracies. Murphy *et al.* (2024) reviewed the success of DL in plant phenomics and discussed best practices, as well as open challenges in this field. Anami *et al.* (2020) developed an automated computer vision system for recognizing and classifying paddy crop stresses, including 2 biotic and 9 abiotic stresses, achieving average stress classification accuracies of 89.12%, 84.44%, and 76.34% using BPNN, SVM, and k-NN classifiers, respectively.

Malvade *et al.* (2022) proposed an automated detection method using pre-trained CNN (CNN) models to identify and classify paddy crop biotic stresses, with ResNet-50 achieving the highest average classification accuracy of 92.61%. Leelavathy *et al.* (2021) developed a DL approach to predict biotic stress in paddy crops using CNNs, achieving an accuracy of 98.33%.

Chowhan *et al.* (2022) discussed the application of remote sensing technology in monitoring and managing various abiotic and biotic stresses in crop plants, highlighting its ability to quantify changes in plant physiological, biochemical, and structural characteristics. Hernanda *et al.* (2024) reviewed the use of thermal imaging for plant stress evaluation, demonstrating its potential as a non-destructive and sustainable method for estimating biotic and abiotic stresses.

Some works are reported on integrated approaches for abiotic stress management and assessment of tolerance using antioxidants and enzymes, (Dutta *et al.*, 2024) (Jin *et al.*, 2024) (Singh *et al.*, 2024) (Rhaman *et al.*, 2024a) (Ali *et al.*, 2024) (Islam *et al.*, 2024) (Rahman *et al.*, 2024b). Again, some are applied remote sensing for this purpose, (Saifullah *et al.*, 2019) (Ramos-Fernández *et al.*, 2024) (Hernández-López *et al.*, 2024) (Gao *et al.*, 2024).

3.2. Nutrient Deficiency Detection

Thomas *et al.* (2020) identified rice genotypes with multiple abiotic stress tolerance, including salinity, drought, and heat stress, using scatter plot analysis to pinpoint promising sources for crop improvement. Bhugra *et al.* (2015) proposed a ML technique to quantify plant stress using leaf color changes and Relative Water Content (RWC), achieving a high correlation coefficient of 0.89.

Xu *et al.* (2020) explored the use of DCNNs for diagnosing nutrient deficiencies in rice plants, with DenseNet-121 performing best (validation accuracy: 98.62%, test accuracy: 97.44%). Taha *et al.* (2022) proposed an integrated approach using color imaging and DCNNs to diagnose nutrient deficiencies in lettuce grown in aquaponics, achieving high accuracy in segmentation (99.1%) and classification (96.5%).

Sathyavani *et al.* (2022) proposed an image acquisition-based classification model using Internet of Things (IoTs) and DenseNet-BC to diagnose nutritional deficiencies in plants, focusing on analyzing leaf texture. Sharma *et al.* (2024) presented a review investigating the application of ML and DL (DL) in rice crop health monitoring and disease/disorder diagnosis, identifying research gaps and providing insights for future research.

Bhabay *et al.* (2022) proposed a model for detecting rice crop pests, diseases, and nutritional deficiencies with high accuracy, finding that InceptionV3 and Inception-ResNetV2 achieved the best validation accuracy of 95.56%. Watchareeruetai *et al.* (2018) proposed a novel image analysis method for identifying nutrient deficiencies in plants based on leaf images, showing superiority over human experts.

Zhai *et al.* (2022) developed a CNN model for automatic rice nitrogen nutrition monitoring, with the GoogleNet model achieving the highest accuracy of 95.7%. Sakthipriya *et al.* (2014) proposed an automated CNN creation approach using genetic algorithms for image classification problems in precision agriculture, achieving high accuracy in soil nutrient deficiency detection. Kavitha *et al.* (2024) proposed a ML approach to identify nutrient deficiencies (NPK) in rice plants using leaf imaging, achieving better accuracy compared to Naive Bayes.

3.3. Vegetation Indices for Stress Monitoring

Ryu *et al.* (2020) investigated the suitability of various vegetation indices for identifying crop responses to elevated air temperatures, heat stress, and herbicide damage in paddy rice, finding that NDVI, MERIS terrestrial chlorophyll index, and photochemical reflectance index were effective in detecting physiological stress.

3.4. Hyperspectral Imaging and Interpretability of Deep Learning Models

Feng *et al.* (2020) utilized HSI and DL techniques to monitor plant phenotypes of 13 okra genotypes under salt stress, demonstrating the effectiveness of combining HSI and DL for high-throughput phenotyping. Nagasubramanian *et al.* (2020) compared popular interpretability methods for DL models in plant stress identification and quantification, concluding that interpretability methods can generate hypotheses and drive scientific insight, despite differences in output explanation maps.

Table 2. ***Summary of state-of-the-art techniques for plant abiotic stress diagnosis using DL and imaging technologies***

Study	Techniques	Focus	Findings
Gao et al. (2020)	DL Techniques	Crop Stress Diagnosis	Reviews recent DL methods for plant stress diagnosis, including classification, object detection, and segmentation.
Gou et al. (2024)	AI and High-Throughput Technologies	Plant Stress and Hormesis	Discusses AI tools and high-resolution imagery for stress detection, with emphasis on ML and DL applications.
Walsh et al. (2024)	AI and Imaging Sensors	Plant Stress Detection	Reviews integration of AI and imaging sensors, focusing on DL algorithms and open-source imaging repositories.

continued on following page

Table 2. Continued

Study	Techniques	Focus	Findings
Goyal et al. (2024)	Custom CNN Framework	Drought Stress in Maize	Achieves high accuracy (98.71%) in detecting drought stress using a custom CNN model compared to other architectures.
Murphy et al. (2024)	DL Phenomics	Crop Improvement	Highlights advantages of DL in non-destructive, high-throughput phenotyping and associated challenges.
Singh et al. (2024)	Stress Signal Transduction	Abiotic Stress and Signal Pathways	Explores various stress signaling pathways and their role in stress adaptation, with insights from recent biotechnological innovations.
Ramos-Fernández et al. (2024)	CWSI and Stomatal Conductance	Water Scarcity and Irrigation	Evaluates water stress using thermal imaging and CWSI under different irrigation regimes, with implications for water management.
Anami et al. (2020)	Automated Computer Vision System	Paddy Crop Stresses	Uses color features to classify 11 stress types in paddy crops with high accuracy.
Malvade et al. (2022)	Pre-trained CNN Models	Biotic Stresses in Paddy Crops	Compares CNN models for detecting biotic stresses, with ResNet-50 achieving the highest accuracy.
Leelavathy et al. (2021)	DL CNN	Biotic Stress in Paddy	Achieves 98.33% accuracy in classifying biotic stress using CNNs.
Chowhan et al. (2022)	Remote Sensing Technology	Abiotic and Biotic Stress Monitoring	Reviews remote sensing techniques for stress monitoring and management in crops.
Hernández-López et al. (2024)	Thermal Imaging	Plant Stress Evaluation	Highlights thermal imaging for non-destructive stress detection and its integration with DL methods.
Bhugra et al. (2015)	ML and Color Features	Plant Stress Quantification	Uses RF to predict stress levels from leaf color features with high correlation.
Xu et al. (2020)	DCNNs (DCNNs)	Nutrient Deficiencies in Rice	Demonstrates high accuracy of DCNNs in diagnosing nutrient deficiencies from leaf images.
Taha et al. (2022)	Color Imaging and DCNNs	Nutrient Deficiencies in Lettuce	Achieves high accuracy in nutrient deficiency classification using DCNNs and color imaging.
Sathyavani et al. (2022)	IoT and DenseNet-BC	Nutritional Deficiencies in Plants	Proposes an IoT-based DenseNet-BC model for diagnosing nutrient deficiencies with improved accuracy.
Sharma et al. (2024)	ML and DL Review	Rice Health Monitoring	Reviews ML and DL applications in rice crop health monitoring, identifying research gaps and trends.

continued on following page

Table 2. Continued

Study	Techniques	Focus	Findings
Bhabay et al. (2022)	TL Algorithms	Pest and Disease Detection	Trains various algorithms to detect rice pests and diseases with high accuracy.
Watchareeruetai et al. (2018)	Image Analysis and CNNs	Nutrient Deficiencies in Plants	Uses CNNs to analyze leaf images and detect nutrient deficiencies with high accuracy.
Zhai et al. (2022)	CNN Models	Nitrogen Nutrition in Rice	Develops a CNN model for automatic monitoring of nitrogen nutrition in rice, achieving high accuracy.
Sakthipriya et al. (2014)	Automated CNN Creation	Soil Nutrient Deficiency	Proposes a genetic algorithm-based CNN approach for efficient soil nutrient deficiency detection.
Kavitha S et al. (2024)	ML and Leaf Imaging	Nutrient Deficiencies in Rice	Uses ML to identify nutrient deficiencies in rice plants from leaf images with high accuracy.
Ryu et al. (2020)	Vegetation Indices	Crop Responses to Stress	Analyzes vegetation indices for detecting heat stress and herbicide damage in rice.
Feng et al. (2020)	HSI and DL	Salt Stress in Okra	Combines HSI and DL to monitor salt stress in okra genotypes.
Nagasubramanian et al. (2020)	Interpretability Methods for DCNNs	Soybean Stress Identification	Compares methods for interpreting DCNN models in stress identification, focusing on feature importance.
Rico-Chávez et al. (2022)	ML and DL in Plant Stress	Hormesis Management	Discusses ML and DL applications in plant stress management and hormesis to enhance crop productivity.
Al-Tamimi et al. (2022)	Phenotyping Imaging Technologies	Abiotic Stress in Crops	Reviews imaging technologies for quantifying crop responses to abiotic stress and selecting resilient genotypes.
Kaur et al. (2021)	Automated Leaf Analysis	Plant Disease Detection	Reviews techniques for automated disease detection based on leaf images, with emphasis on abiotic and biotic stresses.
Watchareeruetai et al. (2018)	Image Analysis and CNNs	Nutrient Deficiencies in Plants	Uses CNNs to analyze leaf images and detect nutrient deficiencies with high accuracy.
Zhai et al. (2022)	CNN Models	Nitrogen Nutrition in Rice	Develops a CNN model for automatic monitoring of nitrogen nutrition in rice, achieving high accuracy.
Sakthipriya et al. (2014)	Automated CNN Creation	Soil Nutrient Deficiency	Proposes a genetic algorithm-based CNN approach for efficient soil nutrient deficiency detection.
Kavitha S et al. (2024)	ML and Leaf Imaging	Nutrient Deficiencies in Rice	Uses ML to identify nutrient deficiencies in rice plants from leaf images with high accuracy.

continued on following page

Table 2. Continued

Study	Techniques	Focus	Findings
Ryu et al. (2020)	Vegetation Indices	Crop Responses to Stress	Analyzes vegetation indices for detecting heat stress and herbicide damage in rice.
Feng et al. (2020)	HSI and DL	Salt Stress in Okra	Combines HSI and DL to monitor salt stress in okra genotypes.
Nagasubramanian et al. (2020)	Interpretability Methods for DCNNs	Soybean Stress Identification	Compares methods for interpreting DCNN models in stress identification, focusing on feature importance.
Rico-Chávez et al. (2022)	ML and DL in Plant Stress	Hormesis Management	Discusses ML and DL applications in plant stress management and hormesis to enhance crop productivity.
Al-Tamimi et al. (2022)	Phenotyping Imaging Technologies	Abiotic Stress in Crops	Reviews imaging technologies for quantifying crop responses to abiotic stress and selecting resilient genotypes.
Kaur et al. (2021)	Automated Leaf Analysis	Plant Disease Detection	Reviews techniques for automated disease detection based on leaf images, with emphasis on abiotic and biotic stresses.

Table 2 provides a comprehensive overview of advanced techniques for diagnosing plant abiotic stress using DL and imaging technologies. It highlights a diverse array of methodologies including custom CNN frameworks, high-resolution imaging, and advanced phenomics to address various stress types such as drought, nutrient deficiencies, and salt stress. Recent studies, such as those by Goyal et al. and Xu et al., demonstrate significant improvements in diagnostic accuracy through specialized models like custom CNNs and DCNNs. Thermal and HSI technologies are emphasized for their role in non-destructive stress assessment and water management, as noted by Ramos-Fernández et al. and Feng et al. Additionally, the integration of AI with genomic and molecular data, as discussed by Singh et al. and Al-Tamimi et al., offers insights into stress signaling pathways and crop resilience. This diverse range of approaches underscores the evolving sophistication in plant stress diagnostics, leveraging both innovative DL models and advanced imaging techniques to enhance crop management and productivity.

3.5. Key findings and advancements include

- DL techniques, such as custom CNN frameworks and TL-based models, have shown high accuracy in detecting and classifying abiotic stresses like drought and nutrient deficiencies.

- Automated detection systems, combining computer vision, ML, and DL, have been developed for the identification and classification of both abiotic and biotic stresses in rice plants.
- The integration of high-throughput phenotyping technologies, like HSI, with DL models has advanced the field of plant stress phenomics, facilitating the selection of stress-resilient genotypes.

While significant progress has been made, ongoing challenges include the need for larger and more diverse datasets, improved interpretability of DL models, and the translation of laboratory findings to practical agricultural applications. Continued interdisciplinary research and the integration of emerging technologies will be crucial for developing comprehensive solutions to address abiotic stresses and ensure sustainable rice production.

4. CONCLUSION

This review highlights significant advancements in image processing technologies for assessing biotic and abiotic stress in rice plants. Key findings include the development of high-resolution imaging systems and sophisticated DL algorithms that enhance the accuracy of stress detection. Innovations such as HSI, thermal imaging, and advanced CNN models have demonstrated their potential in identifying various stress factors with high precision. The integration of these technologies with other agronomic tools, such as IoT and drones, has further refined the capabilities of stress monitoring systems, enabling real-time and non-destructive assessment.

Image processing technologies represent a transformative shift in the way stress assessment is conducted in agriculture. The continued development and adoption of these technologies promise to enhance our understanding of plant responses to stress and improve the management of crops. As these technologies evolve, they are expected to become more accessible and affordable, making them viable for widespread use in diverse agricultural settings. Ultimately, the impact of image processing technologies on stress assessment will likely lead to more resilient and productive agricultural systems, paving the way for future innovations in crop management and agricultural sustainability.

REFERENCES

Aggarwal, M., Khullar, V., Goyal, N., Singh, A., Tolba, A., Thompson, E. B., & Kumar, S. (2023). Pre-trained deep neural network-based features selection supported machine learning for rice leaf disease classification. *Agriculture, 13*(5), 936. DOI: 10.3390/agriculture13050936

Ahad, M. T., Ameen, M., Islam, M. R., Hasan, M. K., & Islam, M. T. (2023). Comparison of CNN-based deep learning architectures for rice diseases classification. *Artificial Intelligence in Agriculture, 9*, 22–35. DOI: 10.1016/j.aiia.2023.07.001

Ahmad, N., Asif, H. M. S., Saleem, G., Younus, M. U., Anwar, S., & Anjum, M. R. (2021). Leaf image-based plant disease identification using color and texture features. *Wireless Personal Communications, 121*(2), 1139–1168. DOI: 10.1007/s11277-021-09054-2

Ajra, H., Nahar, M. K., Sarkar, L., & Islam, M. S. (2020). Disease detection of plant leaf using image processing and CNN with preventive measures. In *2020 Emerging Technology in Computing, Communication and Electronics (ETCCE)* (pp. 1–6). IEEE. DOI: 10.1109/ETCCE51779.2020.9350890

Ali, L. (2024). Influence of abiotic stresses on the production and sustainability of rice crop. *Agrobiological Records, 16*, 99–108. DOI: 10.47278/journal.abr/2024.017

Anami, B. S., Malvade, N. N., & Palaiah, S. (2020). Classification of yield-affecting biotic and abiotic paddy crop stresses using field images. *Information Processing in Agriculture, 7*(2), 272–285. DOI: 10.1016/j.inpa.2019.08.005

Andrianto, H., Faizal, A., & Armandika, F. (2020). Smartphone application for deep learning-based rice plant disease detection. In *2020 International Conference on Information Technology Systems and Innovation (ICITSI)* (pp. xx–xx). IEEE. DOI: 10.1109/ICITSI50517.2020.9264942

Archana, K. S., Srinivasan, S., Bharathi, S. P., Balamurugan, R., Prabakar, T. N., & Britto, A. S. F. (2022). A novel method to improve computational and classification performance of rice plant disease identification. *The Journal of Supercomputing, 78*(xx), 1–21. https://doi.org/xxx. DOI: 10.1007/s11227-021-04245-x

Arora, Y. K., Sharma, M., Vardhan, M., Sharma, D. K., & Batra, R. (2020). A novel approach for detection and classification of rice leaf diseases. In *Applied Computer Vision and Image Processing: Proceedings of ICCET 2020* (Vol. 1). Springer. DOI: 10.1007/978-981-15-4029-5_10

Azim, M. A., Islam, M. K., Rahman, M. M., & Jahan, F. (2021). An effective feature extraction method for rice leaf disease classification. *Telkomnika, 19*(2), 463–470. DOI: 10.12928/telkomnika.v19i2.16488

Balaji, V., Krishnamoorthi, R., Saravanan, S., & Poornima, R. (2023). Deep transfer learning technique for multimodal disease classification in plant images. *Contrast Media & Molecular Imaging, 2023*, 5644727. Advance online publication. DOI: 10.1155/2023/5644727 PMID: 37213211

Bari, B. S.. (2021). A real-time approach of diagnosing rice leaf disease using deep learning-based faster R-CNN framework. *PeerJ. Computer Science, 7*, e432. DOI: 10.7717/peerj-cs.432 PMID: 33954231

Bhabay, Y., & Mane, A. (2022). Rice diseases pests and nutritional deficiency classification using convolutional neural network. *Grenze International Journal of Engineering & Technology (GIJET), 8*(2). [No page numbers available]

Bharanidharan, N.. (2023). *Multiclass paddy disease detection using filter-based feature transformation technique*. IEEE., DOI: 10.1109/ACCESS.2023.3322587

Bhartiya, V. P., Janghel, R. R., & Rathore, Y. K. (2022). Rice leaf disease prediction using machine learning. In *2022 Second International Conference on Power, Control and Computing Technologies (ICPC2T)* (pp. 1–6). IEEE. DOI: 10.1109/ICPC2T53885.2022.9776692

Bhattacharya, S., Mukherjee, A., & Phadikar, S. (2020). A deep learning approach for the classification of rice leaf diseases. In *Intelligence Enabled Research: DoSIER 2019* (pp. 61–69). DOI: 10.1007/978-981-15-2021-1_8

Bhimavarapu, U. (2023). Prediction and classification of rice leaves using the improved PSO clustering and improved CNN. *Multimedia Tools and Applications, 82*(14), 21701–21714. DOI: 10.1007/s11042-023-14631-7

Bhugra, S., Chaudhury, S., & Lall, B. (2015). Use of leaf colour for drought stress analysis in rice. In *2015 Fifth National Conference on Computer Vision, Pattern Recognition, Image Processing and Graphics (NCVPRIPG)* (pp. 1–6). IEEE. DOI: 10.1109/NCVPRIPG.2015.7490060

Bhuyan, P., Saikia, P. A., Barman, S. D., & Deka, P. C. (2023). SE_SPnet: Rice leaf disease prediction using stacked parallel convolutional neural network with squeeze-and-excitation. *Expert Systems: International Journal of Knowledge Engineering and Neural Networks, 40*(7), e13304. DOI: 10.1111/exsy.13304

Bijoy, M. H.. (2024). *Towards sustainable agriculture: A novel approach for rice leaf disease detection using dCNN and enhanced dataset*. IEEE., DOI: 10.1109/ACCESS.2024.3344556

Chaudhari, D. J., & Malathi, K. (2023). Detection and prediction of rice leaf disease using a hybrid CNN-SVM model. *Optical Memory and Neural Networks (Information Optics), 32*(1), 39–57. DOI: 10.3103/S1060992X2301006X

Chowhan, P., & Chakraborty, A. P. (2022). Remote sensing technology—a new dimension in detection, quantification and tracking of abiotic and biotic stresses. In *Plant Stress: Challenges and Management in the New Decade* (pp. 445–457). Springer. https://doi.org/DOI: 10.1007/978-3-030-92336-1_20

Daniya, T., & Vigneshwari, S. (2023). Rice plant leaf disease detection and classification using optimization enabled deep learning. *Journal of Environmental Informatics, 42*(1). Advance online publication. DOI: 10.3808/jei.202300492

•Das, A., Mallick, C., & Dutta, S. (2020). Deep learning-based automated feature engineering for rice leaf disease prediction. In S. De & S. Dey (Eds.), *Computational Intelligence in Pattern Recognition* (pp. 133–141). Springer Singapore. DOI: 10.1007/978-981-15-2449-3_11

Das, S., Dutta, D., & Ghosh, P. (2023). Deep learning analysis of rice blast disease using remote sensing images. *IEEE Geoscience and Remote Sensing Letters, 20*, 1–5. DOI: 10.1109/LGRS.2023.3244324

Dogra, R., Rani, S., Singh, A., Albahar, M. A., Barrera, A. E., & Alkhayyat, A. (2023). Deep learning model for detection of brown spot rice leaf disease with smart agriculture. *Computers & Electrical Engineering, 109*, 108659. DOI: 10.1016/j.compeleceng.2023.108659

Dubey, R. K., & Choubey, D. K. (2024). Reliable detection of blast disease in rice plant using optimized artificial neural network. *Agronomy Journal, 116*(3), 1099–1111. DOI: 10.1002/agj2.21449

Dubey, R. K., & Choubey, D. K. (2024). Reliable detection of blast disease in rice plant using optimized artificial neural network. *Agronomy Journal, 116*(3), 1099–1111. DOI: 10.1002/agj2.21449

Dutta, M., Bhadra, K., Naskar, M., & Chatterjee, P. (2024). Plant signaling and response to abiotic stress. In *Molecular Dynamics of Plant Stress and Its Management* (pp. 211–241). Springer., DOI: 10.1007/978-981-97-1699-9_9

Feng, X., Liu, W., Zhuang, Y., Wei, Y., Wu, Y., & He, Y. (2020). Hyperspectral imaging combined with machine learning as a tool to obtain high-throughput plant salt-stress phenotyping. *The Plant Journal, 101*(6), 1448–1461. DOI: 10.1111/tpj.14597 PMID: 31680357

Firnando, F. M., Muslikh, A. R., & Iriananda, S. W. (2024). Analyzing InceptionV3 and InceptionResNetV2 with data augmentation for rice leaf disease classification. *Journal of Future Artificial Intelligence and Technologies, 1*(1), 1–11. DOI: 10.62411/faith.2024-4

Fukagawa, N. K., & Ziska, L. H. (2019). Rice: Importance for global nutrition. *Journal of Nutritional Science and Vitaminology, 65*(Supplement), S2–S3. DOI: 10.3177/jnsv.65.S2 PMID: 31619630

Ganatra, N., & Patel, A. (2020). A multiclass plant leaf disease detection using image processing and machine learning techniques. *International Journal on Emerging Technologies, 11*(2), 1082–1086.

Gao, C., Wu, X., Huang, Y., & Zhang, W. (2024). The impact of water-saving irrigation on rice growth and comprehensive evaluation of irrigation strategies. *Agronomy (Basel), 14*(7), 1363. DOI: 10.3390/agronomy14071363

Gao, Z., Yu, K., Zhu, Q., Liu, Y., & He, Y. (2020). Deep learning application in plant stress imaging: A review. *AgriEngineering, 2*(3), 368–392. DOI: 10.3390/agriengineering2030029

Ghosal, S., & Sarkar, K. (2020). Rice leaf diseases classification using CNN with transfer learning. In *2020 IEEE Calcutta Conference (Calcon)* (pp. xx–xx). IEEE. DOI: 10.1109/CALCON49167.2020.9106423

Gogoi, M., Barman, S. D., Bhuyan, P., & Saikia, P. A. (2023). Classification and detection of rice diseases using a 3-stage CNN architecture with transfer learning approach. *Agriculture, 13*(8), 1505. DOI: 10.3390/agriculture13081505

Gou, C., Li, S., Zou, D., Xu, X., & Zhang, M. (2024). Machine and deep learning: Artificial intelligence application in biotic and abiotic stress management in plants. *Frontiers in Bioscience (Landmark Edition), 29*(1), 20–45. DOI: 10.31083/j.fbl2901020 PMID: 38287813

Gou, C., Zafar, S., Fatima, Z., Hasnain, Z., Aslam, N., Iqbal, N., Abbas, S., Li, H., Li, J., Chen, B., Ragauskas, A. J., & Abbas, M. (2024). Machine and deep learning: Artificial intelligence application in biotic and abiotic stress management in plants. *Frontiers in Bioscience (Landmark Edition), 29*(1), 20. DOI: 10.31083/j.fbl2901020 PMID: 38287813

Goyal, P., Bhat, M., Yadav, S. K., Gupta, R., & Bansal, R. (2024). A deep learning approach for early detection of drought stress in maize using proximal scale digital images. *Neural Computing & Applications, 36*(4), 1899–1913. DOI: 10.1007/s00521-023-09219-z

Haridasan, A., Thomas, J., & Raj, E. D. (2023). Deep learning system for paddy plant disease detection and classification. *Environmental Monitoring and Assessment, 195*(1), 120. DOI: 10.1007/s10661-022-10656-x PMID: 36399232

Haruna, Y., Qin, S., & Mbyamm Kiki, M. J. (2023). An improved approach to detection of rice leaf disease with GAN-based data augmentation pipeline. *Applied Sciences (Basel, Switzerland), 13*(3), 1346. DOI: 10.3390/app13031346

Hernanda, R. A. P., Doi, R., & Okamoto, M. (2024). Current trends in the use of thermal imagery in assessing plant stresses: A review. *Computers and Electronics in Agriculture, 224*, 109227. DOI: 10.1016/j.compag.2024.109227

Hernández-López, J. A., Andrade, H. J., & Barrios, M. (2024). Agricultural drought assessment in dry zones of Tolima, Colombia, using an approach based on water balance and vegetation water stress. *The Science of the Total Environment, 921*, 171144. DOI: 10.1016/j.scitotenv.2024.171144 PMID: 38401721

Ibrahim, A.-W. S., & Atya, D. B. (2022). Detection of diseases in rice leaf using deep learning and machine learning techniques. *Webology, 19*(1), 1493–1503. DOI: 10.14704/WEB/V19I1/WEB19100

Iqbal, J.. (2023). Early detection and classification of rice brown spot and bacterial blight diseases using digital image processing. *Journal of Computing & Biomedical Informatics, 4*(2), 98–109.

Islam, F., Ali, B., Wang, Y., & Xia, X. (2024). Transcriptomic reprogramming of rice cultivars in response to herbicide, salt and their combined stresses. *Plant Stress, 12*, 100504. DOI: 10.1016/j.stress.2024.100504

Jesie, R. S., Godwin Premi, M. S., & Jarin, T. (2024). Comparative analysis of paddy leaf diseases sensing with a hybrid convolutional neural network model. *Measurement. Sensors, 31*, 100966. DOI: 10.1016/j.measen.2023.100966

Jiang, F., Lu, Y., Chen, Y., Cai, D., & Li, G. (2020). Image recognition of four rice leaf diseases based on deep learning and support vector machine. *Computers and Electronics in Agriculture, 179*, 105824. DOI: 10.1016/j.compag.2020.105824

Jiang, Z., Dong, Z., Jiang, W., & Yang, Y. (2021). Recognition of rice leaf diseases and wheat leaf diseases based on multi-task deep transfer learning. *Computers and Electronics in Agriculture, 186*, 106184. DOI: 10.1016/j.compag.2021.106184

Jin, S., Tang, Q., Li, Y., & Hu, X. (2024). Insights into plant sensory mechanisms under abiotic stresses. *Plants, 13*(14), 1907. DOI: 10.3390/plants13141907 PMID: 39065434

Kartikeyan, P., & Shrivastava, G. (2021). Review on emerging trends in detection of plant diseases using image processing with machine learning. *International Journal of Computer Applications, 975*(8887), 39–48. DOI: 10.5120/ijca2021920990

Kavitha, S. (2024, March). Identification of nutrient deficiency based on leaf image data using machine learning. In *2024 International Conference on Emerging Smart Computing and Informatics (ESCI)* (pp. [insert pages if available]). IEEE. https://doi.org/ [Replace DOI if necessary]DOI: 10.1109/ESCI59138.2024.10565432

Khan, A. R., Akbar, M., Alam, T. M., & Rauf, A. (2023). Automatic diagnosis of rice leaves diseases using hybrid deep learning model. *Journal of Advances in Information Technology, 14*(3), 418–425. DOI: 10.12720/jait.14.3.418-425

Kiratiratanapruk, K., Kositchaiwat, P., Suwanvajokkasikit, K., & Lursinsap, C. (2022). Automatic detection of rice disease in images of various leaf sizes. *arXiv Preprint*, arXiv:2206.07344. https://doi.org//arXiv.2206.07344DOI: 10.48550

Krishnamoorthy, N.. (2021). Rice leaf diseases prediction using deep neural networks with transfer learning. *Environmental Research, 198*, 111275. DOI: 10.1016/j.envres.2021.111275 PMID: 33989629

Kukreja, V., Sharma, R., & Vats, S. (2023). Revolutionizing rice farming: Automated identification and classification of rice leaf blight disease using deep learning. In *2023 Third International Conference on Secure Cyber Computing and Communication (ICSCCC)* (pp. 173–177). IEEE. DOI: 10.1109/ICSCCC58608.2023.10176408

Kulkarni, P., & Shastri, S. (2024). Rice leaf diseases detection using machine learning. *Journal of Scientific Research and Technology, 2024*, 17–22. DOI: 10.61808/jsrt81

Kumar, A., Bhardwaj, A., Sharma, A., & Gupta, A. (2023). Multi-scale feature fusion-based lightweight dual stream transformer for detection of paddy leaf disease. *Environmental Monitoring and Assessment, 195*(9), 1020. DOI: 10.1007/s10661-023-11628-5 PMID: 37548778

Kumar, K., & Kannan, E. (2022). Detection of rice plant disease using AdaBoostSVM classifier. *Agronomy Journal, 114*(4), 2213–2229. DOI: 10.1002/agj2.21070

Kumar, V. S., Jaganathan, M., Viswanathan, A., Umamaheswari, M., & Vignesh, J. (2023). Rice leaf disease detection based on bidirectional feature attention pyramid network with YOLO v5 model. *Environmental Research Communications, 5*(6), 065014. DOI: 10.1088/2515-7620/acdece

Kundu, R., Chauhan, U., & Chauhan, S. P. S. (2022). Plant leaf disease detection using image processing. In *2022 2nd International Conference on Innovative Practices in Technology and Management (ICIPTM)* (Vol. 2, pp. 78–83). IEEE. DOI: 10.1109/ICIPTM54933.2022.9754170

Leelavathy, B., & Kovvur, R. M. R. (2021). Prediction of biotic stress in paddy crop using deep convolutional neural networks. In *Proceedings of International Conference on Computational Intelligence and Data Engineering: ICCIDE 2020* (pp. 361–368). Springer. DOI: 10.1007/978-981-15-8767-2_29

Lu, Y., Tao, X., Jiang, F., Du, J., Li, G., & Liu, Y. (2024). Image recognition of rice leaf diseases using atrous convolutional neural network and improved transfer learning algorithm. *Multimedia Tools and Applications, 83*(5), 12799–12817. DOI: 10.1007/s11042-023-16047-9

Mahadevan, K., Punitha, A., & Suresh, J. (2024). Automatic recognition of rice plant leaf diseases detection using deep neural network with improved threshold neural network. *e-Prime - Advances in Electrical Engineering. Electronics and Energy, 8*, 100534. DOI: 10.1016/j.prime.2024.100534

Mallick, D. K., Ray, R., & Dash, S. R. (2020). Detection and classification of crop diseases from its leaves using image processing. In *Smart Intelligent Computing and Applications:Proceedings of the Third International Conference on Smart Computing and Informatics* (Vol. 1, pp. 539–547). Springer. DOI: 10.1007/978-981-13-9282-5_20

Malvade, N. N., Anami, B. S., & Palaiah, S. (2022). A comparative analysis of paddy crop biotic stress classification using pre-trained deep neural networks. *Artificial Intelligence in Agriculture, 6*, 167–175. DOI: 10.1016/j.aiia.2022.09.001

Mamat, H. S., Zaini, N. A., & Halim, S. A. (2020). Paddy leaf disease recognition system using image processing techniques and support vector machine. *ESTEEM Academic Journal, 16*, 41–50.

Manohar, N., & Gowda, K. J. (2020). Image processing system based identification and classification of leaf disease: A case study on paddy leaf. In *2020 International Conference on Electronics and Sustainable Communication Systems (ICESC)* (pp. xx–xx). IEEE.

Matin, M. M. H., Islam, M. M., Rahman, M. M., & Ahmed, M. (2020). An efficient disease detection technique of rice leaf using AlexNet. *Journal of Computer and Communications, 8*(12), 49–57. DOI: 10.4236/jcc.2020.812005

Mekha, P., & Teeyasuksaet, N. (2021). Image classification of rice leaf diseases using random forest algorithm. In *2021 Joint International Conference on Digital Arts, Media and Technology with ECTI Northern Section Conference on Electrical, Electronics, Computer and Telecommunication Engineering* (pp. xx–xx). IEEE. DOI: 10.1109/ECTIDAMTNCON51128.2021.9425696

Mohammed, L., & Yusoff, Y. (2023). Detection and classification of plant leaf diseases using digital image processing methods: A review. *ASEAN Engineering Journal*, *13*(1), 1–9. DOI: 10.11113/aej.v13.17460

Mohapatra, D., & Das, N. (2023). A precise model for accurate rice disease diagnosis: A transfer learning approach. *Proceedings of the Indian National Science Academy. Part A, Physical Sciences*, *89*(1), 162–171. DOI: 10.1007/s43538-022-00149-3

Murphy, K. M., Singh, D., Zhang, H., & Crisp, P. A. (2024). Deep learning in image-based plant phenotyping. *Annual Review of Plant Biology*, *75*(1), 1–26. DOI: 10.1146/annurev-arplant-070523-042828 PMID: 38382904

Nagasubramanian, K., Jones, S., Singh, A. K., Sarkar, S., Singh, A., & Ganapathysubramanian, B. (2020). Usefulness of interpretability methods to explain deep learning based plant stress phenotyping. *arXiv Preprint*, arXiv:2007.05729. https://arxiv.org/abs/2007.05729

Nalini, S., Krishnaraj, N., Jayasankar, T., Vinothkumar, K., Sagai Francis Britto, A., Subramaniam, K., & Bharatiraja, C. (2021). Paddy leaf disease detection using an optimized deep neural network. *Computers, Materials & Continua*, *68*(1), 1117–1128. DOI: 10.32604/cmc.2021.012431

Nigam, A., Tiwari, A. K., & Pandey, A. (2020). Paddy leaf diseases recognition and classification using PCA and BFO-DNN algorithm by image processing. *Materials Today: Proceedings*, *33*, 4856–4862. DOI: 10.1016/j.matpr.2020.08.397

Ökten, İ., & Yüzgeç, U. (2022). Rice plant disease detection using image processing and probabilistic neural network. In *International Congress of Electrical and Computer Engineering* (pp. xx–xx). Springer International Publishing., https://doi.org/xxx DOI: 10.1007/978-3-031-01984-5_7

Padilla, D. A., Pajes, R. A. I., & De Guzman, J. T. (2020). Detection of corn leaf diseases using convolutional neural network with OpenMP implementation. In *2020 IEEE 12th International Conference on Humanoid, Nanotechnology, Information Technology, Communication and Control, Environment, and Management (HNICEM)* (pp. xx–xx). IEEE. DOI: 10.1109/HNICEM51456.2020.9400004

Pahlawanto, R. D. S., Salsabila, H., & Pratiwi, K. R. (2024). Detection and prediction of rice plant diseases using convolutional neural network (CNN) method. *Journal of Student Research Exploration*, *2*(1), 22–33. DOI: 10.52465/josre.v2i1.254

Pallathadka, H., Chowdhary, C. L., & Ahmad, S. (2022). Application of machine learning techniques in rice leaf disease detection. *Materials Today: Proceedings*, *51*, 2277–2280. DOI: 10.1016/j.matpr.2021.11.398

Pan, J., Wang, T., & Wu, Q. (2023). RiceNet: A two stage machine learning method for rice disease identification. *Biosystems Engineering*, *225*, 25–40. DOI: 10.1016/j.biosystemseng.2022.11.007

Parven, N., Rahman, M. M., Nahar, N., & Rahman, M. M. (2020). Detection and recognition of paddy plant leaf diseases using machine learning technique. *International Journal of Innovative Technology and Exploring Engineering*, *9*(1), 546–551. DOI: 10.35940/ijitee.E2509.039520

Patidar, S., ... (2020). Rice plant disease detection and classification using deep residual learning. In *Machine Learning, Image Processing, Network Security and Data Sciences:Second International Conference, MIND 2020,Silchar, India,July 30–31, 2020, Proceedings, Part I* (Vol. 2, pp. xx–xx). Springer Singapore. DOI: 10.1007/978-981-15-6315-7_23

Patil, P. C., Patil, D. D., & Subramanium, P. (2020). Detection and identification of rice leaf diseases using multiclass SVM. *System*, *7*(06), 1–5.

Peng, J., Gao, H., Zhang, Z., & Chen, Y. (2023). RiceDRA-net: Precise identification of rice leaf diseases with complex backgrounds using a res-attention mechanism. *Applied Sciences (Basel, Switzerland)*, *13*(8), 4928. DOI: 10.3390/app13084928

Petchiammal, M., Rajarajeswari, S., & Bhuvaneswari, T. (2023). Paddy doctor: A visual image dataset for automated paddy disease classification and benchmarking. In *Proceedings of the 6th Joint International Conference on Data Science & Management of Data (10th ACM IKDD CODS and 28th COMAD)* (pp. 17–24). ACM. DOI: 10.1145/3570991.3570994

Poornappriya, T. S., & Gopinath, R. (2022). Rice plant disease identification using artificial intelligence approaches. *International Journal of Electrical Engineering and Technology*, *11*(10), 392–402.

Pothen, M. E., & Pai, M. L. (2020). Detection of rice leaf diseases using image processing. In *2020 Fourth International Conference on Computing Methodologies and Communication (ICCMC)* (pp. xx–xx). IEEE. DOI: 10.1109/ICCMC48092.2020.ICCMC-00080

Rachman, R. K., Mahmud, R., Susilo, E., & Widodo, T. (2024). Enhanced vision transformer and transfer learning approach to improve rice disease recognition. *Journal of Computing Theories and Applications*, *1*(4), 446–460. DOI: 10.62411/jcta.10459

Rahman, C. R., Arko, P. S., Ali, M. E., Khan, M. A. I., Apon, S. H., Nowrin, F., & Hasan, K. (2020). Identification and recognition of rice diseases and pests using convolutional neural networks. *Biosystems Engineering*, *194*, 112–120. DOI: 10.1016/j.biosystemseng.2020.03.020

Rahman, K. S., Rakib, M. R. I., Salehin, M. M., Ali, M. R., & Rahman, A. (2024). Assessment of paddy leaves disease severity level using image processing technique. *Smart Agricultural Technology*, *7*, 100410. DOI: 10.1016/j.atech.2024.100410

Rahman, M. M., Ahmed, S., Akter, S., & Islam, M. R. (2024). Green vanguards: Harnessing the power of plant antioxidants, signal catalysts, and genetic engineering to combat reactive oxygen species under multiple abiotic stresses. *Plant Stress*, *100547*, 100547. Advance online publication. DOI: 10.1016/j.stress.2024.100547

Rajpoot, V., Tiwari, A., & Jalal, A. S. (2023). Automatic early detection of rice leaf diseases using hybrid deep learning and machine learning methods. *Multimedia Tools and Applications*, *82*(23), 36091–36117. DOI: 10.1007/s11042-023-14969-y

Ramesh, S., & Vydeki, D. (2020). Recognition and classification of paddy leaf diseases using optimized deep neural network with Jaya algorithm. *Information Processing in Agriculture*, *7*(2), 249–260. DOI: 10.1016/j.inpa.2019.09.002

Ramos-Fernández, L., Cáceres-Jara, M., Aquino, M. A., & Espinoza, G. (2024). Water stress index and stomatal conductance under different irrigation regimes with thermal sensors in rice fields on the northern coast of Peru. *Remote Sensing (Basel)*, *16*(5), 796. DOI: 10.3390/rs16050796

Rhaman, M. S., Islam, M. R., Jahan, M., & Akter, M. (2024). Phytohormone strigolactone: Involvement in guard cell signaling and abiotic stress tolerance in plants. *Journal of Plant Growth Regulation*, *43*(12), 1–14. DOI: 10.1007/s00344-024-11421-9

Ritharson, P. I., Raimond, K., Mary, X. A., Robert, J. E., & J, A. (2024). DeepRice: A deep learning and deep feature based classification of rice leaf disease subtypes. *Artificial Intelligence in Agriculture*, *11*, 34–49. DOI: 10.1016/j.aiia.2023.11.001

Ryu, J.-H., Jeong, H., & Cho, J. (2020). Performances of vegetation indices on paddy rice at elevated air temperature, heat stress, and herbicide damage. *Remote Sensing (Basel)*, *12*(16), 2654. DOI: 10.3390/rs12162654

Sagarika, G. K., Krishna Prasad, S. J., & Mohana Kumar, S. (2020). Paddy plant disease classification and prediction using convolutional neural network. In *2020 International Conference on Recent Trends on Electronics, Information, Communication & Technology (RTEICT)* (pp. 829–833). IEEE. DOI: 10.1109/RTEICT49044.2020.9315634

Saifullah, M., Iqbal, M., Hussain, M., & Ahmed, S. (2019). Estimation of water stress on rice crop using ecological parameters. *International Journal of Agricultural Sustainable Development, 1*(1), 17–29. DOI: 10.33411/IJASD/20190103

Sakhamuri, S., & Kompalli, V. S. (2020). RETRACTED: An overview on prediction of plant leaves disease using image processing techniques. *IOP Conference Series. Materials Science and Engineering, 981*(2), 022006. DOI: 10.1088/1757-899X/981/2/022024

Sakthipriya, S., & Naresh, R. (2024). Precision agriculture based on convolutional neural network in rice production nutrient management using machine learning genetic algorithm. *Engineering Applications of Artificial Intelligence, 130*, 107682. DOI: 10.1016/j.engappai.2023.107682

Sankareshwaran, S. P., Hariharan, M., & Maheswari, T. (2023). Optimizing rice plant disease detection with crossover boosted artificial hummingbird algorithm based AX-RetinaNet. *Environmental Monitoring and Assessment, 195*(9), 1070. DOI: 10.1007/s10661-023-11612-z PMID: 37610473

Saputra, R. A., Suharyanto, , Wasiyanti, S., Saefudin, D. F., Supriyatna, A., & Wibowo, A. (2020). Rice leaf disease image classifications using KNN based on GLCM fcature extraction. *Journal of Physics: Conference Series, 1641*(1), 012065. DOI: 10.1088/1742-6596/1641/1/012080

Sathyavani, R., JaganMohan, K., & Kalaavathi, B. (2022). Classification of nutrient deficiencies in rice crop using DenseNet-BC. *Materials Today: Proceedings, 56*, 1783–1789. DOI: 10.1016/j.matpr.2021.10.466

Sazzad, T. M. S., Rahman, M. M., Islam, M. M., & Hossain, M. S. (2020). An image processing framework to identify rice blast. In *2020 International Congress on Human-Computer Interaction, Optimization and Robotic Applications (HORA)* (pp. 1–5). IEEE. DOI: 10.1109/HORA49412.2020.9152912

Setiawan, R., Zein, H., Azdy, R. A., & Sulistyowati, S. (2023). Rice leaf disease classification with machine learning: An approach using Nu-SVM. *Indonesian Journal of Data and Science, 4*(3), 136–144. DOI: 10.56705/ijodas.v4i3.114

Sharma, M., Kumar, C. J., & Bhattacharyya, D. K. (2024). Machine/deep learning techniques for disease and nutrient deficiency disorder diagnosis in rice crops: A systematic review. *Biosystems Engineering*, *244*, 77–92. DOI: 10.1016/j.biosystemseng.2024.05.014

Sharma, M., Kumar, C. J., & Deka, A. (2022). Early diagnosis of rice plant disease using machine learning techniques. *Archiv für Phytopathologie und Pflanzenschutz*, *55*(3), 259–283. DOI: 10.1080/03235408.2021.2015866

Sharma, V., Mir, A. A., & Sarwr, A. (2020). Detection of rice disease using Bayes' classifier and minimum distance classifier. *Journal of Multimedia Information System*, *7*(1), 17–24. DOI: 10.33851/JMIS.2020.7.1.17

Shrivastava, G., & Patidar, H. (2022). Rice plant disease identification decision support model using machine learning. *ICTACT Journal on Soft Computing*, *12*(3), 2676–2681. DOI: 10.21917/ijsc.2022.0375

Shrivastava, V. K., & Pradhan, M. K. (2021). Rice plant disease classification using color features: A machine learning paradigm. *Journal of Plant Pathology*, *103*(1), 17–26. DOI: 10.1007/s42161-020-00683-3

Simhadri, C. G., Kondaveeti, H. K., Vatsavayi, V. K., Mitra, A., & Ananthachari, P. (2024). Deep learning for rice leaf disease detection: A systematic literature review on emerging trends, methodologies and techniques. *Information Processing in Agriculture.* Advance online publication. DOI: 10.1016/j.inpa.2024.04.006

Singh, J. P., Pradhan, C., & Das, S. C. (2020). Image processing and machine learning techniques to detect and classify paddy leaf diseases. *RE:view*.

Singh, M., Verma, P., Thakur, R., & Singh, A. (2024). Signaling for abiotic stress tolerance in plants: A review. *Agricultural Reviews (Karnal)*, *45*(2), 1–10. DOI: 10.18805/ag.R-2375

Singh, S. P., Das, B., Pandey, R. P., Singh, H., & Chauhan, D. S. (2023). Custom convolutional neural network for detection and classification of rice plant diseases. *Procedia Computer Science*, *218*, 2026–2040. DOI: 10.1016/j.procs.2023.01.179

Soydan, O., & Taner, A. (2022). Identification of rice leaf blight disease by using image processing techniques. *Anadolu Tarım Bilimleri Dergisi*, *37*(2), 341–360. DOI: 10.7161/omuanajas.1004187

Stephen, A., Punitha, A., & Chandrasekar, A. (2024). Optimal deep generative adversarial network and convolutional neural network for rice leaf disease prediction. *The Visual Computer*, *40*(2), 919–936. DOI: 10.1007/s00371-023-02823-z

Sudhesh, K. M.. (2023). AI based rice leaf disease identification enhanced by dynamic mode decomposition. *Engineering Applications of Artificial Intelligence*, *120*, 105836. https://doi.org/xxx. DOI: 10.1016/j.engappai.2023.105836

Sulistyaningrum, D. R., Rasyida, A., & Setiyono, B. (2020). Rice disease classification based on leaf image using multilevel support vector machine (SVM). *Journal of Physics: Conference Series*, *1490*(1), 012017. DOI: 10.1088/1742-6596/1490/1/012053

Taha, M. F., Saeed, A., Ahmed, H., & El-Sayed, A. (2022). Using deep convolutional neural network for image-based diagnosis of nutrient deficiencies in plants grown in aquaponics. *Chemosensors (Basel, Switzerland)*, *10*(2), 45. DOI: 10.3390/chemosensors10020045

Tejaswini, P., Pooja, G., & Rupa, G. (2022). Rice leaf disease classification using CNN. In *IOP Conference Series: Earth and Environmental Science* (Vol. 1032, No. 1, p. 012037). IOP Publishing. DOI: 10.1088/1755-1315/1032/1/012017

Tholkapiyan, M., Aruna Devi, B., Bhatt, D., Saravana Kumar, E., Kirubakaran, S., & Kumar, R. (2023). Performance analysis of rice plant diseases identification and classification methodology. *Wireless Personal Communications*, *130*(2), 1317–1341. DOI: 10.1007/s11277-023-10333-3

Thomas, T., Joseph, D., Philip, D., & Nair, S. (2020). Identification of rice genotypes for seedling stage multiple abiotic stress tolerance. *Plant Physiology Reports*, *25*(4), 697–706. DOI: 10.1007/s40502-020-00549-x

Tripathy, R., Mandala, J., Pappu, S. R., & Gopisetty, G. K. D. (2024). Optimization based rice leaf disease classification in federated learning. *Multimedia Tools and Applications*, *83*(29), 1–27. DOI: 10.1007/s11042-023-18085-9

Tusher, A. N.. (2022). Automatic recognition of plant leaf diseases using deep learning (multilayer CNN) and image processing. In *International Conference on Image Processing and Capsule Networks* (pp. xx–xx). Springer International Publishing., https://doi.org/xxx DOI: 10.1007/978-3-031-12413-6_11

Tyagi, S., Reddy, S. R. N., Anand, R., & Sabharwal, A. (2024). Enhancing rice crop health: A light weighted CNN-based disease detection system with mobile application integration. *Multimedia Tools and Applications*, *83*(16), 48799–48829. DOI: 10.1007/s11042-023-17449-5

Upadhyay, S. K., & Kumar, A. (2022). A novel approach for rice plant diseases classification with deep convolutional neural network. *International Journal of Information Technology : an Official Journal of Bharati Vidyapeeth's Institute of Computer Applications and Management, 14*(1), 185–199. DOI: 10.1007/s41870-021-00817-5

Verma, S., Kumar, P., & Singh, J. P. (2023). A unified lightweight CNN-based model for disease detection and identification in corn, rice, and wheat. *Journal of the Institution of Electronics and Telecommunication Engineers*, •••, 1–12. DOI: 10.1080/03772063.2023.2169025

Verma, S., Nizam, S., & Verma, P. K. (2013). Biotic and abiotic stress signaling in plants. In Sarwat, M., Ahmad, A., & Abdin, M. (Eds.), *Stress signaling in plants: Genomics and proteomics perspective* (Vol. 1). Springer., DOI: 10.1007/978-1-4614-6372-6_2

Vishnoi, V. K., Kumar, K., & Kumar, B. (2021). Plant disease detection using computational intelligence and image processing. *Journal of Plant Diseases and Protection, 128*(1), 19–53. DOI: 10.1007/s41348-020-00368-0

Wadhawan, R., Garg, M., & Sahani, A. K. (2020). Rice plant leaf disease detection and severity estimation. In *2020 IEEE 15th International Conference on Industrial and Information Systems (ICIIS)* (pp. xx–xx). IEEE. DOI: 10.1109/ICIIS51140.2020.9342653

[] Walsh, J. J., Mangina, E., & Negrão, S. (2024). Advancements in imaging sensors and AI for plant stress detection: A systematic literature review. *Plant Phenomics*, 6, 0153. DOI: 10.34133/plantphenomics.0153

Watchareeruetai, U., Jaroonchon, N., Jongmool, T., & Limthanaporn, N. (2018, March). Identification of plant nutrient deficiencies using convolutional neural networks. In *2018 International Electrical Engineering Congress (iEECON)* (pp. 1–4). IEEE. DOI: 10.1109/IEECON.2018.8712217

Xu, Z., Wang, X., Sun, Y., Wang, J., & Song, H. (2020). Using deep convolutional neural networks for image-based diagnosis of nutrient deficiencies in rice. *Computational Intelligence and Neuroscience, 7307252*, 1–12. Advance online publication. DOI: 10.1155/2020/7307252 PMID: 32952543

Yogeshwari, M., & Thailambal, G. (2023). Automatic feature extraction and detection of plant leaf disease using GLCM features and convolutional neural networks. *Materials Today: Proceedings, 81*, 530–536. DOI: 10.1016/j.matpr.2021.03.700

Zhai, Q., Wang, Y., Guo, Z., Liu, M., Gao, P., & Wang, J. (2022). Rice nitrogen nutrition monitoring classification method based on the convolution neural network model: Direct detection of rice nitrogen nutritional status. *PLoS One*, *17*(11), e0273360. DOI: 10.1371/journal.pone.0273360 PMID: 36413518

Zhang, Z., Gao, H., Peng, J., & Chen, Y. (2023). A high-quality rice leaf disease image data augmentation method based on a dual GAN. *IEEE Access : Practical Innovations, Open Solutions*, *11*, 21176–21191. DOI: 10.1109/ACCESS.2023.3251098

Zhu, J. K. (2002). Salt and drought stress signal transduction in plants. *Annual Review of Plant Biology*, *53*(1), 247–273. DOI: 10.1146/annurev.arplant.53.091401.143329 PMID: 12221975

Chapter 2
Applications of AI In Healthcare, Agriculture, Finance, and Other Industries

P. Selvakumar
https://orcid.org/0000-0002-3650-4548
Department of Science and Humanities, Nehru Institute of Technology, Coimbatore, India

Shabnam
https://orcid.org/0009-0008-6139-0491
Chandigarh University, India

Aditee Huparikar Shah
https://orcid.org/0009-0009-0757-8597
Indira College of Engineering and Management, India

S. Guru Basava Aradhya
https://orcid.org/0000-0002-3335-2681
Chandigarh Group of Colleges, Jhanjeri, India

T. C. Manjunath
https://orcid.org/0000-0003-2545-9160
Rajarajeswari College of Engineering, India

ABSTRACT

Rooted in the idea that machines could mimic cognitive functions associated with the human mind. Today, AI encompasses a variety of technologies and methodol-

DOI: 10.4018/979-8-3693-9846-3.ch002

ogies that empower process information, recognize patterns, and make decisions autonomously or semi-autonomously. language translation, or game playing. These systems operate under a limited set of constraints and excel in their designated areas, often outperforming humans in terms of speed and accuracy. Examples of narrow, its development could revolutionize industries and societal functions. As deep learning architectures continue to evolve, they increasingly enable . Recent advancements in NLP, particularly with transformer models like GPT and BERT, have improved the ability of machines to grasp context and generate coherent text, further blurring the lines between human and machine communication.Computer vision, another critical area of AI, involves teaching machines video analysis, computer vision enables applications ranging from surveillance systems to medical imaging diagnostics.

INTRODUCTION TO ARTIFICIAL INTELLIGENCE (AI): CONCEPTS, TYPES, AND TECHNIQUES

Rooted in the idea that machines could mimic cognitive functions associated with the human mind. Today, AI encompasses a variety of technologies and methodologies that empower process information, recognize patterns, and make decisions autonomously or semi-autonomously. language translation, or game playing. These systems operate under a limited set of constraints and excel in their designated areas, often outperforming humans in terms of speed and accuracy. Examples of narrow, its development could revolutionize industries and societal functions. As deep learning architectures continue to evolve, they increasingly enable. Recent advancements in NLP, particularly with transformer models like GPT and BERT, have improved the ability of machines to grasp context and generate coherent text, further blurring the lines between human and machine communication. Computer vision, another critical area of AI, involves teaching machines video analysis, computer vision enables applications ranging from surveillance systems to medical imaging diagnostics. By employing deep learning techniques, accurately identify and classify objects within images. Moreover, significant and the future of work. From narrow applications that enhance everyday conveniences to the pursuit of general AI that seeks to replicate human cognition, the field continues to grow and evolve. As we navigate the challenges and opportunities presented by AI, it is essential to foster an inclusive dialogue that considers ethical implications and societal impact, ensuring that the advancements in AI benefit humanity how we shape its development will ultimately determine its role in society.

AI Adoption and Implementation: Challenges, Opportunities, and Best Practices

Adoption implementations represent a pivotal shift in how organizations operate across various sectors. While AI offers, improving decision-making, and driving innovation, its integration is not without challenges. Organizations seeking to leverage AI must navigate a complex landscape that includes technological, cultural, and ethical considerations. availability of qualified professionals who can develop, implement, and manage these systems. This skills gap is particularly evident in fields such as data science, machine learning (Acemoglu, et al.,2024)., and AI ethics. Organizations may struggle to find and retain talent, leading to increased competition and rising salaries. To mitigate e that attracts diverse candidates. and maintaining high-quality data. Inaccurate, incomplete. Implementing data management best practices, such as regular audits and the use of data augmentation techniques, can enhance the quality of datasets, enabling more reliable AI models. modern AI solutions. Integrating AI technologies often requires significant investments in infrastructure and a rethinking of workflows. This transition can meet ease the transition.

Ethical considerations are also critical when adopting AI. Concerns about bias, accountability, and transparency are paramount, especially as AI systems. Biased algorithms can perpetuate discrimination, leading to serious ethical and legal repercussions. Organizations must prioritize by implementing rigorous testing procedures and using diverse datasets. Additionally, establishing ethical guidelines and oversight mechanisms can help organizations standards. Organizations that successfully adopt AI can achieve significant operational efficiencies, enhance customer experiences, and drive innovation. For instance, ultimately leading to increased profitability. To capitalize on these opportunities, organizations should adopt best practices in AI implementation. A clear AI strategy is essential for guiding the adoption process. By establishing a structured roadmap, organizations can better allocate resources and measure progress throughout the implementation journey (Acemoglu, et al., 2022). Collaboration is another critical best practice for successful AI adoption. Engaging and end-users can lead to more effective AI solutions. This collaborative approach ensures that diverse perspectives are considered, resulting in systems that address real business challenges. and industry consortia—can provide organizations with access to cutting-edge tools and knowledge. Moreover, organizations should prioritize continuous learning and improvement throughout the AI lifecycle. Regularly monitoring AI system performance against established KPIs allows organizations to identify areas for enhancement and adapt to changing business needs. Implementing feedback loops, where user experiences inform ongoing development, can also drive continuous improvement (Mahadev Madgule, et al.,2023). By remaining agile and responsive, organizations can maximize the value derived from their AI investments In summary, the adoption complex interplay and opportunities. While organizations face hurdles related to talent acquisition, data quality, integration, ranging from increased efficiency to enhanced innovation—are substantial. By embracing best practices such as strategic planning, collaboration, and continuous improvement, organizations can navigate the AI landscape more effectively. Ultimately, successful AI adoption will not only drive

Ethics and Responsible AI: Ensuring Transparency, Accountability, and Fairness

As AI systems become integral to decision-making processes, ensuring transparency, accountability, and fairness is essential. These principles form the foundation of responsible AI, guiding the development and Transparency in AI refers to the clarity with which AI systems operate and make decisions.(Acemoglu, et al.,2011) To foster transparency, organizations should prioritize explainability—developing Techniques such as interpretable machine learning, which offers human-readable explanations for model predictions, can enhance understanding and trust. Additionally, organizations should communicate openly about the data used to train AI systems, the potential biases inherent in those datasets, and the limitations of the models themselves. Accountability is another cornerstone of ethical AI. As AI systems take on greater roles in decision-making (Sugumaran K. M, D., et al.,2024), the question of who is responsible for their outcomes becomes increasingly important. Missteps in AI deployment can result in significant harm, from biased hiring practices to incorrect medical diagnoses. To establish accountability, organizations must implement gover (Babina, Tania, et al.,2024). Furthermore, organizations should adopt robust auditing mechanisms to evaluate AI systems regularly, identifying and mitigating potential biases and inaccuracies before they impact users. Fairness is, which assess and adjust for unfair biases, and using diverse datasets that accurately represent the population. Additionally, organizations can engage with diverse stakeholder groups throughout the AI development process, ensuring that multiple perspectives are considered and that the resulting systems serve the broader community.

The ethical implications of AI extend beyond technical considerations; they also encompass the societal impact of these technologies. As AI systems are deployed in areas such as surveillance, law enforcement, and credit scoring, concerns about privacy and individual rights become paramount. Organizations. Ethical AI practices should involve obtaining informed consent from users, where applicable, and providing mechanisms for users Moreover, fostering organizations is essential for responsible AI development. This involves training employees at all levels on the ethical dilemmas. Organizations can create ethics training programs that equip employees with the knowledge and tools to recognize and address ethical issues in

AI projects. By cultivating an ethical mindset, organizations can better navigate the complexities of AI deployment and promote responsible practices throughout the AI lifecycle. Collaboration across sectors is vital in shaping ethical AI standards and practices. Policymakers, industry leaders, researchers, and civil society must come together to establish guidelines. Computer vision, another critical area of AI, involves teaching machines video analysis, computer vision enables applications ranging from surveillance systems to medical imaging diagnostics. By employing deep learning techniques, accurately identify and classify objects within images. Moreover, significant and the future of work. From narrow applications that enhance everyday conveniences to the pursuit of general AI that seeks to replicate human cognition, the field continues to grow and evolve. As we navigate the challenges and opportunities presented by AI, it is essential to foster an inclusive dialogue that considers ethical implications and societal impact, ensuring that the advancements in AI benefit humanity how we shape its development will ultimately determine its role in society. This collaborative approach can help create a shared understanding of ethical principles, leading to the development of best practices that promote d establishing frameworks that guide ethical AI development globally. In conclusion, as AI continues to permeate various aspects of society, the importance of ethics and responsible, establishing robust governance frameworks, and actively working to eliminate bias, organizations can foster trust and ensure that AI systems serve the common good.

AI in Medical Imaging and Diagnosis: Image Analysis, Detection, and Classification

AI has rapidly transformed the field of medical imaging and diagnosis, introducing advanced techniques for image analysis, detection, and classification that enhance the accuracy and efficiency of medical assessments. In recent years, the integration of artificial intelligence (AI) into radiology and pathology has gained traction, driven by the need for improved diagnostic precision and the ability to manage vast imaging data. utilize often imperceptible to the human eye. assist radiologists in detecting abnormalities (Bai, et al.,2022). For instance, AI algorithms have shown remarkable success in identifying early signs of diseases such as cancer, pneumonia, and fractures. By analyzing thousands of images, these systems can distinguish between normal and abnormal findings with high sensitivity and specificity. This classification. Automated classification systems can categorize images into various (Vijayakumar G, et al., 2024). This classification is particularly valuable in oncological diagnostics, where timely and accurate assessment is critical for patient outcomes. Moreover, AI can assist in risk stratification, helping clinicians determine the best course of treatment based on the classification results. Enhance image quality through techniques such as image reconstruction and denoising. AI algorithms can improve the resolution of images, reduce artifacts, and correct for various distortions, thereby providing clearer and more accurate visualizations. This enhancement is vital in ensuring that the images analyzed by healthcare professionals convey the most precise information possible, ultimately leading to better patient care.AI is also making strides in the realm of personalized medicine. By integrating imaging data with other patient in optimizing resource allocation within healthcare system algorithms (Sugumaran K. M, D., et al.,2024). Variability in imaging techniques, equipment, and patient demographics can affect the generalizability of AI models. Furthermore, there are ethical considerations regarding data privacy and the potential for bias in algorithmic decision-making. alike. Regulatory approval and validation of AI tools are essential for their integration into clinical practice. Organizations working to establish frameworks for the evaluation of AI-based diagnostic tools. Rigorous testing and validation processes will ensure that these technologies meet the safety and efficacy standards required for clinical use. AI researchers is vital to advancing the field. Interdisciplinary teams can work together to develop robust AI models that address specific clinical needs and integrate seamlessly into existing workflows(Babina, et al.,2024). in clinical settings. In conclusion, AI is revolutionizing medical imaging and diagnosis through advanced image analysis, detection, and classification methods. Its potential to improve diagnostic accuracy, enhance image quality, and support personalized medicine is reshaping the landscape of healthcare.

AI-Assisted Clinical Decision Support Systems: Predictive Analytics and Personalized Medicine

AI-assisted clinical decision support systems (CDSS) are revolutionizing healthcare by integrating predictive analytics and personalized medicine to enhance patient care. These. By utilizing machine learning algorithms and advanced analytics, enabling more informed clinical decisions. or hospital readmissions. These predictive models help healthcare providers identify at-risk patients earlier, facilitating timely interventions that can significantly improve outcomes. By prioritizing patients based on their risk profiles, healthcare systems can optimize resource allocation and provide targeted care. Personalized medicine, another pivotal aspect of AI-assisted CDSS, aims to customize healthcare lifestyle factors, and environmental influences most appropriate therapies, thereby maximizing treatment efficacy allowing them to focus on patient interaction and care while still leveraging the latest evidence-based practices. Furthermore, AI-driven CDSS (Caliskan, et al., 2017). As healthcare data evolves and new treatment protocols emerge, these systems can refine their algorithms to stay current with the latest evidence. This adaptability ensures that clinicians receive up-to-date recommendations that reflect the most recent advancements in medical research and practice. Despite the promising advancements in AI-assisted CDSS, several challenges remain. Data t confidentiality while harnessing the benefits of AI. Moreover, underscoring the need for rigorous data management and quality assurance processes. (Cavallo, et al.,2018) Another challenge is the acceptance of AI-driven systems among healthcare providers. There may be resistance to adopting new technologies, particularly if clinicians feel that these systems undermine their clinical judgment or decision-making authority. AI systems will help clinicians leverage these tools confidently and effectively. In summary, AI-assisted clinical decision support systems are transforming the landscape of healthcare through predictive analytics and personalized medicine. By providing evidence-based recommendations tailored to individual patients, these systems enhance the decision-making process and improve patient outcomes (Paramasivam Selvakumar, et al.,2024). CDSS in clinical practice. The future of healthcare will likely be characterized by a more data-driven, personalized approach, where AI serves as a powerful ally for clinicians in their pursuit of optimal patient care.

AI-Driven Patient Engagement and Outcome Optimization: Chatbots, Virtual Assistants, and Wearables

AI-driven patient engagement and outcome optimization are transforming healthcare by harnessing technologies such as chatbots, virtual assistants, and wearables to enhance communication, streamline conditions. Chatbots have medication reminders. By utilizing natural language processing, these chatbots can interact with patients in a conversational manner, making healthcare more accessible and less intimidating. For instance, have medication quire about symptoms or treatment options without the need for direct interaction with a healthcare professional, which can be particularly beneficial for those who may feel hesitant to seek help. This immediacy Virtual assistants, similar to recommendations. For example, virtual assistants can alert patients when they miss a dose of medication or remind them of upcoming appointments. They can also analyze collected data wearable technology has further revolutionized patient engagement by enabling continuous health monitoring. For instance, wearable devices can monitor heart rate and alert both patients and doctors to potential issues, such as arrhythmias or significant changes in physical activity. This proactive monitoring facilitates early intervention, conditions. fosters a collaborative approach between patients and providers. By enabling seamless communication, patients can provide feedback on their health status, share concerns, and report any side effects from treatments. This two-way communication enhances the provider's (Danielsson, et al.,2024), helping providers understand population health trends and improve care delivery strategies. are significant concerns, as sensitive health information is collected and shared through these technologies.

Additionally, by wearables and chatbots must be assured, as misleading information could lead to incorrect assumptions and potential harm. Another challenge lies in ensuring equitable access to improve healthcare access, disparities in technology adoption may exacerbate existing health inequities. Patients from lower socioeconomic backgrounds or those in rural areas may have limited access to the necessary devices or internet connectivity.

Therefore, it is essential for healthcare systems to implement strategies that ensure all patients can benefit from these advancements. Education and training for both patients and healthcare providers are also vital for maximizing the effectiveness of AI-driven engagement tools. Patients need guidance on how to use wearable devices and interact with chatbots or virtual assistants to gain the most benefit. Similarly, healthcare providers should be trained to interpret data from these technologies and integrate it into their clinical practice effectively. By fostering an understanding of how to leverage these tools, both parties can enhance their collaborative efforts in managing health outcomes. In conclusion (Cazzaniga, et al., 2024), AI-driven patient engagement and outcome optimization through chatbots, virtual assistants, and wearables represent a significant shift in healthcare delivery. These technologies empower patients to take an active role in their health, provide continuous monitoring, and facilitate communication with healthcare providers. While challenges related to substantial. As the healthcare landscape

Precision Agriculture with AI: Crop Yield Prediction, Soil Analysis, and Weather Forecasting

AI technologies, agricultural stakeholders can harness vast amounts of data to refine their Crop utilize machine learning algorithms to analyze various factors affecting crop production, such as historical yield data, weather patterns, soil conditions, and crop management practices. By incorporating these diverse datasets. Healthy soil is foundational to productive agriculture, yet traditional soil testing methods can be time-consuming and costly (Ramakrishnan T. et al., 2022) By understanding the unique characteristics of their soil, farmers can improve crop performance and reduce the environmental impact of fertilizers and pesticides. Weather forecasting is another critical component of precision agriculture, as climatic conditions have a direct impact on agricultural productivity. Traditional weather forecasting methods can sometimes be generalized and may not accurately reflect localized conditions on a farm. AI enhances weather forecasting by environmental factors to produce hyper-localized forecasts. This capability allows farmers to anticipate weather events, such as droughts, heavy rainfall, or frost, and adapt their practices accordingly (Eisfeldt, Andrea L., et al., 2023). For example, an accurate prediction of an impending drought can prompt farmers to adjust irrigation schedules proactively, thereby conserving water and protecting crops. Similarly, accurate forecasts of extreme weather can help farmers safeguard their investments by informing them when to take protective measures, such as covering crops or adjusting harvest schedules. environmental footprint and enhance biodiversity. For instance, AI algorithms can help identify the precise amount of fertilizer required for different crops, minimizing excess application that can lead to soil degradation The collection and analysis of agricultural encourage the adoption of AI-driven solutions.(Eisen, Norman, et al.,2023)Moreover, Interdisciplinary partnerships can facilitate knowledge exchange and innovation, leading to the development of tailored AI solutions that meet the specific needs of diverse agricultural contexts. Engaging farmers in the design and testing of these technologies ensures that the solutions are practical and user-friendly, ultimately promoting adoption and effective utilization. In conclusion, the integration of technology in farming not only represents a significant advancement in agricultural productivity but also embodies a commitment to sustainability and responsible stewardship of the environment.

AI-Enabled Livestock Monitoring and Management: Health, Behavior, and Nutrition Analysis

Reshaping the agricultural sector by providing innovative solutions for health, behavior, and nutrition analysis. As the global demand for animal protein rises, efficient and sustainable livestock management becomes increasingly vital. AI technologies facilitate real-time monitoring and analysist practices. Health monitoring. Traditional methods of animal health assessment often rely on visual inspections and routine veterinary checks, which can be time-consuming and may miss early signs of illness. AI-driven systems, equipped with sensors and wearable devices, can continuously monitor vital signs, activity levels, and behavioral patterns Machine learning algorithms analyze this data to identify patterns that may signal illness or distress. Early detection of health problems allows for prompt intervention, reducing the risk of disease spread and minimizing veterinary costs. Behavior analysis is another critical aspect of AI-enabled livestock management. Understanding animal behavior is essential for optimizing welfare and productivity. AI systems can analyze behavioral data, such as feeding patterns, social interactions, and environmental responses, to provide insights into livestock well-being. For instance, deviations in normal behavior, such as reduced feeding or increased restlessness, can indicate stress or discomfort. By monitoring these behavioral changes, farmers can implement strategies to improve animal welfare, such as adjusting housing conditions, providing enrichment, or addressing social dynamics within herds.

Figure 1. Artificial Intelligence (AI) in healthcare

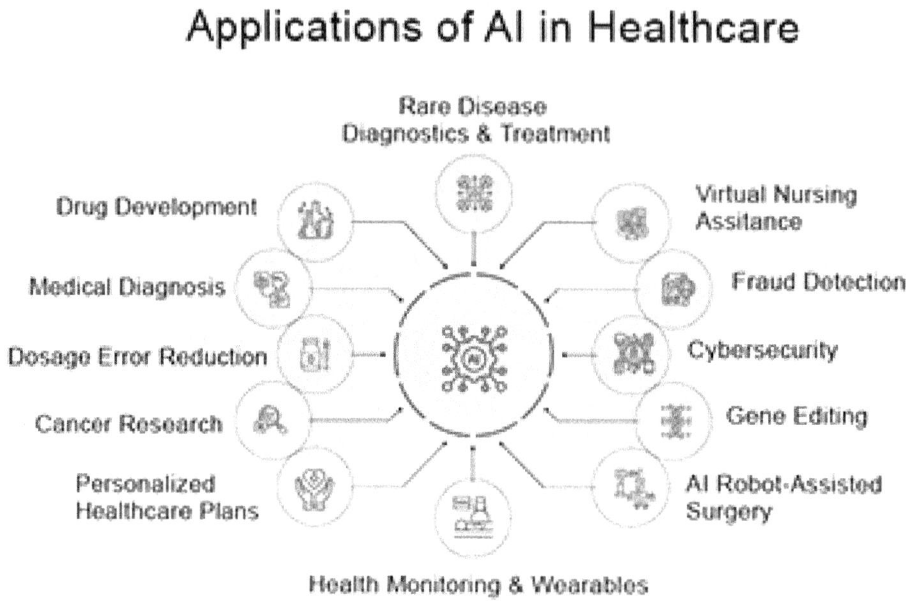

Improved welfare not only enhances the quality of life for livestock but also correlates with better productivity and meat quality. yet traditional feeding practices may not always meet the specific dietary needs of individual animals or groups. AI systems can analyze data on animal weight, age, and health status, along with the nutritional content of available feed. By employing algorithms that consider these variables, farmers can formulate personalized feeding plans that optimize nutrient intake while minimizing waste, (Frey et al., 2017). For example, precision feeding systems can deliver the right amount of, which leads to more efficient feed conversion and improved growth rates. Moreover, AI can assist in predicting future nutritional requirements based on growth trajectories and production goals. on feed intake and animal performance, AI systems can forecast the nutritional needs of livestock at various stages of life. This capability allows farmers to plan feed resources more effectively, reducing costs and ensuring that animals receive adequate nutrition throughout their growth cycles., farmers can optimize breeding strategies, improve herd management practices, and enhance overall operational efficiency. These insights not only support immediate decision-making but also contribute to long-term strategic planning in livestock production. While the benefits of AI-enabled livestock monitoring and management are significant, there are challenges to consider. (Selvakumar, P.,2023) Data privacy and security are critical issues, as sensitive information related to animal health and farm operations is collected and analyzed. Ensuring that data is protected against breaches and misuse is essential for maintaining trust among farmers and stakeholders. Additionally, the interoperating poses a challenge. For effective integration, it is crucial that various AI tools and platforms can communicate and share data seamlessly. Training and education are also vital for the successful adoption of AI technologies in livestock management. Farmers and farmworkers need to within the agricultural sector. Another consideration is the financial investment required for implementing AI-driven systems. While the long-term benefits of improved health and productivity can outweigh initial costs, many farmers may hesitate to invest without clear evidence of return on investment. Demonstrating the tangible benefits of AI technologies through case studies and pilot programs can help build confidence among farmers and encourage adoption.

AI-Driven Farm Automation and Robotics: Autonomous Farming, Harvesting, and Irrigation

AI-driven farm automation and robotics are transforming the agricultural landscape, ushering in an era of autonomous farming, efficient harvesting, and optimized irrigation practices becomes paramount. AI technologies, integrated with robotics and automation, enable farmers to manage resources more effectively, reduce labor costs, and improve crop yields, all while resource scarcity. Autonomous farming agricultural practices (Ganguli, et al., 2022). Through the use of AI-powered machinery, farms can operate with minimal human intervention. Autonomous sophisticated sensors and machine learning algorithms can navigate fields, perform various tasks, and collect data on crop health and soil conditions. For instance, autonomous tractors can plow, seed, and fertilize crops based on real-time data analysis, ensuring that each operation is executed with precision. This level of automation reduces traditional farming methods, allowing farmers to focus on higher-level decision-making and strategic planning. farm automation is in harvesting. Computer vision, another critical area of AI, involves teaching machines video analysis, computer vision enables applications ranging from surveillance systems to medical imaging diagnostics. By employing deep learning techniques, accurately identify and classify objects within images. Moreover, significant and the future of work. From narrow applications that enhance everyday conveniences to the pursuit of general AI that seeks to replicate human cognition, the field continues to grow and evolve. As we navigate the challenges and opportunities presented by AI, it is essential to foster an inclusive dialogue that considers ethical implications and societal impact, ensuring that the advancements in AI benefit humanity how we shape its development will ultimately determine its role in society. Robotics equipped with AI capabilities can identify ripe crops and selectively harvest them, minimizing waste and maximizing efficiency.

For example, robotic harvesters can analyze the color, size, and firmness of fruits and vegetables using computer vision, determining the optimal time for harvest.

This precision not only ensures that crops are picked at their peak ripeness but also reduces the likelihood of damage during the harvesting process. As a result, farmers can increase their yields and enhance the quality of their produce, meeting the growing demands of consumers Water management is a critical aspect of sustainable agriculture, especially in regions facing water scarcity. By optimizing irrigation practices, farmers can not only conserve water resources but also improve crop resilience to drought and other climate-related challenges. can gather extensive data on crop health, soil conditions, and pest infestations. This information is then processed using machine learning algorithms to generate actionable insights for farmers. For instance, AI systems can identify areas of the field that require additional nutrients or pest control measures, enabling farmers to apply treatments precisely where needed. Despite the myriad benefits of AI-driven farm automation, several challenges must be addressed to facilitate widespread adoption. One major barrier is the initial investment required for advanced technologies. Many farmers, particularly those operating on smaller scales, justify purchasing autonomous equipment and robotics. Providing financial incentives, subsidies, or access to affordable financing options can help alleviate this challenge and encourage Additionally, the successful implementation of AI in farm automation requires education and training for farmers and agricultural workers. Understanding how to operate and maintain advanced technologies, interpret data outputs, and integrate AI insights into daily practices is vital for maximizing the benefits of automation. there is a need for collaboration among various stakeholders in the agricultural sector, including farmers, technology developers, researchers, and policymakers. By fostering partnerships and facilitating knowledge exchange, and testing ensures that solutions are practical, diverse agricultural contexts. In conclusion, AI-driven farm automation and robotics represent a significant leap forward in productivity, reduce labor costs, and implement more sustainable practices (Giattino, Charlie, et al., 2023). While challenges related to investment, data security, and education remain, the potential benefits of these innovations are profound. As the agricultural sector faces increasing pressures to meet.

AI in Risk Management and Fraud Detection: Predictive Modeling and Anomaly Detection

Revolutionizing how organizations identify, assess, and mitigate potential threats to their operations. . As the complexity of financial systems and the sophistication of fraud schemes continue to evolve, traditional risk management methods often fall short. AI technologies provide innovative solutions proactively to emerging threats, thereby safeguarding their assets and reputations. Predictive modeling is a key. By evaluating a myriad of factors—including credit history, income levels, and spending behavior—AI systems can predict the likelihood of default, allowing lenders to make informed decisions. This proactive approach not only minimizes financial risk but also enhances customer experience by ensuring that loans are granted to those most likely to succeed. In the realm of insurance, predictive modeling can significantly improve underwriting processes. Insurers can analyze vast datasets to determine risk factors associated with specific policies, allowing them to tailor premiums accordingly. This not only helps in accurately pricing policies but also reduces the incidence of fraudulent claim (Horowitz, et al.,2018) another vital aspect of AI in risk management, focuses on identifying unusual patterns or behaviors. Traditional detection methods often rely on static rules that can be easily circumvented by sophisticated fraudsters.

In contrast, AI algorithms utilize unusual spending spikes or transactions originating from atypical locations. By detecting these anomalies early, organizations can take swift action to investigate and mitigate potential fraud before significant losses occur.AI-driven systems to monitor transactions and user behavior. By analyzing customer purchase patterns and browsing histories, these systems can identify fraudulent activities, such as account takeovers or unauthorized purchases. This proactive detection not only protects the business from losses but also enhances customer trust and loyalty by providing a safer shopping environment. Furthermore, in risk management is not limited to financial fraud; it extends to various sectors, including healthcare and supply chain management. In healthcare, AI algorithms can analyze patient data to detect anomalies that may indicate fraudulent billing practices or unnecessary procedures. In supply chain management, AI can identify irregularities in logistics data that may suggest theft or corruption. This broad applicability highlights the versatility of AI in enhancing risk management across different industries. Despite the advantages of AI in risk management and fraud detection. One significant concern is the quality and availability of data. AI models rely heavily on high-quality, comprehensive datasets to produce accurate fraud detection. adversarial attacks on AI systems. Fraudsters may employ sophisticated tactics to evade detection, necessitating continuous updates and adaptations of AI algorithms. Organizations must implement robust security measures and regularly refine their models to stay ahead of emerging threats. This dynamic approach requires collaboration between data scientists, cybersecurity experts, and business stakeholders to create a comprehensive risk management strategy. Additionally, prioritized. Organizations need to ensure compliance with regulations, such as GDPR, which govern data usage and protection (Huang, et al.,2024). Building trust with customers is essential; therefore, organizations should be transparent about how they collect and utilize data for risk management purposes (Zack, et al., 2023). Engaging stakeholders in discussions about ethical AI practices will foster a more responsible approach to deploying these technologies. Training and skill development within organizations are also crucial for the effective integration of AI in risk management. As businesses continue to navigate increasingly complex environments, embracing AI-driven risk management strategies will be essential for ensuring resilience and long-term success. The integration of advanced technologies in risk management not only fosters a proactive approach to identifying threats but also contributes to a more secure and trustworthy business landscape, ultimately benefiting stakeholders and consumers alike.

AI-Driven Portfolio Optimization and Investment Analysis: Machine Learning and Natural Language Processing

AI-driven portfolio optimization and investment analysis are reshaping the financial landscape by integrating traditional investment strategies often fall short in capturing the nuances of market dynamics. AI technologies provide innovative solutions that enable investors to analyze vast datasets, uncover hidden patterns, and develop more sophisticated investment strategies. At the core of AI-driven portfolio optimization is the application of machine learning algorithms (Klare, Brendan F., et al.,2012). By leveraging techniques(Wheeler, et al.,2023) such as regression analysis, clustering, and reinforcement learning, investors can create models that predict asset performance based on various factors, including market conditions, economic indicators, and historical price movements. For instance, ML algorithms can analyze stock price trends alongside macroeconomic variables to determine how different factors influence market behavior. This predictive capability empowers investors to make informed decisions about asset allocation, ultimately enhancing portfolio performance. By employing AI-driven systems, investors can optimize their portfolios proactively, reducing exposure to risk and capitalizing on emerging opportunities. Traditional financial analysis often relies on structured data, which can limit the scope of insights. NLP techniques, however, can process vast amounts of text, identifying sentiment, key themes, and potential market-moving events. For example, sentiment analysis can gauge market sentiment surrounding specific stocks or sectors, providing investors with a qualitative understanding of public perception. By incorporating this information into their investment strategies, prevailing market sentiment.

Moreover, NLP can enhance risk assessment and management. Discussions, AI systems can detect early signals of potential risks or crises, allowing investors to adjust their portfolios accordingly. For instance, if a negative news story emerges about a company, NLP algorithms can quickly analyze the sentiment and volume of discussions surrounding that event, providing insights into the potential impact on the company's stock price. This real-time monitoring capability enables investors to react swiftly to market developments, mitigating potential losses and safeguarding their investments.AI-driven portfolio optimization also facilitates AI systems can tailor investment portfolios to meet specific needs. For instance, robo-advisors utilize machine learning algorithms to recommend diversified portfolios based on an investor's risk profile, time horizon, and investment objectives. This personalized approach democratizes access to sophisticated investment strategies, allowing individuals to benefit from AI-driven insights that were once the domain of institutional investors. Despite the numerous advantages of AI in portfolio optimization and investment analysis, several challenges remain. Data quality management practices to ensure the reliability of their models. This includes sourcing high-quality financial data, market information, and alternative data sources that provide additional insights into asset performance, (Kogan, et al.,2023). By enhancing model transparency, investors can better understand the factors influencing their investment decisions and feel more confident in the AI-dri). strategies they adopt. Regulatory significant role implementation of AI in finance. is essential to foster trust and credibility in AI-driven investment strategies. Moreover, training and skill development within financial organizations are crucial for the successful integration of AI technologies. Investment professionals AI-driven portfolio optimization and investment analysis are to enhance decision-making and improve investment outcomes. These technologies enable investors to analyze vast datasets, adapt to changing market conditions, and extract insights from unstructured data sources, ultimately leading to more informed and personalized investment strategies (Webb, Michael et al.,2019). While challenges related to data quality, interpretability, paving the way for a more inclusive and efficient financial ecosystem.

AI-Enabled Customer Service and Chatbots in Banking and Finance

AI-enabled customer service and chatbots are transforming the banking and finance sectors by providing efficient, personalized, and round-the-clock support to customers. As the financial landscape becomes increasingly digital, traditional customer service models face challenges in meeting the evolving expectations of consumers who seek quick, accessible, and seamless interactions. This capability allows banks and financial institutions to handle a wide range of customer queries—from balance inquiries and transaction histories to loan applications and investment advice—without the need for human intervention. By automating routine inquiries, chatbots reduce wait times that require personalized attention. chatbots can provide support 24/7, catering to customers' needs at any time. This is particularly valuable for global banking operations, where customers may be located in different time zones. By ensuring that assistance is always available, banks can enhance customer satisfaction and loyalty, as clients appreciate the convenience of accessing services whenever they need them. Personalization is another significant benefit of AI in customer service. Chatbots can leverage recommendations. For example, if a customer frequently inquires about credit card offers, the chatbot can proactively suggest tailored credit products based on the individual's spending habits and financial goals (Lambrecht, et al., 2018). This level of personalization not only enhances the customer experience but also fosters deeper relationships between financial institutions and their clients. Moreover, AI-driven chatbots can analyze customer by employing sentiment analysis techniques, chatbots can gauge whether a customer is satisfied, frustrated, or confused during an interaction. This information can be invaluable for banks, allowing them to identify potential issues in their services concerns. Fraud detection and prevention is another critical application of AI in banking customer service.

Figure 2. Application of AI in agriculture

Chatbots can be programmed transactions in flag suspicious activities based on established patterns. For instance, if a customer makes a transaction that deviates significantly from their typical spending behavior, the chatbot can alert the customer and initiate a verification process. knowing that their financial institutions are vigilant about safeguarding their assets. The on high-impact initiatives that drive business growth. However, the implementation of AI-enabled customer service and chatbots is not without challenges. One major concern is ensuring the accuracy and reliability of chatbot responses. Poorly designed chatbots that fail to understand customer inquiries or provide inaccurate information can lead to frustration and dissatisfaction. To mitigate this risk, banks must invest in robust training and ongoing maintenance of their AI systems, ensuring that chatbots are continuously updated with the latest information and capable of handling a wide range of queries. .Another challenge lies human Data privacy and security are also paramount considerations when implementing Furthermore, training and educating staff to work alongside AI technologies is crucial for successful integration. Employees need to understand how to utilize chatbot insights effectively and handle situations where human intervention is required. This requires ongoing training programs that emphasize the collaborative nature of AI.

AI in Education: Personalized Learning, Intelligent Tutoring Systems, and Automated Grading

AI in education is transforming the learning landscape through significantly enhancing educational outcomes for students across various levels. As traditional educational methods struggle to address the diverse needs of learners, AI technologies offer innovative solutions that adapt to individual learning styles, pacing, and preferences. This evolution is crucial in an increasingly digital world, where the demand for customized education continues to rise. Personalized learning, a cornerstone of AI in education, tailors educational experiences to meet the unique needs of each student. Through data collection and analysis, AI systems can assess a student's strengths, weaknesses, and learning preferences. For instance, adaptive learning platforms utilize algorithms to track students' progress in real-time, providing personalized content and resources that align with their current skill levels (Van Noorden, et al.,2023). This individualized approach understanding. ITS can engage students in interactive dialogues, offering hints, explanations, and encouragement tailored to their responses. For example, in a mathematics tutoring system, or examples until the student achieves mastery. This immediate feedback mechanism not only supports learning but also fosters a growth mindset, encouraging students to persevere through challenges. Moreover, AI-driven intelligent tutoring systems can scale to serve a larger number of students than traditional one-on-one tutoring. By providing support to students who may not have access to personalized instruction due to geographical or financial constraints, ITS democratizes education, making high-quality learning resources available to a broader audience, (McElheran et al.,2024). This accessibility is particularly important in underserved communities, where educational inequalities often persist. Automated grading subjective, especially in subjects requiring detailed feedback, such as writing or essay-based assessments, (Varian, et al.,2019). AI-powered grading systems can analyze student submissions, providing objective assessments based on established criteria. Natural language processing algorithms can evaluate the coherence, structure, and grammar of written work, delivering consistent and unbiased feedback to students.

This not only reduces the workload for educators but also allows for quicker turnaround times, by highlighting these patterns, educators to identify areas where additional support is needed, ultimately enhancing teaching effectiveness and student learning (O'Shaughnessy, et al.,2022). is the quality and bias of the algorithms used in personalized learning and grading. If the underlying data reflects societal biases, there is a risk that AI systems may perpetuate these biases, leading to unequal educational opportunities. Therefore, it is crucial for students. Data privacy and security also Transparency about data usage is essential for building trust with students and parents, fostering a collaborative environment where stakeholders feel confident in the adoption of AI technologies. AI tools into their classrooms. This includes understanding how to interpret AI-generated insights, adapt instructional strategies based on data, and leverage In conclusion, offering innovative solutions that cater to individual student needs and enhance the overall learning experience.

AI in Transportation: Autonomous Vehicles, Route Optimization, and Traffic Management

AI in transportation is, significantly enhancing efficiency, safety, and user experience. As urban populations grow and the demand for mobility increases, traditional transportation systems face unprecedented challenges. AI technologies provide innovative solutions that address these challenges by optimizing resources, improving traffic flow, and paving the way for a new era of autonomous mobility. Moreover, the introduction of AVs promises to alleviate congestion in urban areas. AI systems can coordinate vehicle movements, optimizing traffic patterns and communicating with each other and with traffic infrastructure, AVs can share data about road conditions, accidents, and traffic signals, leading to improved decision-making. This level of connectivity not only enhances safety but also promotes more efficient use of road networks, ultimately reducing travel times and emissions. Route optimization impacts. Advanced algorithms analyze a plethora of data points, including traffic patterns, weather conditions, and road closures, to determine the most efficient routes for vehicles. This capability is particularly valuable for logistics and delivery services, where timely and efficient routing can lead to significant cost savings. For example, companies like UPS and FedEx utilize AI-driven route optimization to minimize fuel consumption and maximize delivery efficiency. By predicting traffic flow and adapting routes in real time, these companies can enhance their operational efficiency and improve customer satisfaction. In public transportation, AI-powered route optimization can enhance service delivery by predicting passenger demand and adjusting schedules accordingly. Such optimization not only I Traffic management is another significant application of AI in transportation. Smart traffic management systems utilize AI to monitor traffic conditions, analyze congestion patterns, and control traffic signals dynamically. This intelligent approach can lead to reduced travel times, lower emissions, and enhanced overall road safety (Obermeyer, et al., 2021). Furthermorees must be addressed to fully realize its benefits requires significant investment in infrastructure. Many existing systems may not be equipped to handle the demands of AI technologies, necessitating upgrades to road networks, communication systems, and traffic management infrastructures. Public and private sectors must collaborate to secure the funding and resources needed for these improvements machine learning, and system management. In conclusion, AI in transportation is paving the way for a future characterized by autonomous vehicles, optimized routing, and intelligent traffic management. These technologies promise to enhance safety, efficiency, and user experience while addressing the challenges of urban congestion and environmental sustainability. While there are challenges related to safety, data privacy, infrastructure investment, and workforce training, the potential benefits of AI in transforming transportation systems are profound. As stakeholders continue to collaborate and innovate, the integration of AI into transportation will play a critical role in shaping the future of mobility.

AI in Manufacturing: Predictive Maintenance, Quality Control, and Supply Chain Optimization

AI in manufacturing is revolutionizing the industry through innovations in predictive maintenance, quality control, and supply chain optimization, significantly enhancing efficiency, productivity, and overall operational effectiveness. As manufacturers face increasing competition and the demand for higher-quality products, AI technologies offer powerful solutions that streamline. In contrast, AI-driven predictive maintenance utilizes sensors and IoT devices to collect data on machinery health, including temperature, vibration, and operational metrics. For example, in a production line for automotive components, Additionally, AI can facilitate continuous improvement in quality control by analyzing historical data to identify trends and root causes of defects. Machine learning algorithms can sift through large datasets from past production runs to uncover patterns that may not be immediately evident to human analysts. Supply chain optimization is another area where AI technologies are driving significant improvements. In today's globalized economy, supply chains are complex and often subject to disruptions due to factors such as demand fluctuations, geopolitical events, and natural disasters. By ensuring that the right amount of product is produced and available at the right time, companies can minimize excess inventory costs and reduce the risk of stockouts. Furthermore, AI can enhance supplier selection and management. By evaluating supplier performance data, 2023) Another significant advantage of AI in supply chain optimization is its ability to enhance logistics and distribution. Machine learning algorithms can analyze traffic patterns, weather conditions, and transportation costs to optimize delivery routes and schedules. For instance, a logistics company could use AI to determine the most efficient route for transporting goods from a manufacturing facility to distribution centers, reducing transportation costs.Despite the many advantages of AI in manufacturing, several challenges must be addressed to fully harness its potential. Data quality is a critical factor in the success of AI initiatives. Accurate and comprehensive.Moreover, the implementation of AI in manufacturing requires a cultural shift within organizations. Many manufacturing environments operate on legacy systems that may not be compatible with modern ensuring minimal disruption to operations. This may involve upgrading infrastructure, adopting cloud-based solutions, and implementing interoperability standards to facilitate data sharing across systems. Finally, ethical considerations related to AI deployment in manufacturing must be taken into account, (Trammell, et al.,2023).

Implementing AI Solutions: Change Management, Talent Acquisition, and Infrastructure Development

Implementing AI solutions in organizations is a multifaceted endeavor that requires strategic focus on change management, talent acquisition, and infrastructure development. As AI technologies continue to advance and permeate various industries, organizations face the imperative to integrate these solutions effectively to enhance operational efficiency, drive innovation, and maintain competitive advantage. However, the transition to an AI-driven environment is not simply a matter of deploying technology. Change management is a critical component of AI implementation. The introduction of AI technologies often disrupts existing workflows, processes, and job roles, leading to resistance among employees who may fear obsolescence or inadequacy in adapting to new systems. To facilitate a smooth transition, organizations must foster a culture of openness and adaptability. This begins with transparent communication about the purpose and benefits of AI integration (Rammer, et al.,2022), highlighting how these technologies AI initiatives, encouraging input and addressing concerns. Creating a shared vision for the future of the organization.Training and support are essential elements of change management. Organizations must invest in comprehensive training programs that equip employees with the necessary skills to work effectively alongside AI technologies. This includes not only technical training on how to use AI tools but also fostering a mindset of continuous learning and adaptation, organizations can enhance employee engagement and commitment, ultimately leading to a more successful AI implementation. Talent acquisition is another crucial aspect of implementing AI solutions, organizations should look for candidates with a blend of technical expertise and domain knowledge. fostering partnerships with educational institutions can help organizations tap into emerging talent and ensure a pipeline of skilled professionals ready to meet the demands of an AI-driven landscape.(Restrepo, et al.,2023) Organizations should implement strategies to ensure inclusivity in their hiring processes, actively seeking candidates from different backgrounds and disciplines. This not only enriches the talent pool but also enhances the organization's ability to develop AI solutions that are equitable and relevant to a broader audience. Infrastructure development is a foundational requirement for successful AI implementation. Organizations must assess their current technological infrastructure and identify areas that need enhancement to support AI initiatives effectively.

This includes investing in robust data management systems, cloud computing capabilities, and advanced analytics tools. High-quality data is essential for training effective AI models; therefore, organizations should establish processes for data collection, cleaning, Collaboration between IT and business units is vital in developing an effective infrastructure for AI. By working together, these teams can ensure that technological solutions align with business goals and address specific operational challenges. Establishing cross-functional teams that include data engineers, software developers, business analysts, and domain experts can enhance communication and collaboration, leading to more effective AI implementations. This collaborative approach can also help identify potential roadblocks early in the process and allow for timely adjustments. Finally, organizations should ensure that they operate transparently and without bias. Developing a framework for ethical AI use can guide organizations in implementing responsible AI solutions. This involves establishing clear guidelines for data usage, algorithmic fairness.

Future of AI: Emerging Trends, Opportunities, and Challenges in Industry Adoption

Transform industries across the globe, driven by emerging trends, burgeoning opportunities, and significant challenges in adoption. As organizations increasingly recognize the potential of AI to enhance operational efficiency, drive innovation, and improve customer experiences, the technology's integration into various sectors is accelerating. This rapid evolution is shaped by I applications that extend far beyond traditional boundaries., which involves algorithms capable of creating content, from text and images to music and video as marketing, entertainment, and education, where personalized content creation can enhance engagement and streamline workflows. Companies are increasingly leveraging generative AI for tasks such as ad copywriting, video production, and even software development, enabling faster turnaround times and reduced costs (Senhadji, et al., 2021). However, the rise of generative AI also raises ethical concerns regarding copyright, misinformation, and the authenticity of content, necessitating careful consideration and regulation. Another significant trend is the growing integration of AI with the Internet of Things (IoT), which creates a synergy that enhances data collection and analysis. As more devices become connected, the amount of data generated is astronomical. AI algorithms can process while AI systems analyze this data to predict maintenance needs and optimize production schedules. The use of AI in automation is another area gaining momentum. Robotic process automation (RPA) combined with AI capabilities allows organizations to automate repetitive tasks that were once labor-intensive.(Shabsigh, et al., 2023) This trend is prevalent in sectors such as finance, where AI systems can process transactions, analyze data, and even manage customer inquiries with minimal human intervention. As businesses look to reduce costs and enhance productivity, the adoption of AI-driven automation is expected to expand, enabling employees AI's role in enhancing decision-making processes is also becoming increasingly significant. With advanced analytics and predictive modeling, AI systems can provide organizations with insights that drive strategic decisions. For example, in healthcare, AI can analyze patient data to identify trends, predict outcomes, and recommend treatment plans tailored to individual needs. In finance, as organizations harness the power of AI to enhance decision-making, the demand for data literacy and analytical skills among employees will rise, prompting a shift in educational and training paradigms Additionally, addressing the ethical implications of AI is crucial for its sustainable adoption. Issues such as algorithmic bias, discrimination, and lack of accountability can undermine the benefits of AI technologies. As AI systems are trained on historical data, they may inadvertently perpetuate existing biases present in that data, leading to unfair outcomes. Organizations must implement rigorous ethical guidelines and frameworks to ensure that AI is developed and deployed responsibly. This includes conducting bias assessments, involving diverse stakeholders in the development process, and establishing accountability mechanisms for AI decisions. Another significant challenge is the skills gap in the workforce.

CONCLUSION

The finance, and beyond illustrate its transformative potential, fundamentally reshaping how organizations operate and deliver value. Predictive analytics can identify at-risk patients, while AI-driven imaging technologies assist in early disease detection, fostering proactive rather than reactive care. In agriculture, AI optimizes crop yields through precision farming, leveraging data from sensors and drones to monitor soil health, weather conditions, and crop performance. This results in sustainable practices that increase productivity while minimizing resource waste, ultimately addressing food security challenges. In finance, AI revolutionizes predict market trends and consumer behaviors, enabling more informed decision-making and tailored financial products. Automated trading systems enhance efficiency in capital markets, while AI-powered chatbots provide customer support, streamlining operations and improving user engagement. Beyond these sectors, industries such as manufacturing benefit from AI-driven predictive maintenance and quality control, which reduce downtime and enhance product consistency. Retailers use AI to optimize inventory management and personalize shopping experiences, driving customer satisfaction and loyalty. As these applications demonstrate, the integration of the journey toward widespread AI adoption is not without challenges. Issues surrounding data privacy, ethical considerations, and the need for a skilled workforce must be addressed to fully realize AI's potential. Organizations must prioritize transparent practices, invest in employee training, and implement robust governance. In conclusion, the future of AI in healthcare, agriculture, finance, and other industries is marked by immense promise and responsibility. By embracing these innovations while navigating the associated complexities, organizations can not only enhance their operational capabilities but also contribute to a more sustainable and equitable future. The ongoing collaboration among industry stakeholders, policymakers, and educators will be crucial in shaping an AI-enabled landscape that prioritizes ethical standards and human-centered values. Ultimately, the successful application of AI across industries will depend on a balanced approach that harmonizes technological advancement with the needs and aspirations of society, paving the way for a transformative impact that benefits all. The rapid advancement of AI technologies has outpaced the development of skills needed to implement and manage these systems effectively. Organizations face difficulties in finding qualified professionals with expertise in AI, machine learning, and data analytics. To bridge this gap, a concerted effort is needed from vocational training, and continuous learning initiatives. By investing in upskilling and reskilling programs, organizations can cultivate a workforce that is equipped to harness the full potential of AI. Infrastructure development is also critical for supporting AI adoption. Many organizations, particularly small and medium-sized enterprises (SMEs), may lack the necessary infrastructure to implement AI solutions effectively. Cloud computing, high-speed internet access, and robust data management.

REFERENCES

Abhilasha, A., Sreenivasulu, A., Manimozhi, T., Kumar, P. S., Selvakumar, P., & Singh, P. (2022, April). The Model of Smart Sensing Device For Sensitive Nano-clusters Modification in Sensing Properties. In 2022 2nd International Conference on Advance Computing and Innovative Technologies in Engineering (ICACITE) (pp. 1043-1047). IEEE.

Acemoglu, D. (2024). Harms of AI. In Bullock, J. B., Chen, Y.-C., Himmelreich, J., Hudson, V. M., Korinek, A., Young, M. M., & Zhang, B. (Eds.), *The Oxford Handbook of AI Governance*. Oxford University Press. DOI: 10.1093/oxfordhb/9780197579329.013.65

Acemoglu, D., Anderson, G. W., Beede, D. N., Buffington, C., Childress, E. E., Dinlersoz, E., Foster, L. S., Goldschlag, N., Haltiwanger, J. C., Kroff, Z., Restrepo, P., & Zolas, N. (2022). "Automation and the Workforce: A Firm-Level View from the 2019 Annual Business Survey". In: Technology, Productivity,and Economic Growth. Ed. by Susantu Basu, Lucy Eldridge, John C. Haltiwanger, and Erich Strassner. Studies in Income and Wealth. University of Chicago Press. DOI: 10.3386/w30659

Acemoglu, D., & Autor, D. (2011). "Chapter 12 – Skills, Tasks and Technologies: Implications for Employment and Earnings". In: Handbook of Labor Economics. Ed. by David Card and Orley Ashenfelter. Vol. 4B. Elsevier, pp. 1043–1171.

Babina, T., Fedyk, A., He, A., & Hodson, J. (2024). Artificial Intelligence, Firm Growth, and Product Innovation. *Journal of Financial Economics, 151*, 151. DOI: 10.1016/j.jfineco.2023.103745

Babina, T., Fedyk, A., He, A. X., & Hodson, J. (2024). Firm Investments in Artificial Intelligence Technologies and Changes in Workforce Composition. In Basu, S., Eldridge, L., Haltiwanger, J. C., & Strassner, E. (Eds.), *Technology, Productivity, and Economic Growth*. National Bureau of Economic Research.

Bai, Y., ... (2022). Constitutional AI: Harmlessness from AI Feedback. Preprint. arXiv. arXiv: 2212.08073.

Caliskan, A., Bryson, J. J., & Narayanan, A. (2017). Semantics Derived Automatically from Language Corpora Contain Human-Like Biases. *Science, 356*(6334), 183–186. DOI: 10.1126/science.aal4230 PMID: 28408601

Cavallo, A. (2018). More Amazon Effects: Online Competition and Pricing Behaviors. Working Paper 25138. National Bureau of Economic Research.

Cazzaniga, M., Jaumotte, F., Li, L., Melina, G., Panton, A. J., Pizzinelli, C., Rockall, E., & Tavares, M. M. (2024). *Gen-AI: Artificial Intelligence and the Future of Work. Staff Discussion Note 2024/001*. International Monetary Fund. DOI: 10.5089/9798400262548.006

Danielsson, J., & Uthemann, A. (2024). On the Use of Artificial Intelligence in Financial Regulations and the Impact on Financial Stability. Working Paper 4604628. SSRN.

Eisen, N., Lee, N. T., Galliher, C., & Katz, J. (2023). AI Can Strengthen U.s.Democracy—and Weaken It. Brookings. url: https://www.brookings.edu/articles/ai-can-strengthen-u-s- democracy-and-weaken-it/

Eisfeldt, A. L., Schubert, G., & Miao, B. Z. (2023). Generative AI and Firm Values. Working Paper 31222. National Bureau of Economic Research.

Ellingrud, K., Sanghvi, S., Dandona, G. S., Madgavkar, A., Chui, M., White, O., & Hasebe, P. (2023). *Generative AI and the Future of Work in America*. McKinsey Global Institute.

Frey, C. B., & Osborne, M. A. (2017). The Future of Employment: How Susceptible AreJobs to Computerisation? *Technological Forecasting and Social Change*, *114*, 254–280. DOI: 10.1016/j.techfore.2016.08.019

Ganguli, D.. (2022). "Predictability and Surprise in Large Generative Models". In: *Proceedings of the 2022 ACM Conference on Fairness, Accountability, and Transparency*. Association for Computing Machinery, pp. 1747–1764. DOI: 10.1145/3531146.3533229

Giattino, C., Mathieu, E., Samborska, V., & Roser, M. (2023). Artificial Intelligence. Our World in Data. url: https://ourworldindata.org/artificial-intelligence. Horowitz, Michael C. (2018). "Artificial Intelligence, International Competition, and the Balance of Power". Texas National Security Review 1(3), pp. 36–57.

Huang, Y. (2024). *"Is the Impact of AI Different from That of IT?" New Technologies, Digitalization, and AI: The Future Is Here. IMF Research Perspectives 26*. Klare, Brendan F., Mark J. Burge, Joshua C. Klontz, Richard W. Vorder Bruegge, and Anil.

Jain, K. (2012). Face Recognition Performance: Role of Demographic Information. *IEEE Transactions on Information Forensics and Security*, *7*(6), 1789–1801. DOI: 10.1109/TIFS.2012.2214212

Kanipriya, M., Sugumaran, D., Pandiyarajan, V., Yogananda, B. S., Selvakumar, P., & Mohan, S. (2024, January). Analysis of Electromyography Signals for Control of Mechanical Prosthesis using Machine Learning Techniques. In 2024 2nd International Conference on Intelligent Data Communication Technologies and Internet of Things (IDCIoT) (pp. 1059-1067). IEEE.

Kogan, L., Papanikolaou, D., Schmidt, L. D. W., & Seegmiller, B. (2023). Technology and Labor Displacement: Evidence from Linking Patents with Worker-Level Data. Working Paper 31846. National Bureau of Economic Research.

Lambrecht, A., & Tucker, C. E. (2018). Algorithmic Bias? An Empirical Study into Apparent Gender-Based Discrimination in the Display of STEM Career Ads. Working Paper 2852260. SSRN. Landemore, Hélène (2023). "Fostering More Inclusive Democracy with AI". Finance &amp. *Development*, *60*(4).

Mahadev Madgule, N. (2023). Mechanical properties and microstructure of activated TIG welded similar joints of Inconel alloys by desirability approaches. *Materials Today: Proceedings*, *77*, 528–533. DOI: 10.1016/j.matpr.2022.12.250

McElheran, K., Li, J. F., Brynjolfsson, E., Kroff, Z., Dinlersoz, E., Foster, L., & Zolas, N. (2024). AI adoption in America: Who, what, and where. *Journal of Economics & Management Strategy*, *33*(2), 375–415.

O'Shaughnessy, M. (2022). One of the Biggest Problems in Regulating AI Is Agreeing on a Definition. Carnegie Endowment. url: https://carnegieendowment.org/2022/10/06/one-of-biggest- problems-in-regulating-ai-is-agreeing-on-definition-pub-88100.

Obermeyer, Z., Nissan, R., Stern, M., Eanef, S., Bembeneck, E. J., & Mullainathan, S. (2021). *Algorithmic Bias Playbook*. The Center for Applied Artificial Intelligence at Chicago Booth.

Pizzinelli, C., Panton, A. J., Ms, M. M. T., Cazzaniga, M., & Li, L. (2023). Labor Market Exposure to AI: Cross-country Differences and Distributional Implications. Working Paper 2023/216. International Monetary Fund.

Ramakrishnan, T., Mohan Gift, M. D., Chitradevi, S., Jegan, R., & Subha Hency Jose, P. Nagaraja, Rajneesh Sharma H.N., Selvakumar P., Sintayehu Mekuria Hailegiorgis, (2022). "Study of Numerous Resins Used in Polymer Matrix Composite Materials" Advances in Materials Science and Engineering, vol. 2022, Article ID 1088926, 8 pages, 2022. DOI: 10.1155/2022/1088926

Rammer, C., Fernández, G. P., & Czarnitzki, D. (2022). Artificial Intelligence and Industrial Innovation: Evidence from German Firm-Level Data. *Research Policy*, *51*(7), 104555. DOI: 10.1016/j.respol.2022.104555

Restrepo, P. (2023). Automation: Theory, Evidence, and Outlook. Working Paper 31910. National Bureau of Economic Research.

Selvakumar, P. (2023). Biomedical applications of polymers in biosensors, cancer vaccines and drug delivery systems. *Biopolimery i Kletka*, *39*(4), 245–256. DOI: 10.7124/bc.000AA0

Selvakumar, P., Mariappan, P., Shunmugam, S., Paul, R., & Jabez, S. (2024).. . *Zeitschrift für Physikalische Chemie*, *2024*. Advance online publication. DOI: 10.1515/zpch-2023-0450

Selvakumar, P., Palanisamy, S. K., Cinthaikinian, S., Palanisamy, V., Mariappan, R., & Selvakumar, P. K. (2024).. . *Zeitschrift für Physikalische Chemie*, *2024*. Advance online publication. DOI: 10.1515/zpch-2023-0406

Senhadji, A., Tieman, A., Gemayel, E., & Benedek, D. (2021). *A Post-Pandemic Assessment of the Sustainable Development Goals. Staff Discussion Note 2021/003*. International Monetary Fund.

Shabsigh, G., & Boukherouaa, E. B. (2023). *Generative Artificial Intelligence in Finance: Risk Considerations. Fintech Note 2023/006*. International Monetary Fund.

Trammell, P., & Korinek, A. (2023). Economic Growth under Transformative AI. Working Paper 31815. National Bureau of Economic Research.

Van Noorden, R., & Perkel, J. M. (2023). AI and Science: What 1,600 Researchers Think. *Nature*, *621*(7980), 672–675. DOI: 10.1038/d41586-023-02980-0 PMID: 37758894

Varian, H. (2019). Artificial Intelligence, Economics, and Industrial Organization. In Agrawal, A., Gans, J., & Goldfarb, A. (Eds.), *The Economics of Artificial Intelligence: An Agenda*. University of Chicago Press. DOI: 10.7208/chicago/9780226613475.003.0016

Webb, M. (2019). The Impact of Artificial Intelligence on the Labor Market. Working Paper 3482150. SSRN.

Zack, T., Lehman, E., Suzgun, M., Rodriguez, J. A., Celi, L. A., Gichoya, J., Ju-rafsky, D., Szolovits, P., Bates, D. W., Abdulnour, R.-E. E., Butte, A. J., & Alsentzer, E. (2023). Coding Inequity: Assessing GPT-4's Potential for Perpetuating Racial and Gender Biases in Healthcare. Preprint. *medRxiv*. DOI: 10.1101/2023.07.13.23292577

Chapter 3
Blockchain Applications for Transparent Fund Management, Donor Accountability, and Intellectual Property Rights in Digital Libraries

Rachit Agarwal
https://orcid.org/0000-0003-0164-5292
Chandigarh University, India

Tanya Kumar
https://orcid.org/0000-0001-7608-9820
Chandigarh Business School of Administration, Chandigarh Group of Colleges, Landran, India

Preeti Sharma
Chandigarh Business School of Administration, Chandigarh Group of Colleges, Landran, India

ABSTRACT

Blockchain technology has emerged as a transformative tool with diverse applications across various sectors, including libraries, where it promises to revolutionize fund management, donor accountability, and intellectual property rights enforcement. This book chapter explores integrating blockchain solutions to enhance transparency, accountability, and rights management within digital libraries. The chapter

DOI: 10.4018/979-8-3693-9846-3.ch003

begins by addressing the challenges faced in traditional fund management and donor accountability practices within libraries. Case studies and examples illustrate successful implementations of blockchain in library settings, highlighting tangible benefits such as increased transparency, enhanced donor trust, and improved rights protection for content creators. Moreover, the chapter explores emerging trends and future directions in blockchain applications for libraries, including interoperability with other technologies, scalability improvements, and regulatory frameworks for blockchain-based DRM solutions.

INTRODUCTION

Blockchain technology, first designed as the foundation of cryptocurrencies, has transformed into a multifaceted and disruptive entity with extensive ramifications across several sectors. Its intrinsic attributes, including decentralization, immutability, and transparency, render it especially adept at tackling significant issues encountered by diverse industries, such as digital libraries (Radovancovici et al., 2022). This research study examines the utilization of blockchain technology in digital libraries, emphasizing its capacity to improve transparency in financial management, strengthen donor responsibility, and protect intellectual property rights. Blockchain technology originated from the necessity for a safe and transparent framework to enable transactions in digital currencies, especially Bitcoin (Rouhani et al., 2021). Blockchain fundamentally constitutes a distributed ledger technology that documents transactions in a tamper-resistant and transparent manner (Khan et al., 2022). Every block in the chain encompasses a compilation of transactions and is interconnected with the preceding block, establishing a chronological and immutable ledger (Beyene et al., 2022). The decentralized structure of this technology guarantees that no one entity possesses control over the entire ledger, hence diminishing the danger of fraud and manipulation. The triumph of blockchain in the financial sector has generated curiosity about its prospective uses in alternative domains (Vaz et al., 2023). Digital libraries, which oversee extensive collections of digital assets and depend on the integrity of their records, might greatly benefit from the distinctive characteristics of blockchain technology. The subsequent sections delineate the principal domains in which blockchain technology might transform digital libraries: financial administration, donor transparency, and intellectual property rights. A fundamental problem encountered by digital libraries is the handling of finances, which frequently entails intricate procedures and numerous parties (Sun & Wang, 2023). Conventional fund management methods may lack transparency, hindering stakeholders' ability to monitor the allocation and expenditure of funds. The absence of transparency may result in inefficiencies, mismanagement, and diminished confidence among contributors and consumers. Blockchain technology can resolve these concerns by offering a transparent and immutable ledger of financial transactions. Every transaction documented on a blockchain is time-stamped and accessible to all network members, guaranteeing that fund administration procedures are transparent and verifiable. Smart contracts are self-executing agreements with provisions encoded in software, enabling the automation and enforcement of fund payments contingent upon predetermined circumstances. This automation diminishes administrative burden and mitigates the likelihood of human mistake. Blockchain's transparent money management system might provide several advantages for digital libraries (Shivsharan et al., 2023). This would allow donors and stakeholders to monitor money distribution in real-time, guaranteeing that donations are utilized efficiently and as intended. The irreversible nature of blockchain records might deter fraud and misuse of funds, ultimately augmenting the legitimacy and reliability of digital libraries. Donor accountability is an essential component of money management for digital libraries. Donors frequently seek guarantees that their gifts are utilized effectively and that the library's financial processes are robust. Conventional reporting and accountability systems might be unwieldy and may not provide the immediacy and transparency desired by funders. Blockchain technology can markedly improve donor accountability by offering a transparent and immutable record of all transactions and activities associated with donations. Donors can employ a blockchain-based system to get real-time data on the utilization of their funds, monitor the development of supported initiatives, and confirm that their contributions have been distributed as pledged. This enhanced openness cultivates trust and confidence among contributors, potentially resulting in heightened involvement and larger donations. Furthermore, the decentralized structure of blockchain guarantees that no individual entity may influence or modify the data. This intrinsic security feature offers an extra degree of confidence for donors, ensuring the accuracy and reliability of the provided information. The administration of intellectual property (IP) rights is a significant domain where blockchain technology may provide considerable enhancements (Caldarelli & Ellul, 2021). Digital libraries frequently oversee a diverse collection of digital assets, encompassing books, research papers, and multimedia material, all of which are governed by intellectual property regulations (Ekpenyong & Owan, 2022). Safeguarding and effectively administering these rights is a multifaceted and sometimes arduous endeavor (Azbeg et al., 2022). Blockchain technology offers a formidable solution for intellectual property rights management by establishing a secure and transparent framework for documenting ownership and usage rights (Sun & Wang, 2023). Every digital asset may be associated with a distinct digital identification on the blockchain, accompanied by information that specifies its ownership and usage conditions (Angara & Saripalle, 2022). This digital record is unalterable and readily accessible for verification by all stakeholders, therefore diminishing the danger of unlawful use or infringement (Jha, 2023). Moreover, blockchain can enable the automation of licensing agreements using smart contracts (Ajibade & Mutula, 2021). These contracts can be configured for automated execution upon the fulfillment of specific criteria, such as royalty payments or compliance with usage terms. This automation optimizes the management of intellectual property rights and guarantees equitable compensation for artists' effort (Beyene et al., 2022). The use of blockchain technology into digital libraries signifies a substantial improvement in the management of money, the assurance of donor responsibility, and the safeguarding of intellectual property rights inside these institutions (Saeed et al., 2024). Although the technology is very novel, its capacity to revolutionize digital libraries is substantial (Vaz et al., 2023).

LITERATURE REVIEW

Panda & Kaur (2023) described the usage of blockchain technology in library as well as information centres. Tawornittayakun & Leelasantitham (2020) developed library information storage system using blockchain technology that can help to solve problems related to data security through management of data collection programs. Blockchain technology can promote digital interactions through transparent and secured mechanism (Lengoatha & Seymour, 2020). Shahzad et al. (2024) identified the effects of blockchain technology on librarians of university in handling library services. Hu (2022) discussed the problems faced by smart libraries in resource collection for storing data as well as disseminating information using blockchain technology. Digital libraries can provide a secure and immutable record of intellectual property rights by documenting ownership and license information on a blockchain (Rouhani et al., 2021). Every digital asset may be associated with a distinct digital identification on the blockchain, accompanied by information that specifies its ownership and usage conditions (Vaz et al., 2023). This solution facilitates straightforward verification of ownership and usage rights, hence minimizing the danger of infringement and illegal access (Beyene et al., 2022). Big data analytics can be used for managing cyber libraries (Wijesekara et al., 2024). Salama et al. (2023) examined the blockchain technology in terms of cybersecurity. (Xiao et al., 2024) thrown a light upon usage of digital copyrights in digital libraries which help the economy to offer digital resources. Bashir & Warraich (2023) discussed upon library and information centre as how they operate and improve their inter-communication by maintaining better privacy in real world. Ekpenyong & Owan (2022) discussed the role of digital research library resources using blockchain technology. Hota et al. (2022) stated upon blockchain mechanism in digital library system. Zhao et al. (2023) discussed upon usage of grid engineering in library. Vubangsi & Al Turjman

(2022) discussed upon usage of blockchain and artificial intelligence techniques for storing real-time database. Lata & Owan (2022) stated the major differences among traditional and moder libraries. Smart libraries can be made by applying centralised storage method by storing digital twin data in their intelligent libraries (Sun & Wang, 2023). Distributed ledger technology led to data management, protection and storage of data (Beyene et al., 2022). Blockchain technology can be used for accessing the digital resources in the libraries (Khan et al., 2022). Azbeg et al. (2022) thrown a light upon IoT and blockchain for digital library management. Jha (2023) proposed the usage of blockchain technology in libraries for maintaining privacy and identification. Authors called it as dependable and fresh technology for transforming library sciences. Vaz et al. (2023) stated the evolution of third generation blockchain technology by making use of networking for promoting education sector. Caldarelli & Ellul (2021) discussed upon the application of blockchain technology in financial sector and how it can have great impact on improvement of transcript and digital libraries. Rouhani et al. (2021) performed detailed analysis by taking into consideration various data-base as per attributes.

TRANSPARENT FUND MANAGEMENT AND DONOR ACCOUNTABILITY IN DIGITAL LIBRARIES

Digital libraries have emerged as essential repository of information, containing a wide range of digital assets, including scholarly publications and multimedia content. As these organizations broaden their activities and adopt new digital frameworks, they encounter considerable problems in conventional money management and donor accountability mcthods (Radovancovici et al., 2022). Challenges include opacity in cash allocation, insufficient traceability in donor monitoring, and challenges in maintaining intellectual property rights underscore the necessity for new solutions (Rawat et al., 2023). This chapter examines how blockchain technology might resolve these issues by offering transparent and immutable transaction records, automating administrative tasks via smart contracts, and improving the administration of intellectual property rights (Beyene et al., 2022). Conventional finance management strategies in libraries sometimes exhibit insufficient openness (Azbeg et al., 2022). The methods for fund allocation can be complex, including several stakeholders such as funders, library administrators, and project managers. This intricacy may result in inefficiencies and ambiguity about the utilization of funds (Caldarelli & Ellul, 2021). Donors may have difficulties in monitoring the effects of their gifts, resulting in diminished trust and possible disengagement (Jha, 2023). Monitoring donations from their source to their ultimate application is a notable difficulty. In traditional systems, donor records are often dispersed over many databases and handwritten logs, complicating the provision of a holistic and real-time perspective on fund utilization (Rouhani et al., 2021). The absence of traceability may lead to anomalies and challenges in reporting to donors, undermining the integrity and accountability of the library's financial procedures (Xiao et al., 2024). Digital libraries oversee an extensive array of digital assets, each potentially governed by distinct intellectual property (IP) regulations and licensing contracts. Effectively managing and enforcing these rights may be a complicated endeavour. Conventional IP management systems sometimes lack the requisite openness to accurately ascertain ownership and usage rights, resulting in possible conflicts and illegitimate exploitation of digital information (Shivsharan et al., 2023). Blockchain technology provides a decentralized ledger system capable of transforming money administration in digital libraries (Vaz et al., 2023).

In the context of fund management, blockchain technology may provide an unalterable record of all financial transactions. Every transaction is documented in a "block" and connected to preceding transactions, creating a "chain" of records (Azbeg et al., 2022). This method guarantees that once a transaction is documented, it cannot be modified or erased, hence offering a dependable and tamper-resistant record of fund distribution. In digital libraries, this implies that each gift may be monitored from the point of receipt until its ultimate application. Donors may obtain real-time data about the utilization of their gifts, therefore augmenting transparency and trust (Sun & Wang, 2023; Panda & Kaur, 2023). Moreover, the permanence of blockchain records aids in the prevention of fraud and mismanagement, guaranteeing that money is utilized as intended (Radovancovici et al., 2022). Smart contracts are a fundamental aspect of blockchain technology that may markedly improve the efficiency of fund management procedures (Lata & Owan, 2022). This automation optimizes the money allocation procedure, ensuring effective distribution of funds in accordance with the established terms (Shivsharan et al., 2023). Moreover, smart contracts can provide instantaneous reporting and auditing. All transactions and contract executions are documented on the blockchain, enabling auditors to obtain a comprehensive and transparent record of fund disbursements. This openness streamlines the auditing process and offers donors and stakeholders a lucid perspective on the utilization of money (Jadhawar, 2024).

The administration of intellectual property rights is another domain where blockchain technology can exert a significant influence. Digital libraries frequently manage a variety of digital content, each accompanied by distinct intellectual property rights and license agreements (Caldarelli & Ellul, 2021). Conventional approaches to manage these rights can be unwieldy and susceptible to inaccuracies, resulting in conflicts and improper material usage. Blockchain technology can provide a more efficient and transparent method for administering intellectual property rights (Adhikari, 2021). Furthermore, the openness of blockchain guarantees that all participants can access identical information, so promoting confidence and clarity in intellectual property management (Jha, 2023). Smart contracts may significantly improve intellectual property rights management by automating licensing agreements and royalty disbursements (Khan et al., 2022). A smart contract might be designed to autonomously disburse royalties to artists upon the access or use of their work (Maidment, 2021; Rawat et al., 2023). This automation streamlines the management of intellectual property rights and guarantees equitable compensation for authors' efforts (Oyelude, 2022).

SAFEGUARDING INTELLECTUAL PROPERTY RIGHTS

Intellectual property (IP) rights are essential for the safeguarding and administration of digital information in libraries. As digital libraries gain prominence in the transmission and preservation of scholarly and creative works, the necessity for efficient and transparent intellectual property management has become paramount (Radovancovici et al., 2022). Conventional intellectual property management systems frequently encounter challenges related to authenticity, enforcement, and effective tracking, rendering producers susceptible to illegal utilization and possible infringement (Khan et al., 2022). Blockchain technology presents a revolutionary solution by delivering a secure, transparent, and immutable framework for the management and enforcement of intellectual property rights (Awalu et al., 2019). This chapter examines the utilization of blockchain for the establishment and enforcement of intellectual property rights in digital libraries, investigating its potential to transform digital rights management (DRM), guarantee equitable remuneration for creators, and safeguard against unauthorized distribution and plagiarism (Sun & Wang, 2023). Digital libraries contain a vast collection of digital information, such as e-books, research articles, multimedia files, and more resources. It is essential to verify that digital information is genuine and accurately credited to its originator (Kasahara et al., 2024). Nevertheless, conventional systems sometimes lack effective means for confirming the origin of digital assets, resulting in problems of misattribution and illicit use (Aurangzeb et al., 2024). Creators anticipate just remuneration for their creations (Vaz et al., 2023). Conventional DRM methods may be ineffective in monitoring the utilization of digital material and guaranteeing that artists receive adequate royalties and acknowledgment (Shivsharan et al., 2023). Digital work is readily repeatable and distributable, hence heightening the danger of unlawful dissemination and plagiarism (Ducrée et al., 2020). Contemporary DRM solutions may insufficiently safeguard against these concerns (He et al., 2022). Overseeing license agreements and usage conditions for digital material may be onerous, particularly when engaging with several parties and diverse rights.

FUTURE OF BLOCKCHAIN TECHNOLOGY IN DIGITAL LIBRARIES

Blockchain technology, characterized by its decentralized and immutable ledger, provides an effective answer to these difficulties. Utilizing blockchain, digital libraries can improve their administration of intellectual property rights. Blockchain offers a secure and immutable record of all transactions and modifications (Khan et al., 2023). Every block in the chain incorporates a hash of the preceding block, establishing an uninterrupted and immutable sequence of records (Jha, 2023). This guarantees that ownership, usage rights, and license agreements are documented in an immutable and tamper-proof way (Hota et al., 2022). In contrast to conventional centralized systems, blockchain functions on a decentralized network of nodes (Caldarelli & Ellul, 2021). This decentralization ensures that no single entity possesses control over the whole system, so diminishing the potential for manipulation or fraud and augmenting the overall security of intellectual property management (Tawornittayakun & Leelasantitham, 2020). Smart contracts are autonomous agreements with the stipulations explicitly encoded. They autonomously implement the stipulations of agreements, including licensing requirements and royalty disbursements, without the necessity for middlemen (Lata & Owan, 2022). This automation optimizes the management of intellectual property rights and guarantees compliance with agreements (Beyene et al., 2022). The transparency of blockchain enables all participants to access identical information (Salama et al., 2023). This function is essential for validating ownership, monitoring usage, and ensuring that all stakeholders are aware of the rules and restrictions related to digital property (Gupta, 2022). Blockchain-based DRM solutions utilize these attributes to tackle the issues of intellectual property management in digital libraries (Khan et al., 2022). Blockchain can document digital signatures and certifications linked to each digital item. These signatures authenticate the material and validate the creator's identity (Vaz et al., 2023). Every transaction, encompassing the generation and alteration of digital material, is documented on the blockchain, establishing a transparent and verifiable record of ownership and provenance (Lund et al., 2019). Every digital item can be allocated a distinct identification, such as a digital fingerprint or hash, which is associated with the blockchain ledger (Shivsharan et al., 2023). This identification facilitates the monitoring of the asset's provenance and any ensuing modifications or applications (Azbeg et al., 2022). Smart contracts can facilitate the automation of license agreement administration. A smart contract can be configured to grant access to digital material just upon receipt of payment or to autonomously allocate royalties to the creator according to use metrics (Radovancovici et al., 2022). This lowers administrative burdens and guarantees the constant enforcement of license requirements (Quan, 2021). Blockchain-based technologies provide real-time monitoring of content utilization. Digital libraries can track compliance with license agreements and detect illicit usage by documenting every instance of access or distribution on the blockchain. The unchangeable characteristics of blockchain records render it challenging for unauthorized individuals to modify or erase information (Cao & Yang, 2019). This function safeguards digital work from illicit dissemination and plagiarism (Sun & Wang, 2023). Blockchain can monitor the dissemination of digital material across several platforms (Dulce & Hurtado, 2021). By documenting each transaction on the blockchain, libraries can identify and rectify occurrences of illicit sharing or replication. Smart contracts can autonomously compute and allocate royalties according to established criteria (Lata & Owan, 2022). For instance, when digital material is viewed or utilized, the smart contract may determine the suitable remuneration for the creator and facilitate the transfer of cash accordingly (Wijesekara et al., 2024). The openness of blockchain enables producers to get comprehensive statistics regarding the utilization of their material and the calculation of royalties. This transparency cultivates trust and guarantees equitable compensation for creators (Ma & Li, 2020).

CASE STUDIES AND EXAMPLES

As blockchain technology advances, other sectors are investigating its possible uses to resolve enduring difficulties. Digital libraries, specifically, can gain from blockchain's potential to improve transparency, accountability, and rights management (Khan et al., 2022). This compilation of case studies and examples demonstrates effective use of blockchain technology in library environments, emphasizing concrete advantages such as heightened transparency, bolstered donor confidence, and strengthened rights protection for content authors (Mohanta et al., 2019).

The University of Melbourne has led the exploration of blockchain applications in academia and research. The institution initiated a project to incorporate blockchain technology into its research funding management system to enhance transparency and accountability (Beyene et al., 2022). The project entailed developing a blockchain-based network for monitoring research awards and funding distributions (Chiu et al., 2021). The platform documents every transaction associated with research funding, from the original contribution to the final allocation (Shivsharan et al., 2023). All transactions are transparent and unalterable, offering a distinct and verifiable record of fund utilization (Vaz et al., 2023). Stakeholders, comprising donors and researchers, possess immediate access to comprehensive information on the allocation of funding (Sun & Wang, 2023). This transparency fosters confidence and guarantees that monies are distributed as planned (Rouhani et al., 2021). Donors may monitor the effects of their gifts, so bolstering their trust in the university's financial processes and promoting more support (Paulavičius et al., 2019). The unalterable characteristics of blockchain records facilitate the auditing process, enhancing the verification of acceptable money utilization (Mohd Pozi et al., 2018). The use of blockchain technology has enhanced accountability and efficiency in the management of research money, hence augmenting the confidence and credibility of the university's financing procedures (Chalkias et al., 2024).

Blockchain technology presents a transformative paradigm for digital libraries, offering solutions to persistent challenges in information management and dissemination. Its decentralized architecture, underpinned by cryptographic principles, enables the creation of immutable and transparent records of digital assets. This is particularly salient in addressing issues of copyright infringement and intellectual property rights, where blockchain can facilitate secure and auditable transactions between creators and users, potentially reducing the estimated $31.8 billion in annual losses attributed to digital content piracy (Verma, 2025). Furthermore, the technology's capacity for enhanced provenance tracking can significantly improve the management of digital archives and special collections, ensuring the authenticity and integrity of historical documents and research data. For instance, a pilot study by the Library of Congress demonstrated the feasibility of using blockchain to manage and preserve digital materials, noting a 40% improvement in tracking efficiency compared to traditional methods. Smart contracts, integral to blockchain systems, can also automate licensing agreements and royalty payments, streamlining access to scholarly resources and potentially increasing revenue for content creators by an estimated 15-20% (Sharma & Singh, 2025).

Ujo Music is a blockchain-based platform aimed at resolving challenges associated with music rights management. The platform seeks to establish a transparent and automated mechanism for the management and enforcement of intellectual property rights within the music business (Caldarelli & Ellul, 2021; Agrawal et al., 2024). Every music track is associated with a distinct digital identification on the blockchain, accompanied by metadata that specifies ownership, license conditions, and usage rights. Smart contracts facilitate the automation of royalty disbursements and licensing arrangements (Sharma & Batth, 2020). The immutable ledger of blockchain guarantees the accuracy and verifiability of ownership and license information, hence diminishing the likelihood of disputes and improper usage (Qian, 2021). Smart contracts autonomously manage royalty disbursements according to established criteria, guaranteeing equitable remuneration for artists' creations (Lengoatha & Seymour, 2020). The platform's decentralized structure enables all stakeholders to access and authenticate rights and payment data, promoting confidence and transparency within the music business (Radovancovici et al., 2022; Aanchal & Prasad, 2021).

Myco is a blockchain platform dedicated to the management and protection of intellectual property rights for digital content providers. The platform offers a mechanism for monitoring the utilization and dissemination of creative works. Myco use blockchain technology to document ownership and license information for digital material, encompassing artwork, literature, and multimedia (Agarwal et al., 2024). The platform's decentralized ledger guarantees that all information is visible and immutable (Jha, 2023). Smart contracts provide automated licensing and payment procedures. Blockchain offers a secure and immutable record of content ownership, mitigating the risk of intellectual property theft and illegal utilization (Azbeg et al., 2022). Automated license agreements and payments optimize the process for artists and users, minimizing administrative burdens and possible conflicts (Wang et al., 2019). All transactions and rights management operations are documented on the blockchain, providing transparent and accessible information for content providers and users (Lamanna & La, 2018). Myco's technology has proven the efficacy of blockchain in safeguarding intellectual property rights and streamlining content administration, providing useful insights for digital libraries seeking to adopt analogous solutions (Pambudi et al., 2021).

A joint initiative among many worldwide digital libraries seeks to investigate the application of blockchain technology for the management of digital collections and the preservation of rare manuscripts. The project established a blockchain-based system for the cataloging and management of rare manuscripts and digital collections (Huo & Pan, 2022). Every object received a digital identification and was connected to a blockchain record that specified its provenance, condition, and access privileges (Lata & Owan, 2022). The system included smart contracts to regulate access rights and process requests for digital copies (Beyene et al., 2022). The unchangeable ledger of blockchain guarantees the precise documentation and preservation of the origin and condition of rare manuscripts (Radovancovici et al., 2022). Smart contracts provide the automation of access rights and reproduction requests, hence enhancing libraries' ability to handle and satisfy user inquiries (Masenya & Chisita, 2022). The decentralized characteristic of blockchain offers substantial protection against manipulation and unwanted entry. The initiative effectively showcased how blockchain technology may improve the administration and preservation of rare digital assets, providing a framework for other libraries aiming to adopt comparable systems (Azbeg et al., 2022).

These case studies demonstrate the revolutionary capacity of blockchain technology in resolving issues encountered by digital libraries (Zhang, 2024). Utilizing blockchain's decentralized ledger, digital libraries may attain more transparency in financial management, bolster donor confidence, and optimize the administration of intellectual property rights (Stana et al., 2021). The effective use of blockchain solutions in many situations underscores the technology's ability to enhance library operations, adhere to legal and ethical norms, and promote a more equal and sustainable environment for digital material distribution (Vaz et al., 2023). Digital libraries are increasingly adopting technology innovations, and blockchain offers a potential opportunity to improve operational efficiency and maintain the integrity of library procedures (Ramya et al., 2024).

Table 1. Comparative highlights for library applications

Feature	Ethereum	Hyperledger Fabric	Hyperledger Sawtooth
Privacy	Public, requires added layers for privacy	Strong privacy features (channels, private data)	Flexible privacy options
Scalability	Improving (Layer-2), historically a concern	High transaction throughput	Designed for scalability
Permissions	Permissionless	Permissioned	Permissioned or public
Smart Contracts	Mature (Solidity)	Chaincode (Go, Java, JavaScript)	Transaction Families (custom logic)
Community & Tools	Large, extensive tooling	Smaller but enterprise-focused	Smaller, evolving ecosystem
Complexity	Relatively easier to get started with public applications	More complex setup for permissioned networks	Moderate complexity
Use Cases in Libraries	Digital asset tracking, rights management	Patron data management, interlibrary loans	Digital archiving, community resource sharing

(Source: Authors' Compilation)

- For applications prioritizing privacy and controlled access to data (e.g., patron records, interlibrary loan systems with sensitive data): Hyperledger Fabric appears to be a strong contender due to its permissioned nature and robust privacy features.
- For applications focusing on transparency and potentially broader access to certain types of library data (e.g., digital asset catalogs, open educational resources) and leveraging a large developer ecosystem: Ethereum could be suitable, provided that appropriate privacy measures are implemented for sensitive data.
- For applications requiring a balance of flexibility in deployment and scalability (e.g., managing diverse library resources and user types): Hyperledger Sawtooth offers an interesting middle ground.

While Ethereum offers a vibrant ecosystem and smart contract capabilities, Hyperledger Fabric's focus on permissioned networks and privacy might make it particularly well-suited for many library applications involving sensitive data.

Hyperledger Sawtooth provides a flexible alternative. Libraries should carefully evaluate their specific requirements and conduct pilot projects to determine the most appropriate blockchain platform for their needs.

ETHICAL CONSIDERATIONS

This chapter part examines the ethical concerns related to blockchain implementation in libraries, emphasizing data privacy, user permission, and the environmental consequences of blockchain networks (Almadani & Hussain, 2023). By analyzing these concerns and implementing responsible deployment procedures that emphasize security, sustainability, and user empowerment (Khan et al., 2022). The transparent and immutable ledger of blockchain is a significant attribute, although it poses distinct issues for data privacy. In a blockchain network, once data is inscribed, it cannot be modified or eradicated (Prakash et al., 2022; Jha, 2023). Libraries manage a diverse range of data, encompassing user information, borrowing records, and intellectual property (Bashir & Warraich, 2022). Protecting sensitive data while utilizing the advantages of blockchain necessitates meticulous deliberation (Kumar & Kaur, 2023). The openness of blockchain may unintentionally reveal user data or sensitive information, resulting in possible privacy violations (Sun & Wang, 2023). To mitigate these issues, libraries must adopt solutions that reconcile blockchain's transparency with the necessity for data protection (Ramya et al., 2024). One method involves employing permissioned blockchains, wherein data access is confined to authorized parties only (Paganelli & Paganelli, 2022). This concept enables libraries to regulate access to data, safeguarding sensitive information while using the benefits of blockchain tcchnology. Another solution is utilizing cryptographic methods, including zero-knowledge proofs (Azbeg et al., 2022). These solutions provide transaction validation without disclosing the underlying data, therefore augmenting privacy while preserving the integrity of the blockchain (Mani & Agarwal, 2022). Furthermore, libraries must develop stringent data governance policies and guarantee that blockchain applications adhere to applicable data protection standards, including the General Data Protection Regulation (GDPR). The use of blockchain technology in libraries requires the collecting and administration of user data (Omame & Alex-Nmecha, 2021). Consequently, acquiring informed permission from consumers is an essential ethical imperative. Users must be well informed regarding the utilization, storage, and dissemination of their data inside the blockchain network (Caldarelli & Ellul, 2021). Users must get detailed information on the recording of their data on the blockchain, the degree of its visibility, and the safeguards used to ensure their privacy (Arsat et al., 2022). Libraries can employ many procedures to guarantee that user permission is acquired and preserved. One strategy is creating accessible consent forms and privacy rules that explicitly delineate the application of blockchain technology and its implications for data privacy (Kösemen et al., 2022). These papers must be composed in comprehensible English and provide consumers with the choice to opt-in or opt-out of blockchain-based data management (Beyene et al., 2022). Furthermore, libraries have to enable users to periodically examine and modify their consent options. This method guarantees that users stay informed and maintain control over their data during their interaction with the library's blockchain technologies (Janardhana & Ganesh, 2023). The ecological ramifications of blockchain networks, especially those utilizing proof-of-work (PoW) consensus processes, represent a considerable ethical issue. Proof of Work necessitates significant computing resources to authenticate transactions and safeguard the network, leading to elevated energy usage and a considerable carbon impact (Lu et al., 2023). This environmental effect may conflict with the sustainability objectives of several libraries and organizations. To reduce environmental effect, libraries might investigate blockchain networks employing more energy-efficient consensus processes, such as proof-of-stake (PoS) or delegated proof-of-stake (DPoS). These options necessitate reduced processing power and, hence, utilize less energy (Jha, 2023). Libraries ought to contemplate the use of blockchain systems engineered for sustainability, including attributes such as carbon offsetting and energy-efficient protocols. Libraries must adopt comprehensive security protocols, encompassing encryption, access restrictions, and periodic security assessments, to protect against possible threats and weaknesses (Mridha et al., 2022). Empowering users is a vital element of healthy blockchain implementation (Azbeg et al., 2022). Libraries must provide users with the knowledge and resources necessary to comprehend and navigate their engagements with blockchain technology. This encompasses supplying instructional materials regarding blockchain's impact on data privacy and user permission, in addition as delivering assistance to users with issues or inquiries (Ramya et al., 2024). Involving stakeholders, such as library personnel, users, and technology vendors, is crucial for the effective and ethical use of blockchain technology (Radovancovici et al., 2022). Libraries have to cultivate open communication with these stakeholders to address concerns, solicit comments, and guarantee that blockchain solutions are conceived and implemented in accordance with ethical norms and community values (Li & Liao, 2022).

A robust framework for ethical implementation, particularly concerning data privacy, rests on several interconnected pillars. Transparency forms the bedrock, demanding clear and accessible communication about data collection, processing, and usage practices. Accountability mechanisms must be established, defining responsibilities and ensuring avenues for redress in case of breaches or misuse. Fairness dictates that data processing should not lead to discriminatory outcomes or undue harm to individuals or groups. Security measures, encompassing both technical and organizational safeguards, are crucial to protect data integrity and confidentiality. Finally, respect for individual autonomy necessitates providing individuals with meaningful choices and control over their data, including the rights to access, rectify, and erase their information. This multi-faceted framework ensures that data privacy is not merely a compliance issue but an integral aspect of ethical technological development and deployment.

EMERGING TRENDS AND FUTURE DIRECTIONS IN BLOCKCHAIN APPLICATIONS FOR LIBRARIES

As blockchain technology advances, its potential uses in digital libraries are extending beyond basic financial management, donor responsibility, and intellectual property rights. This chapter seeks to thoroughly examine contemporary trends and future trajectories in the incorporation of blockchain technology inside library systems (Gatti et al., 2023). The emphasis will be on three principal domains: compatibility with other technologies, enhancements in scalability, and regulatory structures for blockchain-based digital rights management (DRM) systems. The chapter will also include ideas for legislators, library administrators, and stakeholders to address the ethical, legal, and technological difficulties related to blockchain deployment (Panda & Kaur, 2023). A pivotal component for the effective integration of blockchain technology in digital libraries is its capacity to function harmoniously with other developing technologies. As libraries progressively implement sophisticated digital tools and systems, the necessity for compatibility between blockchain and other technologies becomes essential (Sun & Wang, 2023). Cloud computing provides scalable and adaptable resources for the storage and management of digital information. Integrating blockchain with cloud platforms can improve the security and transparency of data storage and transactions (Shahzad et al., 2024). Smart contracts used on a blockchain may automate and authenticate transactions related to cloud-based resources, including access to digital information and the management of user rights (Azbeg et al., 2022). Artificial intelligence technologies, encompassing machine learning and natural language processing, are becoming essential to the operations of digital libraries. Blockchain can augment AI applications by offering a transparent and immutable record of data inputs and decision-making processes (Huang et al., 2022). This collaboration may enhance the precision of AI-generated recommendations, content classification, and user analytics, while maintaining data integrity and traceability (Beyene et al., 2022). The Internet of Things (IoT) facilitates real-time interaction and data sharing among smart equipment in libraries. Blockchain offers a secure and transparent approach for controlling IoT devices, guaranteeing that data produced by these devices is stored with precision and is impervious to tampering (Hu, 2022). IoT sensors might monitor the utilization and status of physical library holdings, while blockchain guarantees that this data is securely recorded and accessible (Radovancovici et al., 2022). As libraries investigate VR and AR technologies for immersive experiences and virtual collections, blockchain can facilitate the management and authentication of virtual assets. Blockchain can provide a secure ledger of ownership and usage rights for digital assets in VR/AR settings, guaranteeing equitable compensation for users and producers while safeguarding material (Hasnain et al., 2023). Scalability continues to pose a considerable barrier for blockchain technology, particularly with large-scale digital libraries with vast holdings and massive user populations. Layer 2 scaling solutions, including state channels and sidechains, provide effective methods for enhancing transaction throughput and minimizing latency. These technologies provide off-chain transactions that are subsequently reconciled on the primary blockchain, reducing congestion and enhancing performance (Sun & Wang, 2023). Addressing technical solutions for digital libraries involves implementing robust systems that ensure efficient storage, retrieval, and management of digital content. Key technical components include cloud-based infrastructure for scalable storage, metadata standards for content organization, and advanced search algorithms for quick access to information. To enhance scalability, microservices architecture and content delivery networks (CDNs) can be integrated, enabling the system to handle increasing user demand and large volumes of data without compromising performance. Additionally, implementing AI-powered tools for automated indexing, personalization, and digital preservation ensures long-term sustainability and user-centric experiences. These solutions collectively support the seamless expansion and adaptability of digital libraries in the evolving digital landscape. This implies that libraries can manage high-volume transactions, including digital content transactions and micro-payments for access, with efficiency (Vaz et al., 2023). The consensus mechanism is an essential element of blockchain networks that dictates the validation and incorporation of transactions into the blockchain (Ali et al., 2024). Novel consensus algorithms, like Proof of Stake (PoS) and Delegated Proof of Stake (DPoS), provide enhanced scalability relative to conventional Proof of Work (PoW) systems. These algorithms can diminish energy usage and enhance transaction speed, rendering them appropriate for extensive library applications. Sharding entails dividing the blockchain network into smaller, more manageable portions known as shards. Each shard independently processes transactions and smart contracts, therefore augmenting the blockchain's total capacity. Implementing sharding in blockchain-based library systems may provide expedited and more efficient transaction processing, including user registrations, content access, and fund transfers (Ojobor, 2023). With the growing integration of blockchain technology into digital libraries, it is imperative to build explicit legislative frameworks to solve legal and ethical issues of digital rights management (DRM). Conventional intellectual property rules may require modification to address the distinctive attributes of blockchain technology. This include resolving matters pertaining to digital ownership, copyright enforcement, and licensing agreements (Jha, 2023, Caldarelli & Ellul, 2021). Regulatory frameworks must delineate the management of personal and sensitive data on the blockchain, ensuring adherence to data protection standards, particularly the General Data Protection Regulation (GDPR) (Chalkias et al., 2024). Libraries must guarantee that blockchain solutions comply with these rules while upholding openness and accountability. The implementation of blockchain technology in libraries must address ethical concerns, including equal access to digital resources, the risk of misuse, and the effects on vulnerable people (Vubangsi & Al Turjman, 2022). Regulatory frameworks must incorporate requirements for ethical behaviors to ensure that blockchain applications foster justice, diversity, and accessibility (Sun & Wang, 2023). Policymakers, library administrators, and stakeholders must address the ethical, legal, and technological dimensions of implementing blockchain technology in library operations. Policymakers must collaborate with library stakeholders to formulate comprehensive policies on the utilization of blockchain technology (Gupta, 2022) (Wijesekara et al., 2024). These regulations must encompass data protection, intellectual property rights, and ethical issues, guaranteeing that blockchain applications conform to legal and ethical requirements. Library managers and personnel have to undergo training on blockchain technology and its ramifications for library operations (Chalkias et al., 2024). Cooperation among libraries, technology providers, and regulatory authorities is crucial for the progression of blockchain solutions (Lu et al., 2023). Stakeholders must participate in continuous discourse to tackle obstacles, exchange best practices, and investigate creative uses of blockchain technology. Ongoing assessment and evaluation of blockchain deployments are essential to guarantee that they achieve their intended objectives and adhere to legislation (Wijesekara et al., 2024). Libraries ought to implement metrics to evaluate the efficacy of blockchain applications and modify them as necessary. Enhancing public knowledge regarding the advantages and ramifications of blockchain technology in libraries might cultivate more comprehension and endorsement. Libraries must engage in transparent communication with users and stakeholders on the ways in which blockchain improves services and safeguards digital assets (Salama et al., 2023).

CONCLUSION

This chapter emphasizes the transformational capacity of blockchain technology in modernizing critical elements of digital library administration, particularly in improving fund management transparency, donor accountability, and the enforcement of intellectual property rights (Khan et al., 2023). The decentralized and irreversible characteristics of blockchain provide a robust answer to the issues encountered by digital libraries, delivering a transparent and secure framework for handling financial transactions, assuring accountability, and safeguarding intellectual property (Azbeg et al., 2022). Blockchain technology offers a revolutionary solution for digital libraries to tackle the issues related to conventional money administration, donor responsibility, and intellectual property rights (Aurangzeb et al., 2024). Utilizing the decentralized and immutable characteristics of blockchain, digital libraries can attain enhanced transparency, efficiency, and security in their operations (Rawat et al., 2023). The use of smart contracts significantly amplifies blockchain's potential by automating administrative functions and facilitating real-time reporting and auditing capabilities (Kasahara et al., 2024). As digital libraries progress and adjust to emerging technological frameworks, blockchain technology presents a viable answer to critical challenges in money management and intellectual property rights management (Sun & Wang, 2023). The appropriate use of blockchain technology may enhance operational efficiency and transparency, thereby benefiting funders, stakeholders, and the wider library community (Oyelude, 2022). This chapter examines the extensive and diverse potential uses of blockchain in digital libraries, facilitating future developments and enhancements in the sector. Blockchain technology offers considerable potential to revolutionize the administration of intellectual property rights in digital libraries (Caldarelli & Ellul, 2021). Utilizing blockchain's immutable ledger, decentralization, and smart contracts, digital libraries can enhance the establishment and enforcement of intellectual property rights with increased transparency, efficiency, and security (Vubangsi & Al Turjman, 2022). The use of blockchain-based DRM systems is a viable solution for guaranteeing equitable credit and remuneration for creators, safeguarding against unlawful distribution and plagiarism, and enabling efficient management of license agreements (Jadhawar, 2024). Despite existing obstacles, the prospective advantages of blockchain for digital libraries render it a tempting domain for continued investigation and advancement (Xiao et al., 2024). As blockchain technology advances, it is expected to assume a progressively significant role in the management of intellectual property rights in the digital era (Azbeg et al., 2022). The implementation of blockchain technology in libraries offers both advantages and obstacles. Although blockchain provides substantial benefits regarding transparency, accountability, and rights management, it also presents critical ethical issues that require meticulous attention (Vaz et al.,

2023). Libraries may responsibly and ethically implement blockchain technology by emphasizing data protection, user consent, and environmental sustainability (Lu et al., 2023). Emphasizing security, fostering user empowerment, and involving stakeholders are essential approaches that will facilitate the realization of blockchain's advantages while alleviating possible hazards and ethical dilemmas (Shkembi et al., 2023). By using blockchain technology thoughtfully and responsibly, libraries may use its disruptive potential while maintaining their dedication to ethical principles and social values (Ajibade & Mutula, 2021). The use of blockchain technology into digital libraries offers several chances to improve transparency, accountability, and rights management (Bashir & Warraich, 2023). By examining future trends, tackling scalability issues, and creating strong regulatory frameworks, libraries may fully leverage blockchain to enhance their operations and services. The guidelines are intended to assist policymakers, administrators, and stakeholders in addressing the intricacies of blockchain deployment and facilitating its effective integration within the library sector (Zhao et al., 2023). The chapter demonstrates how blockchain might enhance transparency in fund administration by integrating theoretical insights with practical implementations, facilitating stakeholders' ability to monitor and authenticate resource distribution (Ekpenyong & Owan, 2022). The improved accountability provided by blockchain alleviates worries over donor contributions, giving real-time transparency in fund utilization and cultivating more confidence among donors (Saeed et al., 2024). Furthermore, blockchain's ability to preserve intellectual property rights guarantees the integrity of digital asset management, safeguarding the interests of creators and rights holders while facilitating licensing and use agreements (Angara & Saripalle, 2022). The chapter discusses significant ethical concerns, highlighting the necessity for regulatory systems that reconcile innovation with legal and ethical norms (Jha, 2023; Hota et al., 2022). This chapter offers a forward-looking perspective on blockchain's role in the evolution of digital library systems by examining emerging trends and future directions, including interoperability with other technologies, enhancements in scalability, and the formulation of comprehensive regulatory guidelines (Azbeg et al., 2022). The primary objective of this chapter is to educate and motivate scholars, practitioners, and policymakers to perceive blockchain not just as a technological novelty but as a driver of significant innovation and sustainability in library administration and digital content stewardship (He et al., 2022; Lata & Owan, 2022). As digital libraries advance in a more digital environment, the adoption of blockchain technology offers a chance to improve operational efficiency, strengthen stakeholder connections, and guarantee the secure and fair administration of digital assets (Radovancovici et al., 2022). By leveraging blockchain technology, the library sector may pioneer innovative solutions that tackle current difficulties and promote future advancement.

REFERENCES

Aanchal, A., & Prasad, P. W. C. (2021, November). Topic: Scoping review of Blockchain based data storage technique in industrial IoT data management. In 2021 6th International Conference on Innovative Technology in Intelligent System and Industrial Applications (CITISIA) (pp. 1-10). IEEE.

Adhikari, S. (2021). Method for Intelligent Information Processing Using Software Reuse Library and Blockchain Architecture. In *AIAA Scitech 2021 Forum* (p. 1919). DOI: 10.2514/6.2021-1919

Agarwal, R., Kumar, T., Singh, A., & Schutte, D. P. (2025). Empowering Indigenous Economies: Financial and Investment Prediction for Sustainable Development. In Irfan, M., Majeed, M., Kassim, S., & Ofori, K. (Eds.), *The Future of Indigenous Inclusivity: Unlocking Finance for Economic Development* (pp. 63–82). IGI Global Scientific Publishing., DOI: 10.4018/979-8-3693-7924-0.ch004

Agrawal, A., Kumar, T., Agarwal, R., & Gupta, A. (2024), "AI-Driven Personalized Risk Management in the Insurance Sector", Taneja, S., Kumar, P., Reepu, Kukreti, M. and Özen, E. (Ed.) Data Alchemy in the Insurance Industry, Emerald Publishing Limited, Leeds, pp. 27-39. https://doi.org/DOI: 10.1108/978-1-83608-582-920241006

Ajibade, P., & Mutula, S. M. (2021). Virtual learning: A disruptive service in academic libraries. *Library Hi Tech News*, *38*(1), 12–13. DOI: 10.1108/LHTN-07-2020-0067

Ali, G., Mijwil, M. M., Buruga, B. A., & Abotaleb, M. (2024). A Comprehensive review on cybersecurity issues and their mitigation measures in FinTech.

Almadani, M. S., & Hussain, F. K. (2023, November). Implementing a Secure Blockchain-Based Wallet System with Multi-Factor Authentication. In 2023 IEEE International Conference on e-Business Engineering (ICEBE) (pp. 23-30). IEEE. DOI: 10.1109/ICEBE59045.2023.00010

Angara, J. S., & Saripalle, R. S. (2022). Towards a virtual water currency for industrial products using blockchain technology. *Water Policy*, *24*(6), 923–941. DOI: 10.2166/wp.2022.285

Arsat, N., Bakar, N. S. A. A., & Yahya, N. (2022, September). Testing in blockchain-based systems: a systematic review. In 2022 10th International Conference on Cyber and IT Service Management (CITSM) (pp. 1-6). IEEE. DOI: 10.1109/CITSM56380.2022.9935846

Aurangzeb, M., Wang, Y., Iqbal, S., Naveed, A., Ahmed, Z., Alenezi, M., & Shouran, M. (2024). Enhancing cybersecurity in smart grids: Deep black box adversarial attacks and quantum voting ensemble models for blockchain privacy-preserving storage. *Energy Reports, 11*, 2493–2515. DOI: 10.1016/j.egyr.2024.02.010

Awalu, I. L., Kook, P. H., & Lim, J. S. (2019, July). Development of a distributed blockchain evoting system. In Proceedings of the 2019 10th International Conference on E-business, Management and Economics (pp. 207-216). DOI: 10.1145/3345035.3345080

Azbeg, K., Ouchetto, O., Andaloussi, S. J., & Fetjah, L. (2022). A taxonomic review of the use of IoT and blockchain in healthcare applications. *Ingénierie et Recherche Biomédicale : IRBM = Biomedical Engineering and Research, 43*(5), 511–519. DOI: 10.1016/j.irbm.2021.05.003

Bashir, F., & Warraich, N. F. (2022). Prospects of semantic web and blockchain technologies in libraries. In *Blockchain and Deep Learning: Future Trends and Enabling Technologies* (pp. 31–45). Springer International Publishing. DOI: 10.1007/978-3-030-95419-2_2

Bashir, F., & Warraich, N. F. (2023). Future libraries' blockchain opportunities and challenges: a systematic literature review and research agenda. Digital Library Perspectives, (ahead-of-print).

Beyene, M., Toussaint, P. A., Thiebes, S., Schlesner, M., Brors, B., & Sunyaev, A. (2022). A scoping review of distributed ledger technology in genomics: Thematic analysis and directions for future research. *Journal of the American Medical Informatics Association : JAMIA, 29*(8), 1433–1444. DOI: 10.1093/jamia/ocac077 PMID: 35595301

Caldarelli, G., & Ellul, J. (2021). Trusted academic transcripts on the blockchain: A systematic literature review. *Applied Sciences (Basel, Switzerland), 11*(4), 1842. DOI: 10.3390/app11041842

Cao, L., & Yang, H. (2019, August). Building virtual digital library based on P2P and Blockchain. In 2019 11th International Conference on Intelligent Human-Machine Systems and Cybernetics (IHMSC) (Vol. 2, pp. 341-345). IEEE.

Chalkias, K. K., Lindstrøm, J., Maram, D., Riva, B., Roy, A., Sonnino, A., & Wang, J. (2024, May). Fastcrypto: Pioneering Cryptography Via Continuous Benchmarking. In Companion of the 15th ACM/SPEC International Conference on Performance Engineering (pp. 227-234).

Chiu, W. Y., Meng, W., & Li, W. (2021, December). LibBlock-towards decentralized library system based on blockchain and IPFS. In 2021 18th International Conference on Privacy, Security and Trust (PST) (pp. 1-9). IEEE. DOI: 10.1109/PST52912.2021.9647821

Ducrée, J., Gravitt, M., Walshe, R., Bartling, S., Etzrodt, M., & Harrington, T. (2020). Open platform concept for blockchain-enabled crowdsourcing of technology development and supply chains. *Frontiers in Blockchain*, *3*, 586525. DOI: 10.3389/fbloc.2020.586525

Dulce, E., & Hurtado, J. (2020, October). The role of the blockchain technology in the elderly care solutions: A systematic mapping study. In *International Workshop on Gerontechnology* (pp. 23-34). Cham: Springer International Publishing.

Ekpenyong, J. A., & Owan, V. J. (2022). Application of Blockchain Technology for Distributed Management of Digital Research Library Holdings. In *Innovative Technologies for Enhancing Knowledge Access in Academic Libraries* (pp. 196–212). IGI Global. DOI: 10.4018/978-1-6684-3364-5.ch012

Gatti, R. R., Singh, C., Agrawal, R., & Serrao, F. J. (Eds.). (2023). *Self-Powered Cyber Physical Systems*. John Wiley & Sons. DOI: 10.1002/9781119842026

Gupta, V. (2022). State of Art of Blockchain Technology in the Educational Sector. *Technology, Business, Innovation, and Entrepreneurship in Industry*, *4*(0), 299–314.

Hasnain, M., Albogamy, F. R., Alamri, S. S., Ghani, I., & Mehboob, B. (2023). The Hyperledger fabric as a Blockchain framework preserves the security of electronic health records. *Frontiers in Public Health*, *11*, 1272787. DOI: 10.3389/fpubh.2023.1272787 PMID: 38089022

He, W., Huang, J., Gao, N., & Chen, H. (2022, January). Data Distribution Scheme for Information Service Based on Blockchain. In 2022 2nd International Conference on Consumer Electronics and Computer Engineering (ICCECE) (pp. 21-25). IEEE. DOI: 10.1109/ICCECE54139.2022.9712811

Hota, P. K., Hota, L., & Dash, P. K. (2022). Blockchain in Digital Libraries: State of the Art, Trends, and Challenges. Machine Learning Adoption in Blockchain-Based Intelligent Manufacturing, 17-32.

Hu, L. (2022, February). Intelligent value added service platform of smart library based on blockchain technology. In *2022 IEEE International Conference on Electrical Engineering, Big Data and Algorithms (EEBDA)* (pp. 1391-1394). IEEE. DOI: 10.1109/EEBDA53927.2022.9744776

Huo, J., & Pan, B. (2022). Study the path planning of intelligent robots and the application of blockchain technology. *Energy Reports*, *8*, 5235–5245. DOI: 10.1016/j.egyr.2022.03.204

Jadhawar, A. A. (2024). Library as a Service (LaaS): Innovating Resource Sharing Among Universities. *Library Progress International*, *44*(1), 87–104.

Jha, S. K. (2023). Application of blockchain technology in libraries and information centers services. *Library Hi Tech News*. Advance online publication. DOI: 10.1108/LHTN-02-2023-0020

JNSS, J., & Ganesh, E. N. (2023). A novel blockchain-based lightweight encryption technique in fog based iot for personal healthcare data application. *International Journal of Intelligent Systems and Applications in Engineering*, *11*(3s), 119–128.

Kasahara, A., Mitchell, J., Yang, J., Cuomo, R. E., McMann, T. J., & Mackey, T. K. (2024). Digital technologies used in clinical trial recruitment and enrollment including application to trial diversity and inclusion: A systematic review. *Digital Health*, *10*, 20552076241242390. DOI: 10.1177/20552076241242390 PMID: 38559578

Khan, A. U., Ma, Z., Li, M., Zhi, L., Hu, W., & Yang, X. (2023). From traditional to emerging technologies in supporting smart libraries. A bibliometric and thematic approach from 2013 to 2022. Library Hi Tech, (ahead-of-print).

Khan, A. U., Zhang, Z., Taleby Ahvanooey, M., & Rafique, W. (2022). Opinion mining towards blockchain technology adoption for accessing digital library resources. *Aslib Journal of Information Management*, *74*(1), 135–157. DOI: 10.1108/AJIM-01-2021-0016

Kösemen, C., Dalkiliç, G., & Öksüzer, Ş. (2022). Performance analysis of lightweight Internet of things devices on blockchainnetworks. *Turkish Journal of Electrical Engineering and Computer Sciences*, *30*(2), 328–343. DOI: 10.3906/elk-2103-110

Kumar, T., & Kaur, S. (2023). Ethical Aspects of Cybersecurity in E-Commerce. In Cybersecurity for Decision Makers (1st ed.). CRC Press. DOI: 10.1201/9781003319887

Lamanna, C., & La Manna, M. (2018). BitView: using blockchain technology to validate and diffuse global usage data for academic publications. In Maturity and Innovation in Digital Libraries: 20th International Conference on Asia-Pacific Digital Libraries, ICADL 2018, Hamilton, New Zealand, November 19-22, 2018 [Springer International Publishing.]. *Proceedings*, *20*, 267–277.

Lata, N., & Owan, V. J. (2022). Contemporary trends and technologies in research libraries: An overview. Innovative technologies for enhancing knowledge access in academic libraries, 40-56.

Lengoatha, L., & Seymour, F. L. (2020, September). Determinant factors of intention to adopt blockchain technology across academic libraries. In Conference of the South African Institute of Computer Scientists and Information Technologists 2020 (pp. 244-250).

Li, F., & Liao, K. (2022, April). Research on multi service collaboration mode of university smart library based on blockchain. In 2022 7th International Conference on Cloud Computing and Big Data Analytics (ICCCBDA) (pp. 217-220). IEEE. DOI: 10.1109/ICCCBDA55098.2022.9778878

Lund, E. H., Jaccheri, L., Li, J., Cico, O., & Bai, X. (2019, May). Blockchain and sustainability: A systematic mapping study. In 2019 IEEE/ACM 2nd International Workshop on Emerging Trends in Software Engineering for Blockchain (WETSEB) (pp. 16-23). IEEE. DOI: 10.1109/WETSEB.2019.00009

Lv, Z., Yang, X., Yu, J., Li, L., Liu, A., & Li, C. (2023, September). Network Traffic Forensics and Traceability Technology Based on Blockchain. In 2023 6th International Conference on Information Communication and Signal Processing (ICICSP) (pp. 717-723). IEEE. DOI: 10.1109/ICICSP59554.2023.10390572

Ma, F., & Li, Z. (2020). Future prospect of library and information science in China in the context of new liberal arts. *Journal of Library Science in China*, *46*(06), 4–15.

Maidment-Otlet, R. (2021). Digital-first approaches and the library brand in a post-pandemic world. In Libraries, Digital Information, and COVID (pp. 103-110). Chandos Publishing. DOI: 10.1016/B978-0-323-88493-8.00016-1

Mani, M., & Agarwal, R. (2022). Payment Banks in India: Insights From the Customers' Perception. [IJABIM]. *International Journal of Asian Business and Information Management*, *13*(1), 1–15. DOI: 10.4018/IJABIM.297851

Masenya, T. M., & Chisita, C. T. (2022). Futurizing library services in a technology-driven dispensation: Reflections on selected academic libraries in Zimbabwe and South Africa. In *Innovative technologies for enhancing knowledge access in academic libraries* (pp. 1–21). IGI Global. DOI: 10.4018/978-1-6684-3364-5.ch001

Mohanta, B. K., Jena, D., Panda, S. S., & Sobhanayak, S. (2019). Blockchain technology: A survey on applications and security privacy challenges. *Internet of Things : Engineering Cyber Physical Human Systems*, *8*, 100107. DOI: 10.1016/j.iot.2019.100107

Mohd Pozi, M. S., Muruti, G., Abu Bakar, A., Jatowt, A., & Kawai, Y. (2018, May). Preserving author editing history using blockchain technology. In *Proceedings of the 18th ACM/IEEE on Joint Conference on Digital Libraries* (pp. 165-168). DOI: 10.1145/3197026.3197068

Mridha, K., Shaw, R. N., Kumar, D., & Ghosh, A. (2022). Driver Drowsiness Alert System Using Real-Time Detection. AI and IoT for Smart City Applications, 65-81.

Ojobor, R. C. (2023). 1 Blockchain Technology for Library Services: Challenges and Opportunities for Libraries from a Nigerian Perspective. *Information Services for a Sustainable Society: Current Developments in an Era of Information Disorder, 183*, 7–23. DOI: 10.1515/9783110772753-003

Omame, I. M., & Alex-Nmecha, J. C. (2021). Application of blockchain in libraries and information centers. In *Handbook of Research on Knowledge and Organization Systems in Library and Information Science* (pp. 384–397). IGI Global. DOI: 10.4018/978-1-7998-7258-0.ch020

Oyelude, A. A. (2022). Trending issues in advancing blockchain technology in libraries, archives and museums. *Library Hi Tech News, 39*(6), 6–7. DOI: 10.1108/LHTN-06-2021-0040

Paganelli, A. L., & Paganelli, A. L. (2023). Blockchain and the Research Libraries: Expanding Interlibrary Loan and Protecting Privacy. In Research Anthology on Convergence of Blockchain, Internet of Things, and Security (pp. 1152-1166). IGI Global.

Pambudi, A., Purnama, S., Ayuninggati, T., Santoso, N. P. L., & Oktariyani, A. (2021, November). Legality on digital document using blockchain technology: an exhaustive study. In *2021 Sixth International Conference on Informatics and Computing (ICIC)* (pp. 1-6). IEEE. DOI: 10.1109/ICIC54025.2021.9632860

Panda, S., & Kaur, N. (2023). Blockchain: A New Technology in Library System and Management. In Handbook of Research on Advancements of Contactless Technology and Service Innovation in Library and Information Science (pp. 211-230). IGI Global. DOI: 10.4018/978-1-6684-7693-2.ch011

Paulavičius, R., Grigaitis, S., Igumenov, A., & Filatovas, E. (2019). A decade of blockchain: Review of the current status, challenges, and future directions. *Informatica (Vilnius), 30*(4), 729–748. DOI: 10.15388/Informatica.2019.227

Prakash, R., Anoop, V. S., & Asharaf, S. (2022). Blockchain technology for cybersecurity: A text mining literature analysis. *International Journal of Information Management Data Insights, 2*(2), 100112. DOI: 10.1016/j.jjimei.2022.100112

Qian, C. (2021, April). Digital copyright management model of university library based on blockchain technology. In *2021 International Conference on Computer, Blockchain and Financial Development (CBFD)* (pp. 508-513). IEEE. DOI: 10.1109/CBFD52659.2021.00108

Quan, R. (2021). National Smart Library System: Opening a New Chapter for the Transformation toward Smart Library. *Journal of Library Science in China, 47*, 4–14.

Radovancovici, M., Galiş, D., & Pungilă, C. (2022, September). A Practical Analysis of Techniques for Minting Genetic Information as NFTs in Blockchain Technology. In 2022 24th International Symposium on Symbolic and Numeric Algorithms for Scientific Computing (SYNASC) (pp. 331-337). IEEE. DOI: 10.1109/SYNASC57785.2022.00060

Ramya, R. S., Ramesh, B. D., Kumudavalli, M. V., Joshi, P., & Venugopal, K. R. (2023, December). Blockchain-Integrated Metaverse for Academia: A Solution for Virtual Classes, Lectures, and Project Fundings. In International Conference on Information and Communication Technology for Competitive Strategies (pp. 387-396). Singapore: Springer Nature Singapore.

Rawat, R., Chakrawarti, R. K., Sarangi, S. K., Patel, J., Bhardwaj, V., Rawat, A., & Rawat, H. (Eds.). (2023). *Quantum Computing in Cybersecurity*. John Wiley & Sons. DOI: 10.1002/9781394167401

Rouhani, S., Belchior, R., Cruz, R. S., & Deters, R. (2021). Distributed attribute-based access control system using permissioned blockchain. *World Wide Web (Bussum), 24*(5), 1–28. DOI: 10.1007/s11280-021-00874-7

Saeed, N., Wen, F., & Afzal, M. Z. (2024). Decentralized peer-to-peer energy trading in microgrids: Leveraging blockchain technology and smart contracts. *Energy Reports, 12*, 1753–1764. DOI: 10.1016/j.egyr.2024.07.053

Salama, R., Al-Turjman, F., Altrjman, C., Kumar, S., & Chaudhary, P. (2023, April). A comprehensive survey of blockchain-powered cybersecurity-a survey. In *2023 International Conference on Computational Intelligence, Communication Technology and Networking (CICTN)* (pp. 774-777). IEEE. DOI: 10.1109/CICTN57981.2023.10141282

Shahzad, K., Khan, S. A., & Iqbal, A. (2024). Effects of blockchain technology (BT) on the university librarians and libraries: a systematic literature review (SLR). Library Hi Tech, (ahead-of-print).

Sharma, S., & Batth, R. S. (2020, June). Blockchain technology for higher education sytem: A mirror review. In *2020 International Conference on Intelligent Engineering and Management (ICIEM)* (pp. 348-353). IEEE. DOI: 10.1109/ICIEM48762.2020.9160274

Sharma, S., & Singh, B. (2025). Blockchain-Based Secure Digital Library System. In *Blockchain Technology for Cyber Defense, Cybersecurity, and Countermeasures: Techniques, Solutions, and Applications* (pp. 195–210). IGI Global.

Shivsharan, N. M., Kambli, S. G., Nikharge, O. V., & Kudatarkar, K. U. (2023, June). Facilitating Cryptocurrency Analysis and Watchlist Management with a Web-Based Platform. In 2023 3rd International Conference on Intelligent Technologies (CONIT) (pp. 1-6). IEEE. DOI: 10.1109/CONIT59222.2023.10205550

Shkembi, K., Kochovski, P., Papaioannou, T. G., Barelle, C., & Stankovski, V. (2023). Semantic Web and blockchain technologies: Convergence, challenges and research trends. *Journal of Web Semantics*, *79*, 100809. DOI: 10.1016/j.websem.2023.100809

Shuiqing, H. U. A. N. G., Xiaoguang, W. A. N. G., Cuijuan, X. I. A., & Ouyang, J. (2022). Advancing the work on ancient classics in the new era and accelerating innovative and intelligent development. Journal of library and information sciences in agriculture, 34(5), 4.

Stana, M. C., Goga, N., Marian, C. V., Popa, R., Vulpe, C. M., & Taslitchi, C. (2021, May). G-Cloud Briefcase-Electronic Archive for Academic Certificates and General Certificates of Education Documents Using Public Private Hyperspace for E-Government Library Services Based on NOSQL Databases. In *2021 IEEE International Black Sea Conference on Communications and Networking (BlackSeaCom)* (pp. 1-5). IEEE. DOI: 10.1109/BlackSeaCom52164.2021.9527826

Sun, J., & Wang, Y. (2023, November). Data Security Mechanism of Digital Library Based on Blockchain. In 2023 International Conference on Ambient Intelligence, Knowledge Informatics and Industrial Electronics (AIKIIE) (pp. 01-06). IEEE. DOI: 10.1109/AIKIIE60097.2023.10390497

Tawornittayakun, J., & Leelasantitham, A. (2020, June). A comparative study of library information storage system through Blockchain technology: A case study of undergraduate researches. In 2020 17th International Conference on Electrical Engineering/Electronics, Computer, Telecommunications and Information Technology (ECTI-CON) (pp. 5-8). IEEE. DOI: 10.1109/ECTI-CON49241.2020.9158239

Vaz, A., Tendulkar, A., & Palaniappan, S. (2023). Tracing the Evolution of Blockchain and Its Implications to Higher Education. In Advancements in Artificial Intelligence, Blockchain Technology, and IoT in Higher Education (pp. 203-225). Apple Academic Press. DOI: 10.1201/9781003300458-11

Verma, R. (2025). The future of academic libraries: Blockchain for secure and transparent information management. In *Driving Socio-Economic Growth With AI and Blockchain* (pp. 401–432). IGI Global. DOI: 10.4018/979-8-3693-8664-4.ch017

Vubangsi, M., & Al-Turjman, F. (2022, August). Design and implementation of a conference attendance monitoring system using blockchain and AI technologies. In *2022 International Conference on Artificial Intelligence in Everything (AIE)* (pp. 197-202). IEEE. DOI: 10.1109/AIE57029.2022.00044

Wang, F., Zhao, H., & Ma, J. Y. (2019). Research and practice progress of data provenance from the perspective of data science. *Journal of Library Science in China*, *45*(5), 79–100.

Wijesekara, D. S. N., & Arachchige, P. (2024). Blockchain and Artificial Intelligence for Big Data Analytics in Networking: Leadingedge Frameworks. *Journal of Engineering Science & Technology Review*, *17*(3).

Xiao, Y., Xiao, R., Zhu, C., & Wei, S. Digital Copyright Management Strategy of Digital Library Based on Blockchain Technology. Journal of Combinatorial Mathematics and Combinatorial Computing, 119, 153–162.

Zhang, J. (2024). Research on Resource Sharing Method of Library and Document Center Under the Multimedia Background. *International Journal of Advanced Computer Science and Applications*, *15*(4). Advance online publication. DOI: 10.14569/IJACSA.2024.0150435

Zhao, C., Shao, J., Zhang, R., Mu, W., & Zhang, J. (2023, May). Analysis of the digital transfer framework of power grid engineering based on the integration of QR code and blockchain. In *Eighth International Symposium on Advances in Electrical, Electronics, and Computer Engineering (ISAEECE 2023)* (Vol. 12704, pp. 98-103). SPIE. DOI: 10.1117/12.2680662

Chapter 4
Breast Cancer Risk Estimation Using Explainable Artificial Intelligence-Based Machine Learning Techniques

Rohit Raj Singh
https://orcid.org/0009-0001-4394-2247
National Institute of Technology, Raipur, India

Bikesh Kumar Singh
National Institute of Technology, Raipur, India

Kushangi Atrey
Rungta College of Engineering and Technology, Bhilai, India

Narendra Kuber Bodhey
https://orcid.org/0000-0003-1736-3329
All India Institute of Medical Sciences, Raipur, India

ABSTRACT

Due to insufficient cancer literacy and lack of understanding of breast cancer risk, Indian women face an higher incidence of breast cancer, regardless of socioeconomic status or education (Gupta et al., 2015). Breast cancer is a common and serious threat to women's physical and emotional well-being. Despite technologi-

DOI: 10.4018/979-8-3693-9846-3.ch004

cal advancements in India, many cancer patients cannot access or afford modern treatment (Consensus document for management of breast cancer, Indian Council of Medical Research). Current computer vision (CV) models focus on image quality, while natural language processing (NLP) models fail to extract key information from mammography reports. This paper proposes an Explainable AI (XAI) risk assessment system for early breast cancer diagnosis using factors like patient history, socioeconomic status, symptoms, and lifestyle. XAI will help highlight significant features, aiding medical decisions. The model will classify a person's breast cancer risk as low or high.

1 INTRODUCTION

Lung cancer used to be the most occurring type of cancer worldwide, but BC has recently surpassed it. The most frequent malignancy in women is BC, also known as the "pink killer," while it is less common in men. Woman's BC peaks between the ages of 45 and 50, whereas man's peak is between the ages of 70 and 75. The average age of diagnosis is between 45 and 55. Patients under the age of 35 make about 10 to 15 percent of the total. The following are some clinical examinations for BC:

(1) Look at the past to determine the patient's BC risk.
(2) Examine the breast's appearance. The breast's shape and size vary depending on the position, so look for any abnormalities in the skin and nipple.
(3) Breast tissue, axillary tissue, and nearby lymph nodes can all be palpated.
(4) Mammography is a crucial imaging test for the diagnosis of BC.

Mammography is a specialized imaging technique used primarily for the prior detection and analysis of BC. It applies low-energy X-rays to produce detailed images of the breast tissue, helping healthcare professionals to identify and analyze abnormalities. The type of mammography includes:

- **Screening Mammography**: This is performed on asymptomatic women to detect BC early, often before symptoms appear. It typically involves two X-ray images of each breast.
- **Diagnostic Mammography**: This is used when abnormalities are detected during screening or when a patient presents with symptoms (e.g., a lump). It provides more detailed images and may involve additional views or techniques.

Artificial intelligence (AI) is increasingly being used to assist radiologists in interpreting mammograms, potentially improving accuracy and reducing workload. Mammography remains a cornerstone in the early detection and classification of BC. Its ability to identify abnormalities before they become symptomatic is crucial for effective treatment. While there are limitations, ongoing advancements in technology and the integration of AI are enhancing the effectiveness of mammography, making it a vital tool in BC screening and diagnosis. Mammography views play a critical role in BC screening by providing comprehensive images of breast tissue. The standard views, including the Craniocaudal and Mediolateral Oblique views, along with additional specialized views, enhance the detection and classification of breast abnormalities. Advances in technology, such as 3D mammography, further improve the effectiveness of BC screening, making it a vital tool in early detection and treatment.

Current BC AI prediction systems fall into two categories: (1) BC prediction with molybdenum mammography. Computer vision (CV) researchers use CNN to train image-centric candidate lesions and classifiers to link feature vectors to candidates to predict genuine lesions. (2) BC prediction using mammograms. Researchers studying natural language processing (NLP), which involves text categorization, use mammography reports as training data. The text classification challenge is separated into feature engineering, explainable AI classification, and standard AI classifier comparison. Common classification models include NB, KNN, DT, SVM, and Random Forest.

XAI refers to a set of methods and steps that makes the end user or human being to understand the result and to conclude the result. The traditional machine learning classification-based AI model operate basically on block boxes, where the logic behind the outcome is unknown. This is one of the major drawbacks of AI specially in the field of healthcare where knowing what feature are affecting the outcome is as important as the outcome itself to have a proper diagnosis and further treatment.

XAI address this concern by making the outcome more transparent and interpretable for the user to understand the outcome more conclusively. The basic aspects of XAI are as follows:

i. Interpretability, by allowing user to know how AI model is working and also by creating relationship between input and output
ii. Transparency, by revealing the internal working of the model, thereby making the user know whether the system is working accurately and ethically
iii. Fairness, ensuring that the model don't work with hidden biases especially in the area of healthcare.

XAI methods can be broadly categorized into two types:

i. Post-hoc interpretability: In this method XAI try to explain already trained models. Examples include feture importance, saliency maps, Local Interpretable Model-Agnostic Explanations (LIME), and Shapley Additive explanations (SHAP).
ii. Intrinsic interpretability: These are models that are inherently interpretable, such as decision trees or linear regression, where the logic of the model is naturally easier to understand.

The integration of AI and XAI in imaging for classification and prediction is a rapidly evolving field with significant implications across various sectors. While AI enhances the capabilities of image analysis, XAI ensures that these advancements are transparent and trustworthy. As technology continues to develop, addressing the challenges associated with data quality, bias, and explainability will be crucial for the successful implementation of AI in imaging. XAI can provide insights into which features (e.g., specific image patterns) are most influential in classification tasks.

Early detection of breast cancer plays a vital role in improving patient outcomes and survival rates. When detected at an early stage, before the cancer has spread, the treatment is often less invasive, and the chances of full recovery are significantly higher. Studies show that the five-year survival rate for localized breast cancer is over 99%, but it drops dramatically for advanced-stage cancers. Mammography, along with newer imaging techniques, is critical in identifying abnormalities in the breast tissue early on, even before symptoms arise. This early detection allows for timely interventions that can greatly improve prognosis and reduce the burden of more aggressive treatments.

In addition to early detection, AI is now enabling personalized medicine for breast cancer patients. By integrating genetic, clinical, and imaging data, AI systems are becoming more adept at predicting which treatments are likely to be most effective for individual patients. This shift toward personalized treatment allows healthcare providers to move away from a "one-size-fits-all" approach, ensuring that therapies are tailored to the specific needs of each patient, (Roscher *et al.*, 2020). For instance, AI models can incorporate genetic markers, such as BRCA1 and BRCA2 mutations, which are associated with a higher risk of breast cancer, to develop personalized risk profiles and treatment plans. This personalized approach not only increases treatment efficacy but also reduces the likelihood of unnecessary treatments, minimizing side effects and improving the patient's overall quality of life.

As AI becomes increasingly integral to breast cancer screening and diagnosis, it is important to address associated ethical and privacy challenges, (Roscher *et al.*, 2020). The effectiveness of AI in healthcare depends on large datasets, often containing sensitive patient information, which raises serious concerns about data privacy and security. Safeguarding patient data and ensuring compliance with regulations like HIPAA and GDPR are vital. Furthermore, AI models must be developed with transparency and accountability to prevent bias in clinical decision-making. For instance, models trained on data from specific populations may not generalize well to diverse patient groups, potentially leading to disparities in care. Addressing these concerns is crucial for the responsible use of AI in breast cancer detection and treatment, (Ak, 2020).

2 PREVIOUS WORK

Dar, Ruey, and Yu (2000) used a self-organizing map (neural network model) and 10-fold cross validation to achieve an 85.6% classification accuracy for the sonographic diagnosis of BC.

Arpita Joshi and Ashish Mehta (2017) makes use of UCI repository contained Wisconsin BC to compare the classification outcomes of KNN (K Nearest Neighbour), Random Forest, SVM (Support Vector Machine) and Decision Tree. Simulations showed that KNN was the top classifier, followed by SVM, Random Forest, and Decision Tree.

Hussein A. Abbass (2002) proposed a pareto-differential evolution algorithm-based evolutionary artificial neural network technique. He made use of a multipropagation artificial neural network, which performs better (98.12%). Better generalization and less expensive computing were obtained by the author. According to Omondiagbe, Veeramani, and Sidhu (2019), A neural network, naive bayes, and support vector machine were tested on the Wisconsin Diagnostic BC (WDBC) dataset. These machine learning methods were used with feature selection and extraction. The simulation results showed that SVM-LDA was preferred over all other strategies because to its longer computation time.

Liu and Ma (2007) conducted experiments on 120 BC images. They discovered that SVM (96.12%) performs better than the Back Propagation Neural Network (93.33%).

Muhammet Fatih Ak (2020) conducted a comprehensive study on BC detection using machine learning and data visualization techniques. In his study, public BC data from Dr. William H. Walberg were analyzed using a variety of machine learning algorithms, including logistic regression, k-nearest neighbors (KNN), support vector machines (SVM), naive Bayes, decision trees, random forests, and rotation forests. The primary objective was to establish a robust predictive model for distinguishing malignant and benign breast tumors. The study demonstrated that logistic regression achieved the highest accuracy (98.1%) compared to other models, highlighting its efficacy in BC diagnosis. Furthermore, data visualization techniques were employed to identify correlations between features and to optimize the predictive performance of machine learning models. This study underscores the potential of machine learning to enhance the accuracy of BC diagnosis and offers insights into the comparative performance of various classification methods in handling clinical datasets.

Sharma, Aggarwal, and Choudhury (2018) conducted a comparative study on BC detection using machine learning algorithms such as Random Forest, k-Nearest Neighbor (kNN), and Naive Bayes. Utilizing the Wisconsin Diagnosis BC dataset, the study evaluated the performance of these algorithms based on key metrics including accuracy, precision, and recall. The results showed that kNN achieved the highest accuracy (95.90%), followed closely by Random Forest (94.74%) and Naive Bayes (94.47%). The study concluded that kNN was the most effective model, outperforming others in accuracy, precision, and F1 score, while Random Forest had the best recall. This work highlights the importance of machine learning models in early BC detection, demonstrating their potential to aid in diagnostic processes.

Nallamala, Mishra, and Koneru (2019) proposed an ensemble voting approach combining Logistic Regression, Neural Networks, and Support Vector Machines (SVM) for BC detection using the Wisconsin BC dataset. Their method achieved an accuracy of 98.50%, demonstrating the effectiveness of combining classifiers to improve diagnostic accuracy. The study highlights the importance of preprocessing and feature extraction in enhancing machine learning performance for medical diagnostics.

Yesuf's (2019) work showed SVM achieving high accuracy (98.8%) in classifying benign and malignant tumors, utilizing advanced image preprocessing, segmentation and feature extraction. These approaches offer valuable insights for our EEG-based depression detection, where feature extraction and classifier performance are crucial. Our study builds on these techniques, integrating RQA, SL, and Coherence to improve mental disorder detection through EEG signals.

The paper by Mohiuddin *et al.* (n.d.) provides a thorough analysis of the application of deep learning techniques for BC detection across various imaging modalities such as mammography, histopathology, and MRI. The authors focus on Convolutional Neural Networks (CNNs) and their effectiveness in feature extraction and classification. Results from various studies reviewed in the paper show that CNN-based approaches achieved high accuracy, with some models like VGG16 reaching 94% accuracy on mammogram images, while others such as U-Net and Autoencoder-based architectures demonstrated improved segmentation and classification performance in histopathology and MRI data. The study also highlights key challenges, including the scarcity of publicly available datasets, high computational costs, and the need for external validations to implement these models in clinical practice.

The paper by Abunasser *et al.* (2022) presents a deep learning-based approach for classifying BC into eight distinct classes using the Xception model. The dataset used, collected from the Kaggle repository, was augmented using Generative

Adversarial Networks (GAN) to improve model performance. The study achieved high accuracy in classifying both benign and malignant tumors, with the customized Xception model achieving a testing accuracy of 97.60%, precision of 97.60%, and F1-Score of 97.58%. This paper highlights the potential of deep learning models, particularly Xception, for accurately classifying multiple types of BC, contributing to early detection and treatment strategies.

The paper by Amethiya *et al.* (2022) provides an extensive review of machine learning (ML) algorithms and biosensors for BC detection. The authors compare various ML models, including support vector machines (SVM), decision trees (DT), artificial neural networks (ANN), and others, applied to datasets such as mammograms, histopathology images, and blood analyses. The study highlights the superior accuracy of hybrid models, particularly the fusion of SVM, Naive Bayes, and decision trees, which achieved an accuracy of 97.31% in classifying BC. The paper also explores biosensors for detecting cancer-related analytes like HER2 and miRNA, emphasizing their importance in improving early diagnosis through electrochemical detection methods.

The paper by Naji *et al.* (2021) explores the use of several machine learning algorithms, including Support Vector Machines (SVM), Random Forest, Logistic Regression, Decision Tree (C4.5), and K-Nearest Neighbors (KNN), for BC prediction using the BC Wisconsin Diagnostic dataset. The study evaluates these models based on accuracy, precision, and sensitivity. Among the classifiers, SVM achieved the highest performance, with an accuracy of 97.2%, outperforming other algorithms. The paper highlights SVM's efficiency in predicting BC, demonstrating its superior precision and area under the curve (AUC) score of 96.6%. The authors conclude that SVM is a highly effective model for BC prediction and diagnosis, though they note the limitation of using a single dataset and suggest applying the methodology to larger datasets in future work.

The integration of Deep Learning (DL) and Case-Based Reasoning system (CBR) in BC detection is explored by Bouzar-Benlabiod *et al.* (2023). The initial step involves enhancing the quality of mammogram images through techniques such as Contrast Limited Adaptive Histogram Equalization (CLAHE) and median filtering. The dataset used in this study is the CBIS-DDSM, which contains a curated set of mammograms with corresponding masks. The segmentation of mammograms is performed using a U-Net architecture, specifically a SE-ResNet-101 model. After segmentation, relevant features are extracted from the Regions of Interest (RoIs) using the Gray-Level Co-occurrence Matrix (GLCM). This method computes texture features that are critical for distinguishing between benign and malignant cases. A wrapper method is employed to select the most informative features, ensuring that only the most relevant attributes are used for classification. The performance of the proposed framework is evaluated using metrics such as accuracy, precision, recall, and F1-score. The results indicate that the hybrid approach outperforms traditional Machine Learning (ML) and DL methods, achieving an accuracy of 86.71% and a recall of 91.34%.

The study by Yan *et al* (2023) introduces an automated BC detection system that employs an ensemble classifier and feature weighting algorithms. Key steps include a novel region extraction technique to isolate the breast region, a pectoral muscle removal algorithm to reduce interference, and an automated ROI segmentation strategy. Textural and morphological features are extracted and the ensemble classifier model combines k-nearest neighbor (KNN), bagging, and Eigen Class algorithms to classify mammograms as normal, benign, or malignant.

The study by Prodan, Paraschiv, and Stanciu (2023) employed a comprehensive approach to analyze mammograms using deep learning techniques. The research utilized the RSNA dataset, containing 28,911 instances of BC. The dataset was split into training and validation sets using a 5-fold stratified cross-validation strategy. The study experimented with various deep learning models, including Convolutional Neural Networks (CNNs) such as ResNet and EfficientNet, as well as Vision Transformer (ViT) architectures. To enhance interpretability, the study employed explainable AI techniques, including class activation maps and centered bounding boxes, to visualize the regions of interest in the mammograms that influenced the model's predictions. The models achieved high accuracy rates, with ResNet18 showing an accuracy of 97% on original images and 94% on processed images addition of synthetic images further improved the performance, with ResNet18 achieving an accuracy of 94% and an AUC of 0.85.

The methodology employed by Anas *et al.* (2024) in recent studies often include preprocessing techniques such as Contrast Limited Adaptive Histogram Equalization (CLAHE) to enhance image quality. Data augmentation strategies are also utilized to expand training datasets, thereby improving model robustness. The combination of these techniques with advanced deep learning architectures has shown to yield superior results compared to traditional approaches. The literature indicates a clear trend towards the integration of AI and deep learning in BC detection, with significant improvements in diagnostic accuracy and efficiency.

In the work reported by Aguerchi *et al.* (2024), the DDSM and MIAS dataset were used. The optimization of CNN parameters involves determining suitable parameters that result in significant accuracy for each task. The Particle Swarm Optimization (PSO) algorithm was employed to search for the optimal hyperparameters, including kernel size, stride, and filter number. The results demonstrated that the PSO-CNN model significantly outperformed existing methods, achieving an accuracy of 98.23% on the DDSM dataset and 97.98% on the MIAS dataset.

The Mammo-Light model is proposed by Raiaan *et al.* (2024), which is a lightweight CNN architecture consisting of 10 layers, including 5 convolutional layers, 3 pooling layers, and 2 dense layers. For the experimentation work, they implemented the CBIS-DDSM and MIAS datasets. The model is designed to minimize the number of parameters while maintaining high accuracy, thus reducing computational time. The results indicate that the model achieves a test accuracy of 99.17% on the CBIS-DDSM dataset and 98.42% on the MIAS dataset, outperforming traditional transfer learning models.

3 PROPOSED EXPERIMENTAL SETUP

Decision Tree, Random Forest, KNN, and SVM were used to detect benign and malignant BC more correctly and efficiently; however, they cannot explain the anticipated output behavior. This study proposed an explainable AI model employing the XGBoost classifier to distinguish benign and malignant BC and explain its behavior. **XGBoost** (Extreme Gradient Boosting) is a highly efficient and scalable machine learning algorithm, particularly well-known for its performance in classification and regression tasks. It is an implementation of the gradient boosting technique but optimized for speed and performance. XGBoost has become one of the most popular algorithms in machine learning, particularly for structured or tabular data, because of its strong predictive power, flexibility, and ability to handle large datasets.

Key Features of XG-Boost:

Optimized for Speed and Performance:

- XGBoost is designed to be highly efficient in terms of both computational resources and execution time. It can handle large datasets with millions of rows and multiple features effectively.
- It uses several optimizations, such as parallelized tree building, efficient memory usage, and hardware optimization.

Regularization:

- XGBoost includes built-in regularization (both L1 and L2 regularization) to avoid overfitting. Overfitting occurs when a model performs well on training data but poorly on new, unseen data. Regularization helps by penalizing overly complex models.

Handling Missing Data:

- XGBoost can manage missing values automatically without requiring pre-processing steps. It learns the best direction to take in a decision tree when a missing value is encountered during training.

Sparsity-Aware:

- XGBoost is designed to handle sparse data effectively. It automatically handles zero values in datasets, which is common in scenarios like text classification where sparse matrices are often used.

Support for Custom Loss Functions:

- While it provides several standard loss functions (like mean squared error for regression and logistic loss for classification), XGBoost allows users to define custom loss functions, giving flexibility for specialized tasks.

Cross-Validation:

- XGBoost provides built-in cross-validation, which makes model tuning easier by allowing users to perform hyperparameter tuning while checking for overfitting and underfitting.

Tree Pruning:

- Unlike traditional boosting algorithms that stop splitting trees when they reach a certain depth, XGBoost uses a technique called "max depth pruning," where it grows trees to a maximum depth and then prunes back based on performance. This helps to avoid overfitting.

Distributed Computing:

- XGBoost supports distributed processing, making it easy to scale training across multiple machines or cloud services.

Advantages of XGBoost:

High Performance:

- XGBoost consistently ranks as one of the top-performing algorithms in machine learning competitions, such as Kaggle. Its ability to efficiently handle large-scale datasets and complex relationships makes it highly competitive.

Flexibility:

- XGBoost supports various objective functions (e.g., regression, classification, ranking), and users can even define their own.

Parallelization:

- The algorithm is designed to be parallelized, allowing it to run faster, especially on large datasets.

Interpretability:

- Although XGBoost builds an ensemble of decision trees, which are often considered "black box" models, the structure of decision trees is relatively easy to interpret compared to other complex models like neural networks.

Handling Imbalanced Data:

- XGBoost has techniques, like controlling the balance of positive and negative weights in binary classification, to handle imbalanced datasets effectively.

Aiming to solve the "black box" issue with machine learning algorithms, XAI techniques visualize and explain model traits and patterns. XGBoost's efficiency in **Explainable AI (XAI)** stems from its unique features and characteristics, making it more interpretable and suitable for producing explainable models compared to other machine learning methods like deep learning. Here's how XGBoost excels in XAI compared to other machine learning algorithms:

Tree-Based Structure: Naturally Interpretable:

- Inherent Interpretability of Decision Trees: XGBoost is based on decision trees, which are relatively interpretable compared to black-box models like deep neural networks. In decision trees, predictions are made through a series of if-then rules, making it easy to trace how a model arrived at a particular decision. This inherent interpretability means that the contribution of each feature to a prediction can be visualized and understood without needing additional explanation tools.

- Feature Importance: XGBoost provides feature importance scores, which rank how much each feature contributes to the model's decisions. This allows users to identify which features have the most influence on the predictions, which can be critical for explaining the model's behavior.

Model-Agnostic XAI Tools Work Well with XGBoost:

- Compatibility with LIME and SHAP: XGBoost works exceptionally well with model-agnostic XAI tools like LIME (Local Interpretable Model-Agnostic Explanations) and SHAP (SHapley Additive exPlanations). These tools can explain individual predictions by approximating the model locally (in the case of LIME) or assigning feature attributions (in the case of SHAP).
- SHAP for Global and Local Interpretations: SHAP values work particularly well with XGBoost because XGBoost's additive tree-based model structure can be interpreted in terms of contributions from each feature, both globally (for the entire model) and locally (for specific predictions). SHAP efficiently calculates the impact of each feature for individual predictions, which is more computationally efficient with XGBoost than with complex models like deep learning.

Handling Missing Values and Sparsity:

- Sparsity-Aware Learning: XGBoost can handle missing values and sparse datasets in a way that allows for better interpretability. When a model encounters missing data during prediction, XGBoost learns the optimal path to take in the tree based on the training data. This decision-making process can be explained and visualized, offering insight into how the model handles real-world data issues, such as incomplete or missing values, without requiring imputation methods.
- Transparency in Handling Missing Data: Unlike algorithms that impute missing values or require pre-processing steps, XGBoost's ability to naturally manage missing values can be explained, improving trust in the model's decisions.

Regularization to Avoid Overfitting

- Regularization for Simplified Models: XGBoost incorporates both L1 (lasso) and L2 (ridge) regularization to avoid overfitting, leading to more generalized and interpretable models. Overfitted models, while potentially performing well on training data, are often difficult to explain since they memorize noise or specific patterns that do not generalize to new data.

- Pruning of Decision Trees: XGBoost applies a pruning technique to prevent the creation of overly complex trees. This results in simpler, more interpretable models that focus on the most important decisions rather than overfitting to noise.

5. Fast and Scalable for Large Datasets

- Efficiency and Scalability: One of XGBoost's key strengths is its ability to handle large-scale datasets efficiently. When applied to big data, traditional algorithms may require simplifications or approximations that make them harder to explain. XGBoost maintains its efficiency and accuracy even with very large datasets, allowing explainable models to be built without compromising on performance.
- Parallel Processing and Distributed Systems: XGBoost's support for parallel processing and distributed computing makes it feasible to use XAI tools like SHAP in a fast and scalable way, even for large datasets.

Comparison to Other Machine Learning Models in XAI:
XGBoost vs. Random Forests:

- While random forests also consist of decision trees and offer interpretability through feature importance, they do not employ boosting, meaning they do not iteratively improve on previous models. XGBoost's boosting process results in better performance while still offering similar levels of interpretability as random forests.

XGBoost vs. Support Vector Machines (SVM):

- SVMs can be challenging to explain, especially when using non-linear kernels. The decision boundaries in high-dimensional spaces are often difficult to visualize and interpret. XGBoost, with its decision tree-based approach, is inherently more interpretable, providing clearer explanations for how features influence predictions.

XGBoost vs. Logistic Regression:

- Logistic regression is often considered more interpretable than XGBoost because it provides linear relationships between features and the target variable. However, logistic regression may not capture complex patterns in the data, reducing its predictive power in complex tasks.
- XGBoost provides a good balance, offering more complex decision-making while still retaining a degree of interpretability. Feature importance scores and decision paths in XGBoost offer more insights into non-linear interactions in the data, which logistic regression might miss.

Other machine learning technique mainly used for comparison with XAI using XGBoost are Naïve Bayes (NV), K Nearest Neighbor (KNN) and Support Vector Machine (SVM). Naive Bayes, K-Nearest Neighbors (KNN), and Support Vector Machine (SVM) are essential algorithms for classification tasks in machine learning. Naive Bayes uses Bayes' theorem to calculate the probability of a class given the features, assuming that the features are conditionally independent. This makes it efficient for large datasets and commonly applied in spam detection, text classification, and sentiment analysis. KNN is a non-parametric, instance-based learning method that classifies data points based on the majority class among its 'k' nearest neighbors in the feature space. It is simple to implement but can be computationally expensive for large datasets and sensitive to the choice of 'k' and distance metrics.SVM, on the other hand, aims to find a hyperplane in an N-dimensional space (N being the number of features) that distinctly classifies data points. It uses support vectors to define this hyperplane and maximizes the margin between different classes. SVM can also handle non-linearly separable data using kernel functions like polynomial or radial basis function (RBF), mapping input data to higher dimensions. It is robust to overfitting in high-dimensional spaces and is widely used in image recognition, bioinformatics, and text categorization.

Figure 1. Flow of design model

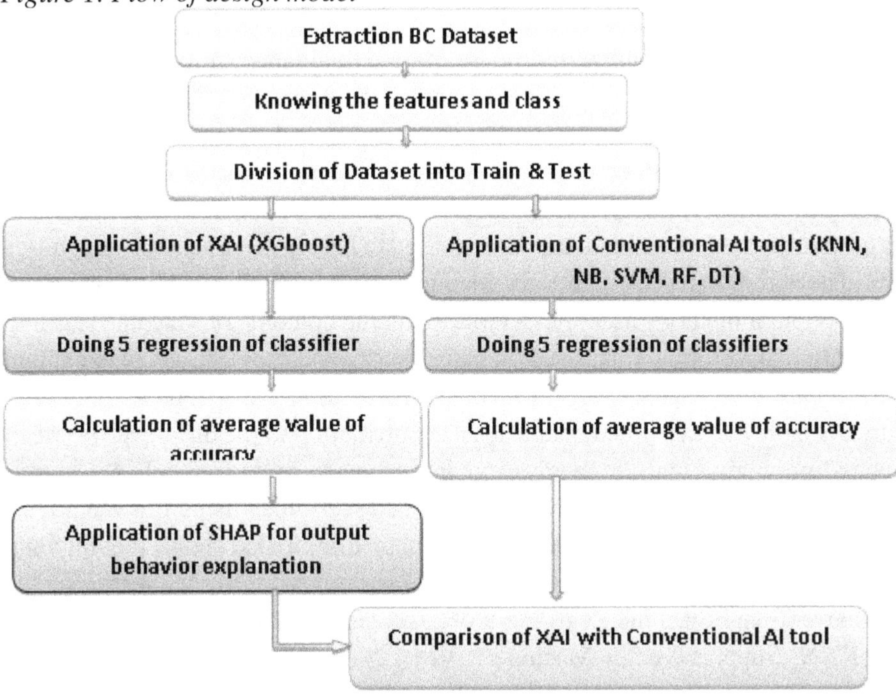

These justifications can aid medical professionals in comprehending the model's logic and locating any biases or mistakes. Data used for this work was obtained from the BC Surveillance Consortium (BCSC): https://www.bcsc-research.org/. The BCSC has 2,392,998 screening mammograms from women who are part of this risk estimation dataset, (Barlow *et al.,* 2006).

Data was preprocessed before being subjected to ML algorithms using XAI and classifiers like XGBoost in addition to other conventional ML classifiers like Naive Bayes, Support Vector Machines, and KNN for BC detection features. Performance parameters like accuracy, specificity, sensitivity, and f1score are used to evaluate machine learning algorithms. In the proposed study, the system was design to make use of fivefold cross validation on each classifier and average out the accuracy, precision, sensitivity, and f1score as shown in Figure 1.

3.1 Machine Learning Techniques

The proposed methodology makes use of an explainable AI machine learning tool and comparing it with conventional machine learning tools. XAI Explained AI attempts to simplify sophisticated, black-box AI models and system outputs. Roscher *et al.* (2020) used unique concepts from descriptive term definitions to identify transparency, interpretability, and explainability. SHAP is one of several analytical tools used in the study to explain the outcome and to help in more accurate diagnoses of BC. The acronym SHAP stands for Shapley Additive Explanations, it is one of the most popular methods used in XAI to explain the outcome of Machine learning classification algorithms. It is based on game theory and provides a way to explain the contribution of each feature to the model's predictions. SHAP assigns each feature a Shapley value, which quantifies its contribution to the difference between the actual prediction and the average prediction made by the model. A feature's impact on the value of the target variable can be ascertained using this method. It can provide both global insights and local explanation, it also shows that on what features the model is most reliable thereby giving high score to such features.

In conventional machine Learning tools, we were making use of Naive Bayes (NB), KNN, Support Vector Machine (SVM). Naive Bayes (NB) A supervised machine learning algorithm called Naive Bayes classifier classifies text. Like other generative learning algorithms, it simulates input distribution within a class or category. Support vector machines (SVM) SVMs are supervised machine learning methods that use decision planes to set boundaries between class data points in high-dimensional space. KNN KNN simply records all occurrences and categorizes data using similarity. It is used in statistical estimation.

4 RESULT

Traditional method for detecting BC is large scale analysis, feature extraction and classification, also the traditional machine leaning classification-based AI model don't give the information about the factors which are responsible for making decision whether the particular tumor is cancerous or not. This paper deals address this shortcoming by making use of Machine learning Classification based XAI model.

Figure 2. Shap values impact on model output (left) and summary plot shows the feature importance of each feature in the model (right)

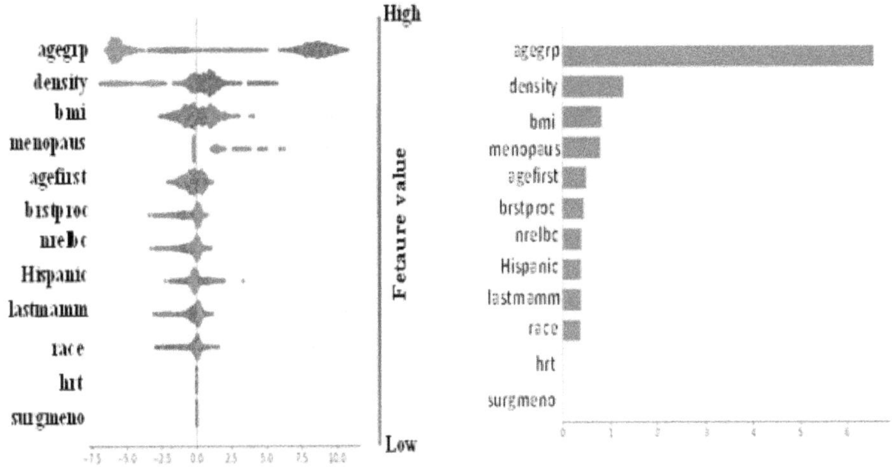

Figure 3. Waterfall plot (left) and decision plot (right)

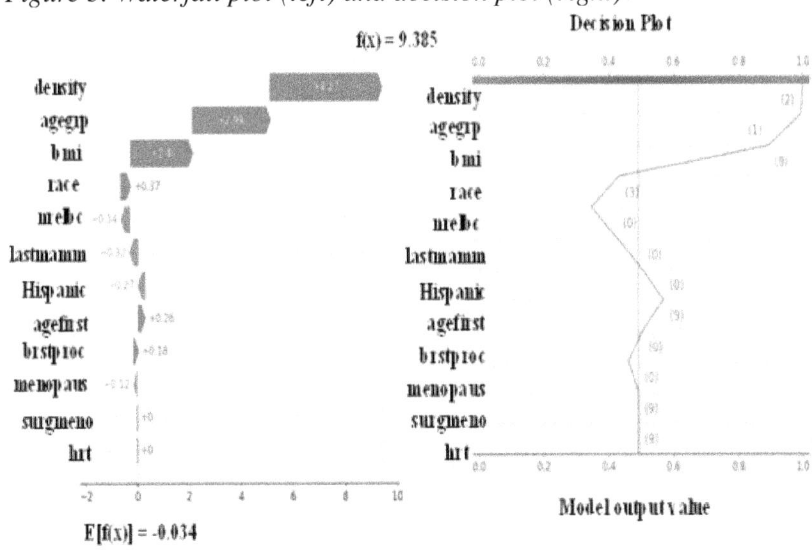

Figure 4. Force plot

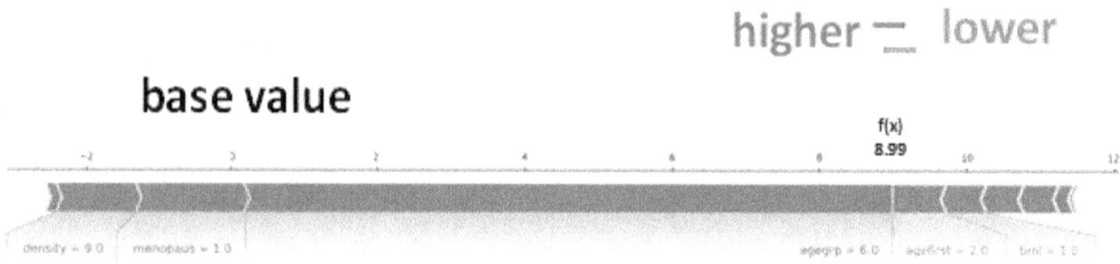

Figure 2 (left) shows the label "0" summary plot. On the Y-axis, feature names were ranked by importance. The SHAP value on the X-axis shows the log odds change. The feature values of each graph point are red for high values and blue for low values. Data rows from the original dataset are represented by each point. The "agegrp" feature is typically high with a negative SHAP value. **Figure 2** (right) shows the summary plot using SHAP and the testing set. The summary plot indicates model feature relevance. Agegrp, density, and bmi significantly affect the outcome. **Figure 3** (left) shows Waterfall plots provide explanations for particular forecasts; hence, they call for one row of an explanation object. A waterfall plot shows how each feature's +ve (red) or -ve (blue) contribution changes the value from the expected model output over the background dataset to the model output for this prediction in each row. **Figure 3** (right) shows the presentation of the decision plot. It maps cumulative SHAP values for each prediction to show model judgment. Each decision plot line indicates how strongly each feature contributed to a model prediction, describing what feature values drove it. **Figure 4** force plots show how each characteristic affects prediction for a given observation. Positive SHAP values are on the left and negative on the right in this plot, as though competing. The predicted observation value is indicated. SHAP force plots reveal which features most affected the model's prediction for a particular observation.

Table 1. Table shows the comparison between conventional AI tools and XAI

Parameters Classifiers	Sensitivity	Specificity	Accuracy	F1Score
NB	88.19	86.47	93.27	88.19
KNN	94.36	93.97	96.94	94.36
SVM	96.65	96.5	98.26	96.65
XAI(XGB)	98.81	98.78	99.29	98.81

In the above table, we can see the parameters like sensitivity, specificity, accuracy, and f1 score, which are the fundamental standards for determining the overall quality of any classifier for making decisions. This shows that XAI has better numbers as compared to conventional AI tools used for the detection of BC.

Figure 5. ROC curve

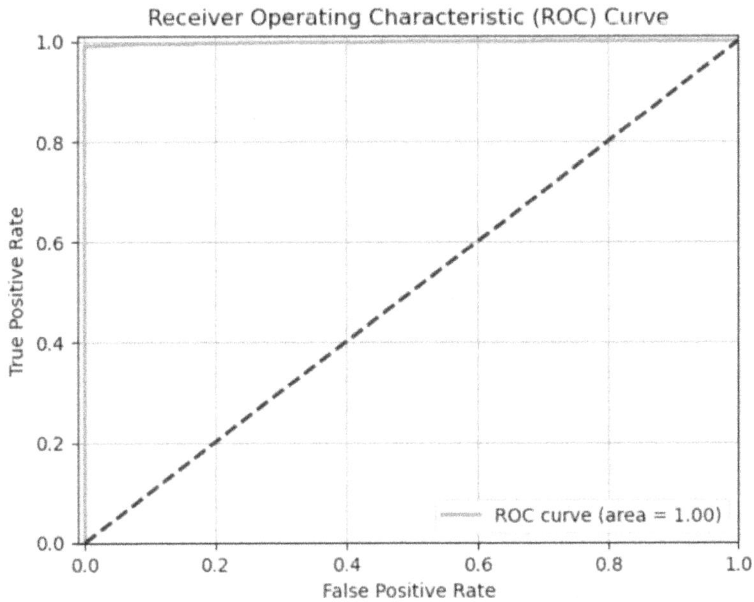

Figure 5 shows the ROC curve from the XGBoost classifier closely fits the top-left corner of the plot, indicating good classification performance. This shows a high true positive rate and low false positive rate across thresholds. The model's AUC of 0.9972 shows its good discrimination between cancer and non-cancer classes. A high AUC score shows the model can reliably distinguish the two classes in clinical screening, reducing misclassification risk. The sharp ROC curve increase, and early saturation demon-strate the model's robustness and confidence in predictions.

5 DISCUSSION

The present study demonstrates the significant advantages of using Explainable Artificial Intelligence (XAI) over conventional machine learning models for breast cancer (BC) risk estimation. The XGBoost model, paired with SHAP values, provides not only high predictive accuracy but also interpretability, making it more suitable for clinical decision-making. This combination of performance and transparency addresses the "black-box" issue often associated with traditional AI models, where decisions lack interpretability. Compared to conventional classifiers like Naive Bayes, KNN, and SVM, the XAI approach using XGBoost demonstrated superior metrics across sensitivity, specificity, accuracy, and F1-score. The ability of XAI to highlight feature importance enables a deeper understanding of the factors contributing to the predictions, such as age group, breast density, and BMI. This is crucial in healthcare, where understanding the reasons behind a prediction can guide personalized treatment plans.

SHAP values provided insights into both global and local feature importance, allowing clinicians to see which factors were most influential across the dataset and for individual predictions. For example, features like breast density and menopausal status consistently showed high SHAP values, indicating their strong influence on BC risk prediction. The use of SHAP also helped identify any potential biases in the model, contributing to more ethical and accurate decision-making. The results emphasize that incorporating explainable models in healthcare AI can enhance trust and facilitate more informed clinical decisions. The study lays the groundwork for future work integrating more complex patient data, including genetic information and lifestyle factors, which may further improve prediction accuracy and interpretability.

However, this study is with some limitations as the dataset used was obtained from a single source (BCSC), and although it is demographically diverse, external validation on datasets from different geo-graphic and healthcare backgrounds was not performed, which may affect the model's generalizability.

6 CONCLUSION

The purpose of this study aims to assess XAI BC prediction algorithms. Based on their strong performance in prior studies, AI machine learning techniques NB, KNN and SVM were selected for comparison with XAI using XGBoost for prediction of cancer and survival. Using XAI, we were able to comprehend the algorithm's forecast and produce a precise cancer prognosis. The output of the model was examined using SHAP, a collection of widely used, explainable AI methods. 'Black-box' AI models and system outputs that were more complex were transformed into easily understandable representations or naturally explicable models. The major objective of this research is to provide accurate disease forecasts based on symptoms.

The use of SHAP values allowed us to explore feature influence across individuals, which could help sur-face potential biases. The proposed XGBoost-based model is lightweight and computationally efficient, making it suitable for large-scale deployment in clinical environments. Since it operates on structured tabular data, it can be integrated into existing hospital information systems.

The model developed in this study was trained using demographic and tumor dimensional features from the BCSC dataset, which contains over 2 million mammograms. However, our analysis utilized a subset of this dataset consisting only of structured tabular data (i.e., demographic and clinical features), and not the raw image data.

The model training was conducted on a standard desktop workstation with the following specifications:

- Processor: Intel Core i7 (12th Gen)
- RAM: 16 GB
- GPU: None (model training was CPU-based)
- Operating System: Windows 10

Given the nature of the dataset and the model, the training time was relatively low. On average:

- Model training time: Approximately 10-15 seconds per run.
- Cross-validation (10-fold): Completed in under 5 minutes.

The computational cost was minimal, with no requirement for high-end GPUs or cloud resources. This suggests that the proposed approach is feasible for real-world deployment, especially in clinical settings with limited computational infrastructure.

7 FUTURE SCOPE

1) We intend to include clinical diagnosis and patient feedback into the model learning process in the future.
2) Although computer vision was not used in this article, we want to utilize it in the future to compare the two approaches and determine which is more accurate.
3) While this study focused solely on the BCSC dataset to establish a baseline and maintain data consistency, we plan to extend our future work by integrating additional publicly available datasets such as:
    - The Wisconsin Breast Cancer Dataset (WBCD)
    - Breast Cancer Coimbra Dataset
    - Indian Breast Cancer Dataset

This integration will allow us to evaluate cross-dataset generalizability and improve the model's adaptability to varied clinical environments.

4) To enhance clinical relevance, future work will involve real-world validation through pilot studies in collaboration with healthcare institutions, enabling prospective testing and assessment in diverse clinical settings. This will help evaluate the model's generalizability and trustworthiness in actual screening workflows.
5) To facilitate clinical adoption, future work will focus on developing an intuitive visual interface that allows clinicians to easily interact with the model and interpret its predictions.
6) As of now this study is based on BCSC standard dataset but in near plans are there to prepare our own dataset and for the same a set of AI ethical guidelines will be prepare focusing on patients consent.

ACKNOWLEDGEMENTS

This study was funded by Indian council of medical research (ICMR) India (Grant No. R.11013/15/2021-GIA/HR). The authors of this article are extremely thankful to them. "Data collection and sharing were supported by the National Cancer Institute-funded BC Surveillance Consortium (HHSN261201100031C). You can learn more about the BCSC at: https://www.bcsc-research.org/." The funding agencies had no role in the design, conduct, interpretation or writing of the study.

REFERENCES

Abbass, H. A. (2002). An evolutionary artificial neural networks approach for breast cancer diagnosis. *Artificial Intelligence in Medicine*, *25*(3), 265–281. DOI: 10.1016/S0933-3657(02)00028-3 PMID: 12069763

Abunasser, B. S., Al-Hiealy, M. R. J., Zaqout, I. S., & Abu-Naser, S. S. (2022). Breast cancer detection and classification using deep learning Xception algorithm. *International Journal of Advanced Computer Science and Applications*, *13*(7). Advance online publication. DOI: 10.14569/IJACSA.2022.0130729

Aguerchi, K., Jabrane, Y., Habba, M., & El Hassani, A. H. (2024). A CNN hyper-parameters optimization based on particle swarm optimization for mammography breast cancer classification. *Journal of Imaging*, *10*(2), 30. DOI: 10.3390/jimaging10020030 PMID: 38392079

Aiswarya, J. S. S. N. S., Sowbhagya Sree, P. S., Hemanjali, K., Pranavi, P., Babu, V., & Kumar, S. (2022). A review: Various challenges and causes based on cervical cancer. *2022 5th International Conference on Contemporary Computing and Informatics (IC3I)*, 1452–1457. https://doi.org/DOI: 10.1109/IC3I57145.2022.10002142

Ak, M. F. (2020). A comparative analysis of breast cancer detection and diagnosis using data visualization and machine learning applications. *Health Care*, *8*(2), 111. DOI: 10.3390/healthcare8020111PMC PMID: 32357391

Al-Tam, R. M., & Narangale, S. M. (2021). Breast cancer detection and diagnosis using machine learning: A survey. *Journal of Scientific Research*, *65*(5), 265–285. DOI: 10.37398/JSR.2021.650532

Amethiya, Y., Pipariya, P., Patel, S., & Shah, M. (2022). Comparative analysis of breast cancer detection using machine learning and biosensors. *Intelligent Medicine*, *2*(2), 69–81. DOI: 10.1016/j.imed.2021.08.004

Anas, M., Ul Haq, I., Husnain, G., Ali, S., & Jaffery, F. (2024). Advancing breast cancer detection: Enhancing YOLOv5 network for accurate classification in mammogram images.

Anastasi, G., Franchini, M., Pieroni, S., Buzzi, M., Buzzi, M. C., Leporini, B., & Molinaro, S. (2024). Machine learning techniques in breast cancer preventive diagnosis: A review. *Multimedia Tools and Applications*. https://doi.org/ *(Add DOI if available)*DOI: 10.1007/s11042-024-16809-9

Barlow, W. E., White, E., Ballard-Barbash, R., Vacek, P. M., Titus-Ernstoff, L., Carney, P. A., Tice, J. A., Buist, D. S. M., Geller, B. M., Rosenberg, R., Yankaskas, B. C., & Kerlikowske, K. (2006). Prospective breast cancer risk prediction model for women undergoing screening mammography. *Journal of the National Cancer Institute, 98*(17), 1204–1214. DOI: 10.1093/jnci/djj331 PMID: 16954473

Bouzar-Benlabiod, L., Harrar, K., Yamoun, L., Khodja, M. Y., & Akhloufi, M. A. (2023). A novel breast cancer detection architecture based on a CNN-CBR system for mammogram classification. *Computers in Biology and Medicine, 163*, 107133. DOI: 10.1016/j.compbiomed.2023.107133 PMID: 37327756

Choudhari, G., Desai, R., Dagale, P., & Dashetwar, I. (2023). Breast cancer prediction using various machine learning algorithms and their comparative analysis. In *Proceedings of the 2023 IEEE Pune Section International Conference (PuneCon)* (pp. 1–6). DOI: 10.1109/PuneCon58714.2023.10450082

Dar, C. R., Ruey, C. F., & Yu, H. L. (2000). Breast cancer diagnosis using self-organizing map for sonography. *Ultrasound in Medicine & Biology, 26*(3), 405–411. DOI: 10.1016/S0301-5629(99)00156-8 PMID: 10773370

Dar, R. A., Rasool, M., & Assad, A. (2022). Breast cancer detection using deep learning: Datasets, methods, and challenges ahead. *Computers in Biology and Medicine, 149*, 106073. DOI: 10.1016/j.compbiomed.2022.106073 PMID: 36103745

Ghose, P., Uddin, M. A., Islam, M. M., Islam, M., & Acharjee, U. K. (2022). A breast cancer detection model using a tuned SVM classifier. *2022 25th International Conference on Computer and Information Technology (ICCIT)*, 102–107. https://doi.org/DOI: 10.1109/ICCIT57492.2022.10003819

Gupta, A., Shridhar, K., & Dhillon, P. K. (2015). A review of breast cancer awareness among women in India: Cancer literate or awareness deficit? [Research Online]. *European Journal of Cancer, 51*(14), 2058–2066. DOI: 10.1016/j.ejca.2015.07.008 PMID: 26232859

Jony, A. I., & Arnob, A. K. B. (2024). Deep learning paradigms for breast cancer diagnosis: A comparative study on Wisconsin diagnostic dataset. *Malaysian Journal of Science and Advanced Technology, 4*(1), 109–117. DOI: 10.56532/mjsat.v4i2.245

) Joshi, A., & Mehta, A. (2017). Comparative analysis of various machine learning techniques for diagnosis of breast cancer. *International Journal on Emerging Technologies (Special Issue NCETST-2017), 8*(1), 522–526.

Karmakar, R., Chatterjee, S., Das, A. K., & Mandal, A. (2023). BCPUML: Breast cancer prediction using machine learning approach—A performance analysis. *SN Computer Science*, *4*(4), 377. Advance online publication. DOI: 10.1007/s42979-023-01825-x

Liu, J., & Ma, W. (2007). An effective recognition method of breast cancer based on PCA and SVM algorithm. In *Proceedings of the International Conference on Management of e-Commerce and e-Government* (pp. 57–64). Springer. DOI: 10.1007/978-3-540-77413-6_8

Maiti, S., Juneja, S., Goyal, R., & Chaudhry, N. (2023). Reviewing the landscape of deep learning approaches for breast cancer detection. In *Proceedings of the 2023 International Conference on Computing, Communication, and Intelligent Systems (ICCCIS)* (pp. 193–199). DOI: 10.1109/ICCCIS60361.2023.10425219

Meenalochinia, G., & Ramkumar, S. (2021). Survey of machine learning algorithms for breast cancer detection using mammogram images. *Materials Today: Proceedings*, *37*(2), 2738–2743. DOI: 10.1016/j.matpr.2020.08.543

Mudaliar, N. K., Redekar, S., Kadam, N. M., & Ramachandran, J. (2023). Machine learning-based prediction model for cancer detection. *2023 2nd International Conference on Applied Artificial Intelligence and Computing (ICAAIC)*, 383–388. https://doi.org/DOI: 10.1109/ICAAIC57855.2023.10193819

Naji, M. A., El Filali, S., Aarika, K., Benlahmar, E. L. H., Abdelouhahid, R. A., & Debauche, O. (2021). Machine learning algorithms for breast cancer prediction and diagnosis. *Procedia Computer Science*, *191*, 487–492. DOI: 10.1016/j.procs.2021.07.062

Nallamala, S. H., Mishra, P., & Koneru, S. V. (2019). Breast cancer detection using machine learning way. *International Journal of Recent Technology and Engineering*, *8*(2–3), 1402–1405.

) Omondiagbe, D. A., Veeramani, S., & Sidhu, A. S. (2019). Machine learning classification techniques for breast cancer diagnosis. [Publication details needed].

Prodan, M., Paraschiv, E., & Stanciu, A. (2023). Applying deep learning methods for mammography analysis and breast cancer detection. *Applied Sciences (Basel, Switzerland)*, *13*(7), 4272. DOI: 10.3390/app13074272

Raiaan, M. A. K., Fahad, N. M., Mukta, M. S. H., & Shatabda, S. (2024). MammoLight: A lightweight convolutional neural network for diagnosing breast cancer from mammography images. *Biomedical Signal Processing and Control*, *94*, 106279. DOI: 10.1016/j.bspc.2024.106279

Roscher, R., Bohn, B., Duarte, M. F., & Garcke, J. (2020). Explainable machine learning for scientific insights and discoveries. *IEEE Access : Practical Innovations, Open Solutions*, *8*, 42200–42216. DOI: 10.1109/ACCESS.2020.2976199

Rovshenov, A., & Peker, S. (2022). Performance comparison of different machine learning techniques for early prediction of breast cancer using Wisconsin Breast Cancer Dataset. *2022 3rd International Informatics and Software Engineering Conference (IISEC)*, 1–6. https://doi.org/DOI: 10.1109/IISEC56805.2022.10009384

Sandhu, J. K., Kaur, A., & Kaushal, C. (2022). Analysis of breast cancer in early stage by using machine learning algorithms: A review. *2022 IEEE International Conference on Current Development in Engineering and Technology (CCET)*, 1–7. https://doi.org/DOI: 10.1109/CCET56127.2022.00010

Shabbir, A., Ali, R. H., Shabbir, M. Z., Abideen, Z. U., Khan, T. A., Ijaz, A. Z., Ali, N., & Abu Bakar, M. (2023). Genetic algorithm-based feature selection for accurate breast cancer classification. In *Proceedings of the 2023 International Conference on IT and Industrial Technologies (ICIT)* (pp. 1–6). DOI: 10.1109/ICIT59216.2023.10335827

Sharma, S., Aggarwal, A., & Choudhury, T. (2018). Breast cancer detection using machine learning algorithms. In *Proceedings of the 2018 International Conference on Computational Techniques, Electronics and Mechanical Systems (CTEMS)* (pp. 114–118). IEEE. DOI: 10.1109/CTEMS.2018.8769187

Sneha, S., Sharma, B. S., Ruchitha, H. K., Kodipalli, A., Rao, T., & Rohini, B. R. (2023). Analysis of various machine learning classifiers for prediction and classification of breast cancer. In *Proceedings of the 2023 International Conference on Computational Intelligence for Information, Security and Communication Applications (CIISCA)* (pp. 285–290). DOI: 10.1109/CIISCA59740.2023.00062

Srinidhi, B., & Bhargavi, M. S. (2023). An XAI approach to predictive analytics of pancreatic cancer. In *Proceedings of the 2023 International Conference on Information Technology (ICIT)*.

Yan, F., Huang, H., Pedrycz, W., & Hirota, K. (2023). Automated breast cancer detection in mammography using ensemble classifier and feature weighting algorithms. *Expert Systems with Applications*, *227*, 120282. DOI: 10.1016/j.eswa.2023.120282

Yesuf, S. H. (2019). Breast cancer detection using machine learning techniques. *International Journal of Advanced Research in Computer Science*, *10*(5), 27–33. DOI: 10.26483/ijarcs.v10i5.6464

Chapter 5
IoT Applications in Agriculture, Healthcare, and Industry:
IoT Security and Privacy Challenges

P. Selvakumar
https://orcid.org/0000-0002-3650-4548
Department of Science and Humanities, Nehru Institute of Technology, Coimbatore, India

Shivprakash Varimani
https://orcid.org/0009-0000-1946-1809
Chandigarh Group of Colleges, Jhanjeri, India

Amitava Kar
Gopal Narayan Singh University, India

Rajkumar Mandal
https://orcid.org/0009-0005-1754-0569
Gopal Narayan Singh University, India

T. C. Manjunath
https://orcid.org/0000-0003-2545-9160
Rajarajeswari College of Engineering, India

ABSTRACT

The architecture of IoT can be categorized into several layers: data collection through sensors that detect environmental conditions, user inputs, or device states.

DOI: 10.4018/979-8-3693-9846-3.ch005

improving urban living conditions and resource efficiency. In agriculture, IoT solutions enhance crop monitoring and irrigation management, enabling farmers to maximize yield while minimizing water usage. Transportation systems benefit from IoT through connected vehicles that provide real-time traffic updates and enhance safety features, ultimately leading to more efficient logistics and reduced congestion. proliferation of particularly regarding security With billions of connected devices. This shift is particularly beneficial for applications requiring immediate feedback, In conclusion,, enabling automation, and fostering innovation. As we continue to explore and harness the capabilities of IoT, it is essential to prioritize responsible implementation practices that address these challenges while maximizing the benefits for society as a whole.

INTRODUCTION TO INTERNET OF THINGS (IOT): CONCEPTS, ARCHITECTURE, AND APPLICATIONS

The architecture of IoT can be categorized into several layers: data collection through sensors that detect environmental conditions, user inputs, or device states. improving urban living conditions and resource efficiency. In agriculture, IoT solutions enhance crop monitoring and irrigation management, enabling farmers to maximize yield while minimizing water usage. Transportation systems benefit from IoT through connected vehicles that provide real-time traffic updates and enhance safety features, ultimately leading to more efficient logistics and reduced congestion. proliferation of particularly regarding security With billions of connected devices. This shift is particularly beneficial for applications requiring immediate feedback, In conclusion,, enabling automation, and fostering innovation. As we continue to explore and harness the capabilities of IoT, it is essential to prioritize responsible implementation practices that address these challenges while maximizing the benefits for society as a whole. The future of IoT is not just about connecting devices; it's about creating smarter, more sustainable systems that enhance our daily lives and drive progress in numerous fields. As industries and consumers increasingly. which can communicate autonomously and enhance both efficiency and user experience across multiple sectors. Devices encompass sensors, actuators, and smart appliances that collect and act on data. Extension services and agricultural organizations can play a vital role in providing training programs that help farmers understand the benefits of precision agriculture and how to implement IoT solutions effectively. integrate robotics, machine learning, and artificial intelligence. For example, autonomous tractors and drones may become standard tools for crop monitoring and management, allowing farmers to scale their operations while maintaining precision and efficiency. Additionally, the growing emphasis on sustainability and environmental stewardship will likely drive further innovation in precision agriculture, as farmers.

IOT ECOSYSTEM AND TECHNOLOGIES: DEVICES, NETWORKS, AND PLATFORMS

Communication networks, data processing platforms, and applications. Central to this ecosystem are the devices themselves, which include a wide variety of sensors, actuators, smart appliances, and embedded systems. the necessary hardware to collect data from their environments and cooling systems to optimize energy usage and comfort. Other include smart wearables that track health metrics, industrial machines that report performance statistics, and agricultural sensors that provide real-time soil moisture data. The communication networks serve as the backbone of the IoT ecosystem, facilitating, Wi-Fi offers high-speed connectivity ideal for smart home devices, while LoRaWAN excels in low-power, long-range applications suitable for remote sensor networks in agriculture or environmental monitoring surgeries. Data processing platforms are another critical component of the IoT ecosystem. These platforms aggregate (Mahadev Madgule, et al.,2023). Cloud computing has traditionally played a significant role in this aspect, allowing for scalable storage and advanced analytical capabilities. However, edge computing is increasingly gaining traction as a complementary approach. adjustments to optimize operations. This hybrid approach to data processing enhances overall system efficiency and responsiveness. The role of software platforms cannot be overstated, as they provide the necessary tools for managing, analyzing, and visualizing complex. The industry is making strides toward developing universal standards that facilitate interoperability. Initiatives improved patient outcomes. Smart city initiatives leverage IoT technology to enhance urban infrastructure, optimize traffic flow, and improve energy efficiency through smart grid applications. In agriculture, precision farming techniques use IoT data to monitor crop health, soil conditions, and weather patterns, helping farmers make informed decisions that increase Similarly, in transportation, IoT applications facilitate fleet management, enabling real-time tracking of vehicles and efficient routing based on current traffic conditions(Agiollo, A., et al.,2021)..As IoT technology The rise of artificial intelligence is enabling more sophisticated data analysis, allowing for greater automation and smarter decision-making. Moreover, the increasing adoption of 5G technology is expected to enhance to environmental conservation efforts. The interplay between these components creates a dynamic environment that drives innovation and efficiency across various sectors. While challenges such as security, privacy, and interoperability remain (Ahmad, R., et al.,2021), ongoing advancements in technology and standardization efforts promise to unlock the full potential of IoT. As industries and consumers increasingly adopt IoT solutions, the future.

IOT SECURITY AND PRIVACY: OVERVIEW OF CHALLENGES AND CONCERNS

As billions of IoT devices come online, they create an intricate network where vulnerabilities can be exploited, leading to data breaches, unauthorized access, and even physical harm. Another major concern is the issue of data metrics. The aggregation of such data raises significant privacy issues, particularly regarding how it is stored, processed, and shared. Many consumers are unaware of the extent of data collection performed by their devices, and they may not fully understand the implications of this data being shared with third parties, including advertisers and data brokers. Moreover, the lack of transparency regarding data handling practices can erode consumer trust in IoT technologies, which is essential for widespread adoption. In addition to device-level vulnerabilities and privacy concerns, (Sugumaran K. M, D. et al., 2024), transmitted over networks can be intercepted if not adequately encrypted. Techniques such as being communicated, leaconcerns ((nauthorized modiD., et al. or access. Furthermore, the diversity of communication protocols and standards used across different IoT devices complicates the establishment of consistent security measures.

This distributed architecture complicates the implementation of uniform security policies and makes it challenging to conduct comprehensive security audits. Moreover, were physical and legal requirements. This disparity can leave consumers vulnerable and hinder the development of effective security strategies within the industry (Aldabbas, H., et al.,2021) can educating consumers importance of changing default passwords, enabling two-factor authentication, and understanding privacy settings. Clear communication about data collection practices and user consent can empower consumers to make informed choices regarding their IoT devices. Another effective strategy is to design IoT systems with security in mind from the outset, a concept known as "security by design." This approach emphasizes integrating security. By considering security as a fundamental aspect rather than an afterthought, manufacturers can create more resilient IoT devices that better withstand potential attacks.it also poses considerable security and privacy proactively addressed. From device vulnerabilities and data privacy concerns to regulatory compliance and the complexities of managing decentralized systems, the landscape of IoT security is multifaceted. By implementing robust security measures, educating consumers, designing with security in mind, and fostering collaboration among stakeholders (Balevi, E., et al.,2018)., the industry can create a safer and more trustworthy IoT environment. Ensuring the and realizing their full potential in enhancing our daily lives and transforming various sectors.

PRECISION AGRICULTURE WITH IOT: CROP MONITORING, YIELD OPTIMIZATION, AND WEATHER FORECASTING

Precision agriculture, empowered by tionizing the farming landscape by enhancing crop monitoring, optimizing yields, and improving weather forecasting. This innovative approach harnesses a network of connected devices their agricultural practices, allowing for more efficient and sustainable farming methods. decisions that significantly enhance productivity while minimizing resource use, ultimately contributing At the core of precision agriculture is crop monitoring, which involves the continuous observation of various factors(Sugumaran K. M, D., et al.,2024) For instance, if soil moisture levels drop below optimal thresholds, farmers can receive alerts, enabling them to irrigate precisely when needed, thus conserving water and ensuring crops receive adequate hydration. Additionally, satellite imagery and drones equipped with multispectral cameras can helping identify areas that may require special attention, such as pest infestations or nutrient deficiencies. Yield optimization is another significant benefit of integrating IoT into agriculture. For example, precision planting technologies utilize data to determine the ideal planting density and configuration for different crop varieties, leading to improved competition among plants and application can also be optimized through variable rate technology (VRT), which adjusts the amount of fertilizer applied based on specific soil and crop needs. By applying resources more effectively, Weather forecasting plays a pivotal role in precision agriculture, as weather conditions have a profound influence on crop performance. IoT-enabled weather stations can meteorological data, including levels (Vijayakumar G, et al., 2024). This data can be integrated with predictive analytics to provide accurate forecasts tailored to specific locations and crop types. Armed with this information, farmers can make informed decisions regarding planting planting and harvesting decisions.

This predictive capability not only helps in optimizing current crop performance but also aids in long-term strategic planning, ensuring that farmers can adapt to changing environmental conditions and market demands. (Bårli, E. M., et al.,2021).; they can also contribute to broader agricultural sustainability. By reducing over-reliance on water, fertilizers, and pesticides, IoT-driven practices can mitigate the negative environmental impacts often associated with traditional farming methods. For instance, precision irrigation techniques that use IoT data can lead to significant water savings, crucial in regions facing water scarcity. Similarly, optimized fertilizer application can decrease runoff, reducing pollution in nearby waterways and improving soil health over time. (Chemodanov, D., et al., 2020). However, the implementation of IoT in precision agriculture is not without its challenges. One major concern is the cost associated with deploying advanced technologies, including sensors, drones, and data analytics platforms. Small-scale farmers may find it difficult to invest in these technologies, which could exacerbate existing inequalities in agricultural productivity. Furthermore, the effectiveness of IoT solutions relies heavily on connectivity; regions with poor internet access may struggle to fully leverage the benefits of precision agriculture. To address these issues, (Paramasivam Selvakumar, et al.,2024) stakeholders must work towards creating affordable solutions and improving connectivity in rural areas, ensuring that all farmers can, and cybersecurity measures and transparent data-sharing policies is essential to building trust in IoT technologies among farmers. Extension services and agricultural organizations can play a vital role in providing training programs that help farmers understand the benefits of precision agriculture and how to implement IoT solutions effectively. integrate robotics, machine learning, and artificial intelligence. For example, autonomous tractors and drones may become standard tools for crop monitoring and management, allowing farmers to scale their operations while maintaining precision and efficiency. Additionally, the growing emphasis on sustainability and environmental stewardship will likely drive further innovation in precision agriculture, as farmers.

LIVESTOCK MONITORING AND MANAGEMENT WITH IOT: HEALTH, BEHAVIOR, AND NUTRITION ANALYSIS

Livestock monitoring and management have been significantly transformed by enhancing the health, behavior, and nutrition analysis of animals. For instance, a sudden change in an animal's temperature or behavior may indicate the onset of disease, allowing farmers to take prompt action, such as isolating of diseases within herds, ultimately leading to lower veterinary costs and increased productivity. Behavior monitoring is another critical aspect of IoT-enabled livestock management. Understanding animal behavior is essential for improving welfare and productivity. Sensors can track movement patterns, feeding habits, and hierarchies' farmers (Chowdhury, R. R et al.,2022)). Nutrition analysis, facilitated by IoT, allows farmers to optimize feed efficiency and improve livestock health. Smart feeders equipped with sensors can measure the amount of feed consumed by each animal, providing valuable data on dietary intake and nutritional needs. By analyzing this data, farmers requirements of different animals, accounting for factors such as age, weight, and reproductive status. This Ramakrishnan T., et al.,2022) The use of IoT in livestock management also enhances traceability and compliance with food safety standards.

Each animal can be assigned a unique identifier linked to its health and nutrition data, creating a digital record that can be accessed and audited by stakeholders in the supply chain. This level of traceability not only ensures transparency for consumers but also helps farmers comply with regulatory requirements related to animal welfare and food safety.

In the event of an outbreak or contamination issue, having detailed records allows for swift action. In addition to health and nutrition, IoT technologies can assist in environmental monitoring, ensuring that livestock farming weather patterns, providing farmers with insights needed to manage grazing and reduce overgrazing. By analyzing this data, farmers can implement rotational grazing practices that enhance soil health and prevent land degradation. Furthermore, by optimizing governments and agricultural organizations can provide support through grants or subsidies that encourage the adoption of IoT solutions among all farmers, promoting equity in the agricultural sector. Farmers must be assured that their data will be protected through robust cybersecurity measures and transparent data-sharing agreements. Implementing best practices in data management and involving stakeholders in discussions about data ownership will be vital in building trust in IoT technologies. Moreover, training and education are crucial for the effective use of IoT in livestock management. Farmers must be equipped with the skills devices and integrate these insights into their management practices. Extension services and agricultural education providing the necessary training, ensuring that farmers understand how to leverage IoT technologies effectively. Looking to enabling predictive analytics that forecast health issues (Danielis, P., et al.,2020), optimize nutrition, and improve overall herd management. Additionally, advancements in drone technology may provide aerial monitoring capabilities, offering proactively manage livestock health, optimizc fccding strategies, and ensure sustainable farming practices.

IOT-ENABLED FARM AUTOMATION AND ROBOTICS: AUTONOMOUS FARMING, HARVESTING, AND IRRIGATION

IoT-enabled farm automation and robotics are revolutionizing agriculture by enhancing efficiency, precision, and productivity across various farming operations. As the global ntly reduce labor costs and optimize resource use (Selvakumar, P.,2023). These tractors use real-time data from sensors to adjust their operations machines analyze various factors such as fruit color, size, and texture to determine the optimal harvesting time, thereby maximizing yield quality. By automating this labor-intensive process, farmers can reduce harvest time and labor costs while minimizing damage to crops. Moreove(Doan, T. V. T., et al.,2023).r, IoT-enabled harvesters can communicate with other farm machinery, creating a cohesive ecosystem where data is shared across devices, further enhancing operational efficiency. Irrigation management is critical in agriculture, especially in regions where water resources are limited. IoT technology allows enabling farmers to apply water only when necessary. This not only conserves water but also promotes healthier crops by preventing over- or under-irrigation. Furthermore, IoT-enabled irrigation systems can be programmed to operate autonomously, adjusting water delivery based on predictive analytics that factor in weather forecasts and soil conditions. The integration of drones into farm automation is also gaining traction. areas quickly and take corrective actions before issues escalate. This aerial data complements ground-based sensors, creating a comprehensive monitoring system that enhances decision-making and operational effectiveness.The benefits of IoT-enabled farm automation extend beyond operational optimizing resource use—such as water, fertilizers, and pesticides—automated systems can reduce the environmental impact of farming. For example, precision application technologies ensure that inputs are applied only where needed, minimizing the gap between large agribusinesses and smaller operations.

To mitigate this, governments and agricultural organizations can offer financial assistance, grants, or subsidies to encourage the adoption of IoT technologies among all farmers, fostering inclusivity in agricultural innovation. Connectivity issues, particularly in rural areas, also pose a challenge. IoT systems rely heavily on reliable internet access to function effectively. In regions where connectivity is limited, farmers may struggle to implement automated solutions. Investing in rural broadband infrastructure and exploring alternative communication methods, such as satellite-based systems, can help bridge this gap, ensuring that all farmers can access the benefits of IoT technologies. Data security and privacy. Farmers must be assured that their data will be protected from unauthorized access and misuse. Establishing robust cybersecurity measures and transparent data-sharing policies will be vital in building trust among users of IoT technologies in agriculture. Farmers need to understand how to operate new technologies, interpret data effectively, and integrate insights into their management practices. Extension services and agricultural education programs can provide the necessary support, ensuring that farmers are equipped to leverage IoT solutions to their fullest potential. Looking toward the future, the potential of IoT-enabled farm automation and robotics is vast. anticipate further innovations that enhance the capabilities of autonomous systems. For instance, advancements in artificial intelligence and greater transparency regarding the origins of their food. In conclusion, IoT-enabled farm automation and robotics represent a paradigm shift in agriculture, allowing for autonomous farming, efficient harvesting, and optimized irrigation.

IOT IN HEALTHCARE: PATIENT MONITORING, TELEMEDICINE, AND WEARABLE DEVICES

Revolutionizing patient care, enhancing monitoring, facilitating telemedicine, and enabling the use of wearable devices, ultimately leading to more efficient and personalized healthcare solutions. As the demand for better health outcomes grows alongside the challenges posed by aging populations and rising healthcare costs, IoT technologies provide innovative solutions that improve patient engagement, that significantly improves clinical decision-making. Telemedicine has also gained significant traction through IoT technologies (Fatorachian, H., et al.,2021)., providing a viable alternative to traditional in-person consultations. With the advent of high-speed internet and smart devices, healthcare providers can conduct remote consultations, monitor patients' conditions, and deliver care more efficiently. conferencing, mobile apps, and digital platforms, ensuring that patients have access to medical advice without the barriers of distance smartwatches, and specialized medical wearables, empower individuals to take control of their health by monitoring physical activity, sleep patterns, and vital signs. For instance, devices like the Apple Watch or Fitbit can track heart rate variability, detect irregularities, and provide insights into overall health and fitness levels. More advanced wearables, designed for specific medical conditions. Moreover, IoT-enabled wearables numerous advantages of IoT in healthcare, there are significant challenges that must be addressed to ensure the successful implementation and adoption of these technologies. One major concern is data security and privacy. The sensitive nature of health information makes it a prime target for cyberattacks (Fatorachian, H., et al.,2021). Standardization of protocols and collaborative efforts among manufacturers, healthcare providers, and regulatory bodies can help facilitate interoperability, ensuring that data can be shared and utilized effectively across different platforms. Training and education are also critical components for the knowledge and skills to effectively use IoT devices and interpret the data generated. This includes understanding how to engage patients in their care, guiding them on how to use wearables, and ensuring they are aware of the importance of monitoring their health. This predictive capability can empower healthcare systems to implement preventive measures and allocate resources more effectively.

IOT-ENABLED CLINICAL DECISION SUPPORT SYSTEMS: PREDICTIVE ANALYTICS AND PERSONALIZED MEDICINE

The advent of IoT-enabled clinical decision support systems (CDSS) is transforming healthcare by integrating predictive analytics and personalized medicine, enhancing the quality of care while improving patient outcomes. real-time, Predictive analytics, a cornerstone of IoT-enabled CDSS, real-time patient data, identifying patterns that can predict future health outcomes (Haq, T. U., et al.,2021). This capability allows healthcare providers to adopt a proactive approach, implementing preventive measures and lifestyle modifications that can significantly alter a patient's health trajectory. For example, if a predictive model indicates a high risk of heart disease in a patient, the healthcare team can prioritize interventions such as dietary changes, exercise programs, or regular monitoring to mitigate that risk. Furthermore, IoT-enabled CDSS enhances. In contrast, personalized medicine focuses on understanding these differences to deliver more precise treatments. IoT devices, such as wearables and smart health monitors, continuously collect data on vital signs, medication adherence, and lifestyle habits, feeding this information into CDSS for comprehensive analysis. This enables the system to offer personalized treatment options that are most likely to be effective for a particular patient, thus optimizing therapeutic outcomes. An excellent example of the application of IoT-enabled CDSS is in oncology. Cancer treatment often involves complex decision-making processes that require consideration of numerous factors, including tumor genetics, including genomic sequencing and real-time health monitoring, CDSS can assist oncologists in selecting the most appropriate therapies, dosing regimens, and monitoring protocols tailored to individual patients. This level of personalization can lead to better responses to treatment, reduced side effects, and improved overall survival rates. Additionally, IoT-enabled CDSS can facilitate collaborative care among multidisciplinary must be equipped with the knowledge and skills to interpret predictive analytics and understand the implications for patient care. Educational initiatives and ongoing training programs can help healthcare providers gain proficiency in utilizing these advanced systems, enhancing their ability to make data-driven decisions that improve patient outcomes. Looking forward, the potential for IoT-enabled CDSS in healthcare is vast. These tools will not only improve the accuracy of predictions but also enhance the system's ability to learn from new data over time, further refining personalized treatment recommendations. Moreover, the integration of genomics and in addition, (Hassani, H. L., et al., (2021). the expansion of remote monitoring capabilities through IoT devices will enhance the utility of CDSS. By continuously tracking patients' vital signs and health status, these systems can provide real-time alerts and recommendations, allowing for timely interventions.

IOT-DRIVEN MEDICAL RESEARCH AND INNOVATION: DATA ANALYTICS AND ARTIFICIAL INTELLIGENCE

This wealth of information allows researchers and healthcare professionals to identify trends, monitor disease progression, and evaluate treatment efficacy in real-time (Imteaj, A., et al.,2021)., fundamentally transforming the research paradigm and leading to more informed decisions in patient care-scale, real-world evidence studies. Traditional clinical trials often face challenges such as limited participant diversity, lengthy timelines, and high costs, which can hinder the generalizability of findings. In contrast, IoT technologies enable researchers to gather data from diverse populations over extended periods, capturing a more comprehensive picture of patient experiences and treatment responses. For instance, wearable devices can monitor a participant's vital signs and activity levels outside the controlled environment of a clinical trial, analyzing data from thousands of patients with chronic conditions, researchers can uncover factors that contribute to disease exacerbation or treatment failure. These insights can inform clinical guidelines and lead to the development of targeted therapies Artificial intelligence, when combined with IoT data, takes the potential for medical research to new heights. AI algorithms can analyze a patient's genomic data alongside real-time monitoring from wearable devices to recommend personalized treatment plans based on individual tumor characteristics and patient health status. Collaboration among researchers, healthcare providers, and technology developers is essential for maximizing the benefits of IoT-driven medical research. Interdisciplinary partnerships facilitate the sharing of knowledge. By working together, stakeholders can develop more robust IoT infrastructures, standardize data collection methods, and ensure compliance with ethical and regulatory guidelines. Such collaboration is crucial for building trust among patients, as they must be assured that their data will be used responsibly and ethically. Despite the transformative Interoperability among IoT devices and health information systems is another critical challenge. Many IoT devices operate on diverse platforms and protocols, complicating data integration and analysis. Establishing common standards for data formats and communication protocols will be essential for creating cohesive ecosystems that facilitate seamless data sharing among devices and systems. Efforts to promote interoperability will enable researchers to harness the full potential of IoT data, ultimately leading to more robust research findings and improved patient care and interpret the data generated. Educational initiatives and training programs can help build expertise in data analytics.

IOT IN MANUFACTURING: PREDICTIVE MAINTENANCE, QUALITY CONTROL, AND SUPPLY CHAIN OPTIMIZATION

As industries strive for greater productivity and reduced operational costs, IoT technologies are proving invaluable in enabling manufacturers to monitor processes in real time, make informed decisions, and ultimately drive continuous improvement. By integrating connected devices, sensors, and advanced analytics, manufacturers can collect and analyze vast amounts of data, leading to insights that enhance production capabilities and streamline workflows. In contrast, IoT-enabled predictive maintenance uses real-time data from sensors installed on machinery to monitor equipment health continuously. By analyzing data on vibration, temperature, pressure, and other operational parameters, manufacturers can identify early warning signs of potential failures. Quality control is another critical area where IoT technologies are making a significant impact. Ensuring product quality is paramount in manufacturing. By continuously analyzing this data, manufacturers can detect deviations from predefined quality standards immediately. For instance, if a temperature fluctuation is detected during the curing process of a composite material, the system can trigger corrective actions to ensure that product specifications are met. This real-time monitoring not only enhances product quality but also improves compliance with industry regulations, as manufacturers can provide verifiable data on production processes. manufacturing landscape is increasingly globalized, with complex supply chains that involve multiple suppliers, logistics providers, and distribution channels.

This visibility allows for better inventory management, as manufacturers can avoid stockouts or overstock situations by knowing exactly when to reorder materials or products. Furthermore, real-time data on shipment status can help manufacturers respond swiftly to delays or disruptions, ensuring that production schedules are maintained and customer commitments are met. The integration of IoT in manufacturing also paves the way for enhanced data analytics and machine learning capabilities. By collecting and analyzing data. For instance, by analyzing historical data, manufacturers can identify peak production times and allocate resources accordingly to maximize output (Khan, M. A., et al.,2018). As manufacturing systems become increasingly interconnected, the risk of cyberattacks escalates. A breach in the system can lead to production downtime, loss of sensitive data, and substantial financial losses. Manufacturers the manufacturing requires a cultural shift within organizations. Employees at all levels must understand the value of IoT technologies and be trained to leverage them effectively. This involves fostering a mindset of continuous improvement and encouraging collaboration between AI with IoT data can lead to more sophisticated predictive maintenance models that take into account not only equipment performance but also external factors such as weather conditions and market demand. This level of intelligence will enable manufacturers to optimize their operations dynamically, adapting to changing circumstances in real time. With its ultra-low latency and high-speed connectivity, 5G will enable manufacturers to connect a larger number of devices and sensors in real time.

IOT-ENABLED ENERGY MANAGEMENT AND SMART GRIDS: ENERGY EFFICIENCY AND RENEWABLE ENERGY

As global demand for energy continues to rise alongside the pressing need to combat climate change, IoT technologies are playing a pivotal role in enhancing energy efficiency, optimizing renewable energy sources, and creating resilient energy systems. By leveraging connected devices, sensors, and advanced analytics, stakeholders can monitor energy consumption patterns in real time, facilitate demand response initiatives, and ensure the seamless integration of renewable energy into the grid, ultimately paving the way for a more sustainable energy future. Smart meters, connected appliances, and environmental sensors provide real-time insights into energy usage across residential, commercial, and industrial sectors. (Khan, M. A., et al., 2018) Moreover, IoT technologies enable electricity demand, a utility can signal smart appliances to temporarily reduce energy consumption, alleviating pressure on the grid. This not only helps maintain grid proactive approach to grid management ultimately leads to more reliable energy delivery and improved service for consumers. Aspect of IoT-enabled energy management. As the world shifts towards cleaner energy solutions, the ability to manage the variability of renewable sources becomes increasingly important. IoT technologies facilitate the real-time monitoring of renewable energy generation and consumption, enabling utilities to optimize the use of these resources. For example, smart inverters can adjust the flow of electricity from solar panels based on grid demand, ensuring that excess energy is stored or redirected appropriately.

Figure 1. Impact of IoT in agriculture

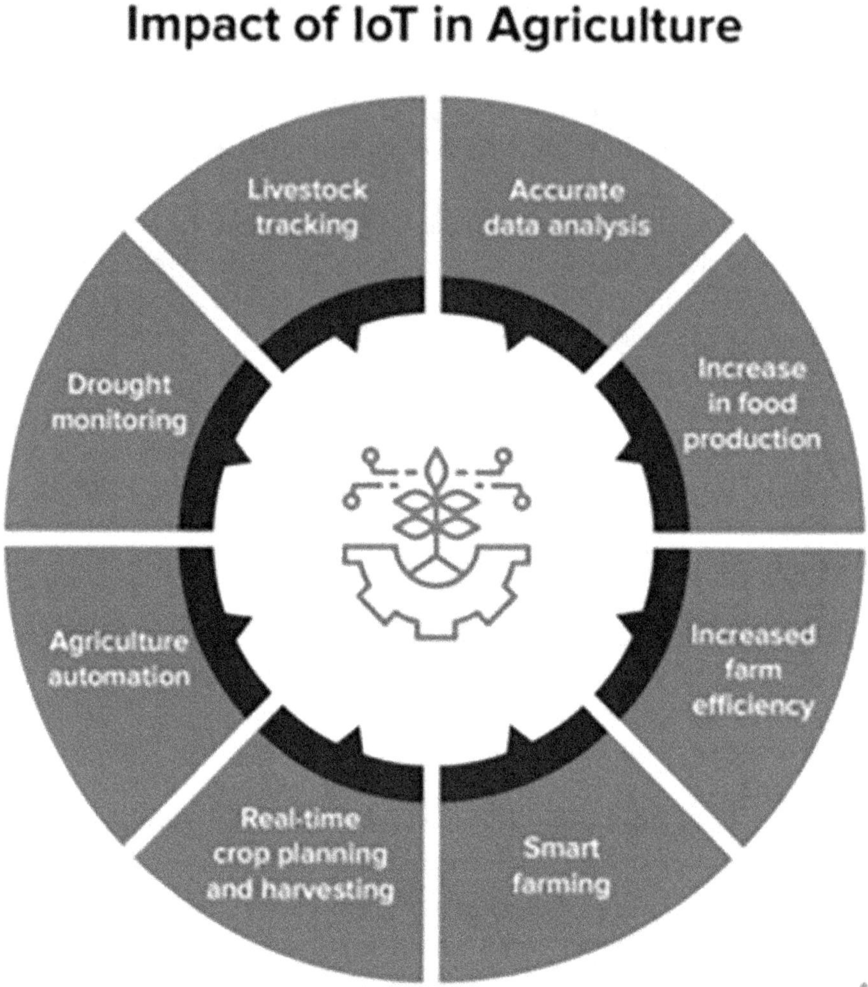

This capability not only maximizes of low demand and discharged during peak periods, providing valuable grid services while optimizing energy costs for consumers. This distributed approach building owners can monitor and control energy usage, implement automated demand response strategies, and achieve sustainability certifications such as LEED (Leadership in Energy and Environmental Design). The collection of vast amounts of energy usage data raises questions about how this information is stored, used, and protected. Utilities and technology providers must prioritize robust cybersecurity measures to safeguard sensitive consumer data from cyber threats. This includes implementing encryption, secure access controls

(Liang, X., et al.,2021)., and continuous monitoring for potential vulnerabilities. compliance with relevant regulations can help build trust in IoT-enabled energy management systems. Interoperability is the successful implementation of IoT in energy management. The energy sector consists of a diverse range of devices and systems, often operating on different communication protocols. Establishing common standards and ensuring compatibility among various technologies will be crucial for facilitating seamless integration and data sharing. can energy systems. implement demand response strategies and manage distributed energy resources. Educational programs and training initiatives can empower professionals in the energy sector to embrace new technologies and adapt to changing industry dynamics. renewable energy integration. Moreover, the integration of artificial intelligence (AI) create resilient energy systems. While challenges such as data privacy, interoperability, and training must be addressed.

IOT-DRIVEN TRANSPORTATION SYSTEMS: AUTONOMOUS VEHICLES, ROUTE OPTIMIZATION, AND TRAFFIC MANAGEMENT

Route optimization and traffic management. As urban populations swell and demand for efficient mobility solutions increases, IoT technologies offer transformative potential for creating smarter, safer, and more efficient transportation networks. By leveraging connected devices, real-time data analytics, and advanced algorithms. Autonomous vehicles (AVs) are at the forefront of IoT-driven transportation innovations, (Mahmoud, R., et al.,2015). By continuously analyzing data from various sources, including traffic signals and road conditions, these systems can optimize driving patterns, reduce fuel consumption, and lower emissions. For instance, adaptive cruise control systems can adjust a Route optimization is another critical application of IoT in transportation (Ni, J., et al.,2017).. Advanced algorithms can analyze this data to identify the fastest or most efficient routes, taking into account current traffic conditions, construction activities, and even weather forecasts. For example, ride-sharing platforms use IoT data to match passengers with drivers efficiently, optimizing routes in real time to minimize wait times and reduce congestion. Traffic management systems also benefit significantly from IoT technologies. Traditional traffic management methods often rely

In contrast, IoT-enabled traffic management systems utilize a network of connected sensors and cameras to monitor traffic flow continuously. This real-time data allows traffic management centers or rerouting traffic as needed. For example, during peak hours or after an accident, traffic signals can be adjusted to prioritize routes with the highest congestion, ensuring smoother traffic flow and reducing delays. By providing real-time information on public transit schedules, routes, and occupancy levels, IoT technologies can encourage more people to use public transportation options. For instance, mobile applications robust cybersecurity measures, including encryption, secure communication protocols, and continuous monitoring of systems for vulnerabilities. Interoperability is another critical challenge facing IoT-driven transportation systems. The transportation ecosystem comprises a diverse range of devices, platforms, and technologies, often operating on different communication standards. Establishing common protocols and frameworks is essential for facilitating seamless data sharing and integration across various systems. Collaborative efforts among transportation agencies, technology. Transportation authorities must embrace innovation and be willing to invest in new technologies that enhance efficiency and safety. Additionally, public education campaigns can help raise awareness about the benefits of IoT in transportation, encouraging citizens to adopt new mobility solutions and participate in smart transportation initiatives. systems is promising. Evolution, enabling faster and more reliable communication between vehicles, infrastructure, and cloud-based systems. This enhanced connectivity will facilitate real-time data exchange, supporting the infrastructure investments and traffic management strategies.

IOT SECURITY THREATS AND VULNERABILITIES: DEVICE, NETWORK, AND DATA SECURITY

The rapid expansion of and even jeopardize public safety. Understanding the various dimensions of IoT security threats is critical for developing effective strategies to mitigate risks and ensure the integrity and privacy of IoT systems (Pacheco, J., et al.,2020). At the device level, IoT security vulnerabilities often stem from inadequate security measures during the design and manufacturing phases. hardcoded credentials that users fail to change, allowing attackers to gain unauthorized access easily. Additionally, the lack of regular firmware updates means that known vulnerabilities remain unpatched, providing a pathway for Network security to overwhelm networks. Data security is equally important in the context of IoT. health records, and financial details. This data is a prime target for cybercriminals who seek to exploit it for identity theft, fraud, or corporate espionage. Ensuring data security involves implementing encryption protocols both in transit and at rest, ensuring that data is unreadable to unauthorized. Another significant concern in IoT security is the of standardized security protocols can lead to inconsistencies in how devices communicate and authenticate each other. This variability increases the potential for misconfigurations and vulnerabilities. For instance, if one device in a network uses outdated encryption methods while others use more secure protocols, it creates a weak link that attackers can exploit. Industry-wide collaboration is essential for developing common security standards that promote interoperability while ensuring robust protection against threats. Moreover, the lifecycle of IoT devices poses challenges for security. Unlike traditional IT assets, IoT devices often remain in operation for several years lifespan. The role of users in IoT security cannot be overstated. For instance, a cyberattack on an industrial control system could disrupt As the IoT landscape. For example, AI-driven security systems can monitor device behavior and establish baselines for normal activity, enabling them to detect unusual patterns indicative of a cyberattack. However, the use of AI in security also presents challenges, including the potential for adversarial attacks that seek to manipulate AI algorithms (Pisner, D. A., et al.,2020). Therefore, looking to the future, the development of regulatory frameworks and industry standards will be crucial in addressing IoT security challenges. Governments and industry bodies must collaborate to establish guidelines that mandate robust security practices for IoT manufacturers and users. Regulations that promote transparency in device security features, data protection measures, and incident reporting can help create a safer IoT ecosystems. In conclusion, IoT security threats and vulnerabilities present significant challenges across device, network, and data security domains. The interconnected nature of IoT devices creates a complex landscape where a single vulnerability can lead to widespread consequences. To address these challenges, stakeholders must prioritize security throughout the entire IoT lifecycle.

IOT PRIVACY CONCERNS AND DATA PROTECTION: DATA OWNERSHIP, CONSENT, AND ANONYMITY

It has also raised significant privacy concerns regarding data protection, ownership, consent, and anonymity. The interconnectedness of IoT devices not only amplifies the scale of data generated but also complicates the management of privacy rights, making it imperative for stakeholders to address these concerns comprehensively. At the heart of IoT privacy concerns is the issue of data ownership (Rana, A., et al.,2024). Unlike traditional data generation, where individuals often retain control over their information, IoT complicates this dynamic. When devices collect data, it raises questions about who owns that data—the user, the manufacturer, or the service provider. For instance, data generated by a fitness tracker can provide insights into a user's health, but the company behind the device often retains ownership rights. This ambiguity can lead to conflicts regarding how data is used, shared, and monetized. Users may be unaware that their data is being sold or shared with third parties for purposes beyond their original intent. Therefore, establishing clear policies on data ownership is critical to empower users and ensure transparency. Consent is another pivotal aspect of privacy in the IoT ecosystem. In many cases, users must agree to terms and conditions to use IoT devices, but these agreements are often lengthy and complex, making it challenging for users to fully understand what they are consenting to. This lack of clarity can lead to uninformed consent, where users unknowingly permit extensive data in understandable language. information about a person's lifestyle, habits, and preferences. While some IoT systems may implement anonymization techniques to protect user identities, the effectiveness of these methods is often debated. Skilled adversaries may be able to re-identify anonymized data through sophisticated techniques, potentially leading to privacy breaches.

Therefore, implementing robust anonymization methods that ensure data cannot be traced back to individuals is vital for preserving user privacy (Riahi, A. et al.,2013), secure communication protocols, and regular software. Regulatory frameworks play a crucial role in addressing IoT privacy concerns. As the landscape of connected devices continues to evolve, governments and regulatory bodies must adapt existing privacy laws or create new regulations to account for the unique challenges posed by IoT. For instance, the General Data Protection Regulation (GDPR) associated with their devices. Educational initiatives can help users understand how their data is collected, what it is used for, and how they can protect their privacy. Manufacturers can play a proactive role by providing resources that educate consumers about best practices for device security, can foster a more informed and vigilant consumer base that actively participates in managing their privacy. As IoT technologies continue to advance, emerging trends such as edge computing and decentralized data storage offer potential solutions for enhancing privacy and data protection. is generated, reducing the need to transmit sensitive information to centralized servers. This approach can help minimize data exposure during transmission and limit the amount of personal data that is stored in the cloud. Similarly, decentralized storage solutions can empower users to maintain control over their data by allowing them to store information on their devices rather than relying on third-party services. Also, an essential consideration. Companies must balance the benefits of data-driven insights with the need to respect user privacy. Ethical data practices involve and respect for user autonomy. Organizations should establish ethical guidelines for data collection and usage, ensuring that users' rights are upheld and that their data is used responsibly. This ethical approach can help build trust between companies and consumers, fostering a positive relationship that encourages user engagement and loyalty.

IOT SECURITY STANDARDS AND REGULATIONS: COMPLIANCE, CERTIFICATION, AND BEST PRACTICES

Creating new opportunities for innovation. However, this proliferation of connected devices has also raised considerable security concerns, necessitating the establishment of robust IoT security standards and regulations (Salem, O., et al.,2021). Similarly, the Internet Engineering Task Force (IETF) has been working on protocols and standards, such as the Constrained Application Protocol (CoAP), to enhance security in IoT communications. in Europe impose strict requirements on. Furthermore, the California Consumer Privacy Act (CCPA) has also introduced data privacy regulations that impact IoT manufacturers and service providers. Organizations must remain vigilant in understanding and implementing compliance measures to adhere to evolving regulations across different jurisdictions. Independent certification bodies assess IoT devices and systems against established security standards, ensuring they meet minimum security requirements before they enter the market. The IoT Security Foundation (IoTSF) and the Trusted Computing Group (TCG) are examples of organizations that provide certification frameworks for IoT devices. These programs often include rigorous testing and validation processes, encompassing aspects such as device authentication, secure communication protocols, and data protection measures. By obtaining certifications, manufacturers can demonstrate their commitment to security, fostering consumer trust and encouraging adoption. Best practices for IoT security encompass a broad range of strategies aimed at minimizing vulnerabilities and enhancing overall security posture. One critical best practice is to implement a robust security-by-design approach during the development phase of IoT devices. This approach involves embedding security features into the design process from the outset, address potential vulnerabilities before products are deployed, software to protect against newly discovered threats. using virtual private networks (VPNs) and firewalls can help protect data in transit and enhance overall network security. Data encryption is another crucial best practice (Salem, O., et al.,2021). However, the deployment of AI in security must be carefully managed, as adversaries can also employ AI techniques to launch sophisticated attacks. Continuous monitoring and improvement of security measures are necessary to stay ahead of evolving threats. necessitating a coordinated approach to establishing regulations and best practices. Industry alliances and partnerships can foster knowledge sharing, encourage the adoption of standardized security measures, and promote responsible innovation. Organizations must actively engage with regulators, standards bodies, and industry consortia to contribute to the development of effective security frameworks, underscores the need for stringent security standards. A successful breach in these areas could have dire consequences, affecting public safety and national security. Consequently, it is imperative tailored to high-risk environments, ensuring that adequate protections are in place. In conclusion, IoT security standards and regulations are critical for addressing the vulnerabilities associated with connected devices. Compliance, certification, and best practices play essential roles in safeguarding.

IMPLEMENTING IOT SOLUTIONS: CHANGE MANAGEMENT, TALENT ACQUISITION, AND INFRASTRUCTURE DEVELOPMENT

Implementing talent acquisition, they must navigate the complexities associated with this transformation, (Zhang, Z. K., et al. (2014). A successful IoT implementation requires not only the integration of new technologies but also a holistic approach that addresses the human, organizational, and technical aspects of change. Change management is critical in ensuring that all stakeholders are aligned and that the transition to particularly if employees feel threatened by the changes or are unsure of how these technologies will impact their roles. Effective change management strategies begin with clear communication about the benefits of IoT implementations and how they align with the organization's goals. Leadership must engage employees at all levels, fostering a culture of collaboration and openness to change. This includes conducting workshops and training sessions to familiarize staff with new systems and processes, thereby reducing anxiety and promoting acceptance of the IoT initiatives. Moreover, that allow for testing and feedback before full-scale implementation. These pilot projects can serve as valuable learning opportunities (Webb, J., et al.,2018), helping organizations identify teams and ease the transition process. Talent acquisition plays, data analytics, cybersecurity, and system integration. Hiring strategies should focus on attracting individuals with expertise in these areas, whether through direct recruitment, partnerships with educational institutions, (Zankl, A., et al.,2021)., or reskilling existing employees. The demand for professionals with IoT expertise is growing, leading to increased competition among organizations to secure top talent. Companies can enhance their attractiveness by offering Additionally, fostering a culture of innovation and continuous learning can help organizations retain talent and encourage employees to stay abreast of evolving technologies. Furthermore, organizations should consider forming strategic partnerships with technology providers, startups, and academic institutions to access specialized knowledge and resources. Collaborations can help fill skill gaps, facilitate knowledge sharing, and accelerate the implementation process. Such partnerships may also enable organizations to stay informed about emerging. Infrastructure development is a critical component of successful IoT implementation. As organizations deploy IoT solutions, they must abilities. Cloud computing evaluates their cloud strategies, considering factors such as data sovereignty, compliance, and integration with existing systems. Selecting the right cloud service In addition to cloud solutions, businesses must also focus on creating a cohesive ecosystem for their IoT devices. This includes implementing standardized communication protocols and ensuring interoperability among devices from different manufacturers. Security must be a fundamental consideration in both infrastructure development and IoT implementation. As organizations integrate connected devices, they become increasingly vulnerable to cyber threats. To foster a successful IoT implementation, organizations must also create a data governance framework that defines stored, accessed, and used. This framework should establish policies for data quality, privacy, and compliance.

FUTURE OF IOT: EMERGING TRENDS, OPPORTUNITIES, AND CHALLENGES IN IOT ADOPTION AND INNOVATION

The future of the Internet of Things (IoT) is poised for transformative growth, driven by emerging trends, new opportunities, and the challenges that accompany widespread adoption and innovation. As industries increasingly recognize the potential of IoT to enhance operational efficiency, improve decision-making, and foster new business models, the landscape is evolving rapidly. However, navigating this complex terrain requires a thorough understanding of the driving forces behind IoT advancements and the barriers that organizations must overcome.

latency but also minimizes bandwidth costs and enhances data security by limiting the amount of data transmitted to centralized servers. The adoption of edge computing will become increasingly vital in sectors such as autonomous vehicles, these advancements empower IoT devices to make intelligent decisions based on data analysis, improving automation and operational effectiveness. For instance, predictive analytics can be employed to forecast equipment failures in manufacturing settings, enabling organizations to implement maintenance strategies proactively. Similarly, in smart homes, AI-driven IoT devices can learn user preferences to optimize energy consumption and enhance comfort. The synergy between IoT and AI presents vast opportunities for innovation across various industries, from agriculture to healthcare, by creating smarter, more responsive systems that enhance productivity and user experiences.5G technology is another where real-time data exchange is essential. Despite the numerous s opportunities presented by IoT, organizations must also navigate several challenges in their adoption journey. Security remains a paramount concern, as the increasing number of connected devices creates a larger attack surface for cyber threats. Many IoT devices lack robust security features, making them vulnerable to attacks that can compromise sensitive data and disrupt operations. To mitigate these risks, organizations must prioritize the implementation and utilized, and organizations must implement transparent data practices to build trust. This includes obtaining informed consent from users, providing clear privacy policies, and implementing data governance frameworks that define data usage and sharing protocols. The diverse range of devices and platforms, often developed by different manufacturers, can lead to compatibility issues that hinder seamless the skills gap presents another barrier to effective IoT adoption. Organizations often struggle to find programs to upskill their existing workforce while also fostering partnerships with educational institutions to cultivate a talent pipeline equipped to meet the demands of the evolving IoT landscape. Sustainability is an increasingly important consideration in the future of IoT.

CONCLUSION

However, as these technologies proliferate, they bring forth significant security and privacy challenges that must be addressed to fully realize their potential. privacy and trust. Moreover, regulatory compliance, such as adherence to HIPAA in the United States, adds another layer of complexity in ensuring that data is handled appropriately. Similarly, in industrial applications, IoT technologies drive automation, predictive maintenance, and supply chain optimization, which are essential for maintaining competitiveness in a fast-paced global market. security technologies such as encryption, secure authentication methods, and continuous monitoring to protect against potential threats. Furthermore, fostering a culture of security awareness among employees and stakeholders is critical for mitigating risks associated with human error and negligence. guidelines that promote best practices in data management and security. These regulations should not only safeguard individual privacy rights but also encourage innovation by providing a stable framework within which organizations can operate. while safeguarding the interests of individuals and communities. In doing so, they will not only enhance operational efficiencies but also contribute to a safer, more secure digital landscape that fosters innovation across all sectors. Ultimately, the success of IoT initiatives hinges on a collaborative effort to develop robust security measures, effective regulatory frameworks, and an informed user base, ensuring that the transformative power of IoT is harnessed responsibly and ethically. As organizations seek to reduce. Smart sensors and connected devices can optimize resource usage, reduce waste, and improve energy efficiency across various sectors. By integrating meeting, the growing consumer demand for environmentally responsible practices. The democratization of IoT technology is another trend that will shape its future. As IoT solutions become more accessible, smaller organizations and startups are increasingly adopting IoT technologies to drive innovation and efficiency. Cloud-based IoT platforms and pre-built solutions are enabling companies without extensive technical resources to implement IoT applications, leveling the playing field. This democratization fosters a diverse range of applications and uses cases across industries, driving competition and accelerating innovation. Looking ahead, the among participants. This integration can prove particularly beneficial in supply chain management, where tracking the provenance of goods and verifying authenticity are critical. By leveraging blockchain, organizations can create secure and transparent ecosystems that facilitate seamless data sharing while maintaining privacy. Interoperability, and the skills gap will be critical for successful IoT adoption.

REFERENCES

Abhilasha, A., Sreenivasulu, A., & Manimozhi, T. KUMAR, P. S., Selvakumar, P., & Singh, P. (2022, April). The Model of Smart Sensing Device For Sensitive Nano-clusters Modification in Sensing Properties. In 2022 2nd International Conference on Advance Computing and Innovative Technologies in Engineering (ICACITE) (pp. 1043-1047). IEEE.

Agiollo, A., Conti, M., Kaliyar, P., Lin, T. N., & Pajola, L. (2021). DETONAR: Detection of routing attacks in RPL-based IoT. *IEEE Transactions on Network and Service Management*, *18*(2), 1178–1190. DOI: 10.1109/TNSM.2021.3075496

Ahmad, R., & Alsmadi, I. (2021). Machine learning approaches to IoT security: A systematic literature review. *Internet of Things : Engineering Cyber Physical Human Systems*, *14*, 100365. DOI: 10.1016/j.iot.2021.100365

Aldabbas, H., & Amin, R. (2021). A novel mechanism to handle address spoofing attacks in SDN based IoT. *Cluster Computing*, *24*(4), 3011–3026. DOI: 10.1007/s10586-021-03309-0

Balevi, E., & Gitlin, R. D. (2018). Optimizing the number of fog nodes for cloud-fog-thing networks. *IEEE Access : Practical Innovations, Open Solutions*, *6*, 11173–11183. DOI: 10.1109/ACCESS.2018.2808598

Bårli, E. M., Yazidi, A., Viedma, E. H., & Haugerud, H. (2021). DoS and DDoS mitigation using variational autoencoders. *Computer Networks*, *199*, 108399. DOI: 10.1016/j.comnet.2021.108399

Chemodanov, D., Calyam, P., & Palaniappan, K. (2020). "Fog computing to enable geospatial video analytics for disaster-incident situational awareness" in Fog Comput. *Theory into Practice*, •••, 473–503.

Chowdhury, R. R., & Abas, P. E. (2022). A survey on device fingerprinting approach for resource-constraint IoT devices: Comparative study and research challenges. *Internet of Things : Engineering Cyber Physical Human Systems*, *20*, 100632. DOI: 10.1016/j.iot.2022.100632

Danielis, P., Beckmann, M., & Skodzik, J. (2020). "An ISO-compliant test procedure for technical risk analyses of IoT systems based on STRIDE" in 2020 IEEE 44th Annual Computers,Software, and Applications Conference (COMPSAC) (IEEE), 499–504.

Doan, T. V. T., Messai, M. L., Gavin, G., & Darmont, J. (2023). A survey on implementations of homomorphic encryption schemes. *The Journal of Supercomputing*, *79*(13), 1–42. DOI: 10.1007/s11227-023-05233-z

Fatorachian, H., & Kazemi, H. (2021). Impact of industry 4.0 on supply chain performance. *Production Planning and Control*, *32*(1), 63–81. DOI: 10.1080/09537287.2020.1712487

Ferrara, P., Mandal, A. K., Cortesi, A., & Spoto, F. (2021). Static analysis for discovering IoT vulnerabilities. *International Journal of Software Tools for Technology Transfer*, *23*(1), 71–88. DOI: 10.1007/s10009-020-00592-x

Goudarzi, S., Anisi, M. H., Soleymani, S. A., Ayob, M., & Zeadally, S. (2021). An IoT-based prediction technique for efficient energy consumption in buildings. *IEEE Transactions on Green Communications and Networking*, *5*(4), 2076–2088. DOI: 10.1109/TGCN.2021.3091388

Haq, T. U., Shah, T., Siddiqui, G. F., Iqbal, M. Z., Hameed, I. A., & Jamil, H. (2021). Improved twofish algorithm: A digital image enciphering application. *IEEE Access : Practical Innovations, Open Solutions*, *9*, 76518–76530. DOI: 10.1109/ACCESS.2021.3081792

Hassani, H. L., Bahnasse, A., Martin, E., Roland, C., Bouattane, O., & Diouri, M. E. M. (2021). Vulnerability and security risk assessment in a IIoT environment in compliance with standard IEC 62443. *Procedia Computer Science*, *191*, 33–40. DOI: 10.1016/j.procs.2021.07.008

Imteaj, A., Thakker, U., Wang, S., Li, J., & Amini, M. H. (2021). A survey on federated learning for resource-constrained IoT devices. *IEEE Internet of Things Journal*, *9*(1), 1–24. DOI: 10.1109/JIOT.2021.3095077

Jazzar, M., & Hamad, M. (2022). "An analysis study of IoT and dos attack perspective" in Proceedings of international conference on intelligent cyber-physical systems: ICPS 2021 (Singapore: Springer Nature Singapore), 127–142.

Kanipriya, M., Sugumaran, D., Pandiyarajan, V., Yogananda, B. S., Selvakumar, P., & Mohan, S. (2024, January). Analysis of Electromyography Signals for Control of Mechanical Prosthesis using Machine Learning Techniques. In 2024 2nd International Conference on Intelligent Data Communication Technologies and Internet of Things (IDCIoT) (pp. 1059-1067). IEEE.

Karuppannan, P. G., Paramasivam, S., Mariappan, P., Shunmugam, S., & Richard Paul, S. J. (2024). Removal of chemicals from effluent using photobioreactor technology to improve environmental and health impacts&quot. *Zeitschrift für Physikalische Chemie*, *238*(7), 1295–1302. Advance online publication. DOI: 10.1515/zpch-2023-0450

Khan, M. A., & Salah, K. (2018). IoT security: Review, blockchain solutions, and open challenges. *Future Generation Computer Systems*, *82*, 395–411. DOI: 10.1016/j.future.2017.11.022

Khurshid, A., Alsaaidi, R., Aslam, M., & Raza, S. (2022). EU cybersecurity act and IoT certification: Landscape, perspective and a proposed template scheme. *IEEE Access : Practical Innovations, Open Solutions*, *10*, 129932–129948. DOI: 10.1109/ACCESS.2022.3225973

Liang, X., & Kim, Y. (2021). "A survey on security attacks and solutions in the IoT network"in 2021 IEEE 11th Annual Computing and Communication Workshop and Conference (CCWC) (IEEE), 853–859.

Madgule, M., Vinayaka, N., Sonkhaskar, Y. M., Patel, D. R., Karthikeyan, R., & Selvakumar, P. (2023). Mechanical properties and microstructure of activated TIG welded similar joints of Inconel alloys by desirability approaches. *Materials Today: Proceedings*, *77*, 528–533.

Mahmoud, R., Yousuf, T., Aloul, F., & Zualkernan, I. (2015). "Internet of things (IoT) security: current status, challenges and prospective measures" in 2015 10th International Conference for Internet Technology and Secured Transactions (ICITST) (IEEE), 336–341.

Makarenko, I., Semushin, S., Suhai, S., Kazmi, S. A., Oracevic, A., & Hussain, R. (2020). "A comparative analysis of cryptographic algorithms in the internet of things" in 2020 InternationalScientific and Technical Conference Modern Computer Network Technologies(MoNeTeC) (IEEE), 1–8.

Ni, J., Zhang, K., Lin, X., & Shen, X. (2017). Securing fog computing for internet of things applications: Challenges and solutions. *IEEE Communications Surveys and Tutorials*, *20*(1), 601–628. DOI: 10.1109/COMST.2017.2762345

Pacheco, J., Benitez, V. H., Felix-Herran, L. C., & Satam, P. (2020). Artificial neural networks- based intrusion detection system for internet of things fog nodes. *IEEE Access : Practical Innovations, Open Solutions*, *8*, 73907–73918. DOI: 10.1109/ACCESS.2020.2988055

Paramasivam, S., Palanisamy, S. K., Cinthaikinian, S., Palanisamy, V., Mariappan, R., & Selvakumar, P. K. (2025). Biosensors and its diverse applications in healthcare systems. *Zeitschrift für Physikalische Chemie*, *239*(1), 37–43.

Ramakrishnan, T., Mohan Gift, M. D., Chitradevi, S., Jegan, R., Hency Jose, P. S., Nagaraja, H. N., & Hailegiorgis, S. M. (2022). Study of numerous resins used in polymer matrix composite materials. *Advances in Materials Science and Engineering*, *2022*(1), 1088926.

Rana, A., Gupta, S., & Gupta, B. (2024). A comprehensive framework for quantitative risk assessment of organizational networks using FAIR-modified attack trees. *Frontiers of Computer Science*, *6*, 1304288. DOI: 10.3389/fcomp.2024.1304288

Riahi, A., Challal, Y., Natalizio, E., Chtourou, Z., & Bouabdallah, A. (2013). "A systemic approach for IoT security" in 2013 IEEE International Conference on Distributed Computing in Sensor Systems (IEEE), 351–355.

Salem, O., Alsubhi, K., Shaafi, A., Gheryani, M., Mehaoua, A., & Boutaba, R. (2021). Man-in- the-middle attack mitigation in internet of medical things. *IEEE Transactions on Industrial Informatics*, *18*(3), 2053–2062. DOI: 10.1109/TII.2021.3089462

Selvakumar, P. (2023). Biomedical applications of polymers in biosensors, cancer vaccines and drug delivery systems. *Biopolimery i Kletka*, *39*(4), 245–256. DOI: 10.7124/bc.000AA0

Verma, A., & Ranga, V. (2020). Machine learning based intrusion detection systems for IoT applications. *Wireless Personal Communications*, *111*(4), 2287–2310. DOI: 10.1007/s11277-019-06986-8

Webb, J., & Hume, D. (2018). "Campus IoT collaboration and governance using the NIST cybersecurity framework" in Living in the internet of things: cybersecurity of the IoT- 2018 (IET), 1–7.

Zankl, A., Seuschek, H., Irazoqui, G., & Gulmezoglu, B. (2021). "Side-channel attacks in the internet of things: threats and challenges" in Research anthology on artificial intelligenceapplications in security (IGI Global), 2058–2090.

Zhang, Z. K., Cho, M. C. Y., Wang, C. W., Hsu, C. W., Chen, C. K., & Shieh, S. (2014). *"IoT security: ongoing challenges and research opportunities" in 2014 IEEE 7th InternationalConference on Service-Oriented Computing andApplications*. IEEE.

Chapter 6
IoT-Based Drip Irrigation Sequential Data Classification Using SVM

Ida Ayu Devian Branitasandhini Putra
https://orcid.org/0009-0002-3421-005X
University of Mataram, Indonesia

Wirarama Wedashwara
Independent Researcher, Indonesia

I. Komang Damar Jaya
https://orcid.org/0000-0001-9194-344X
Independent Researcher, Indonesia

Ariyan Zubaidi
https://orcid.org/0000-0002-2139-9749
University of Mataram, Indonesia

ABSTRACT

The use of IoT in drip irrigation has been widely applied in previous studies in agriculture to overcome water management problems well in agriculture by utilizing certain sensors in IoT. The method used in this study to classify IoT-based drip irrigation data is the Support Vector Machine (SVM) method. The SVM method is used to find the best classification and distinguish be-tween the two classes in the training data. From the results of the data using the SVM method, the original data before diff has an accuracy of 0.8334 and after diff is done on the data it results in an increase in accuracy to 0.9127. It can be concluded that the data after diff has greater accuracy. Cloudy weather results in faster watering and longer drying. With the rbf graph, it can be concluded that the higher the temperature, the more

DOI: 10.4018/979-8-3693-9846-3.ch006

watering is required and will result in high soil moisture, while if the temperature de-creases, the soil will dry out.

1 INTRODUCTION

Irrigation in agriculture or plantations is an activity related to the provision or distribution of water to plants to maintain and increase soil moisture according to plant needs, (Adibi, Sholichin, & Prayogo, 2024). Regulating the amount of water provided and needed by plants is very important to ensure their needs are met. If the water supply is excessive, this can cause rot in plants. In addition, the amount of water provided is also influenced by temperature and rainfall factors. There are two irrigation techniques that can be used, namely, providing irrigation water with surface irrigation techniques and pressurized irrigation. The most widely used surface irrigation is drip irrigation.

Irrigation is a crucial component in the process of plant cultivation. Traditional irrigation methods used by farmers are often labor-intensive (Hassan *et al.*, 2022; S *et al.*, 2021). In agriculture, the availability and regulation of water play a significant role, as improper water management, especially over-irrigation, can lead to plant root. Common irrigation methods include surface irrigation, sub-surface irrigation, sprinkler irrigation, and drip irrigation (Krishnan *et al.*, 2020; Umam *et al.*,2023).

One widely used technique is drip irrigation, which delivers water in small droplets directly to the base of plants. This method allows for efficient water usage, making it especially suitable in areas with limited water resources. So, this method is good for use in sandy soil conditions, limited water supply, and dry climate (Adibi *et al.*, 2024).

One of the irrigation methods is the drip irrigation method. Drip irrigation method or irrigation water delivery method by delivering droplets near plants, this method uses limited water availability to be managed efficiently. So, this method is good for use in sandy soil conditions, limited water supply, and dry climate (Sangeetha *et al.*, 2022; Umam *et al.*, 2023).

This drip irrigation can help when the weather is difficult to predict due to global climate change and changes in rainfall patterns so that it can increase water availability. When implementing a drip irrigation system, not only pay attention to weather conditions but also plant conditions. The condition of the plant in question is the acceleration of plant roots when absorbing air, soil moisture, and the duration of the soil from after being watered until it dries again, which is done to determine the capacity data of the plant roots. Watering that is carried out by paying attention to the weather and plant conditions will result in plants having the opportunity to experience a lack of air or excess air (Sangeetha *et al.*, 2022; Umam *et al.*, 2023).

Several studies have discussed the success of using drip irrigation systems. One of them is a study using the drip irrigation method, which involves providing air in small and continuous volumes. The results of the study are increasing the efficiency of water use in agricultural land, especially in areas with limited water availability. By using the drip irrigation method, air can be effectively absorbed by plant roots, reducing air loss through evaporation, and maintaining optimal soil air condition for plant growth (Enjellina, 2021).

This study explores the application of IoT in drip irrigation systems. The data analyzed is collected from IoT-enabled drip irrigation setups. These systems use sensors to monitor soil moisture, temperature, air humidity, and light intensity. (Glória *et al.,* 2020; Halim *et al.,* 2023; Hutabarat *et al.,* n.d..).

Furthermore, research conducted by Faikul Umam, *et al.* (2023) aims to overcome challenges in plant cultivation, such as a climate that does not support conditions and less intensive cultivation processes. The stability of the drip irrigation system can be further optimized by integrating decision-making methods such as fuzzy logic. This study focuses on collecting temperature and humidity data, which is then processed using fuzzy logic to estimate the optimal watering time.

The research conducted by Desy Enjellina D aims to determine the effect of the duration and interval of the drip irrigation system on the growth and yield of pak choi and to determine the duration and interval of the drip irrigation system that is efficient in the use of water on pak choi plants. The results of her research are that the duration and interval of the drip irrigation system have a very significant effect on the growth of pak choi plants, both in plant height, number of leaves, and yield of pak choi plants (Enjellina, 2021). Several studies show the success of using drip irrigation systems in water utilization in agriculture.

Thc usc of IoT in drip irrigation has been widely implemented in previous agricultural studies to effectively address water management challenges through the use of specific IoT sensors. (Ghosh *et al.,* 2016; IoT-Enabled Smart Drip Irrigation System Using ESP32, t.t.; Jain, 2023; Kavianand *et al.,* 2016; Ndunagu *et al.,* 2022). The results show that drip irrigation systems utilizing IoT are highly efficient and manage water resources and can be used to increase agricultural productivity.

The data used in this study were taken from the drip irrigation system in previous studies. The drip irrigation system is used in one of the farmer's gardens in Amor-Amor Hamlet, West Lombok Regency, West Nusa Tenggara. Amor-Amor Hamlet was chosen as the research location because it has characteristics that support the development of a solar-based drip irrigation system. Many hamlets experience drought, one of which is Amor-Amor Hamlet; therefore, Amor-Amor Hamlet was chosen as the research location. Amor-Amor Hamlet has the potential to be developed into productive agricultural land for various food crops and horticultural agricultural commodities. The dry land in Amor-Amor Hamlet is the reason that Amor-Amor

Hamlet can have this potential; farmers are trying to develop the dry land. Limited water availability and low average rainfall, drip irrigation is the most appropriate method that can be used. Drip irrigation is not only relevant as a solution but also important to improve food and agricultural security in Amor-Amor Hamlet. Therefore, farmers in Amor-Amor Hamlet are trying to develop the dry land. Drip irrigation is an appropriate method to use in this area due to limited water availability and low average rainfall (Wedashwara *et al.,* 2020, 2021, 2023).

In previous studies conducted in Amor-Amor Hamlet, the drip irrigation system consisted of nodes equipped with Arduino Nano microcontrollers and nRF24L01 modules for data communication. The nRF24L01 modules, powered by 3.3V and connected to various digital inputs, transmitted data only to adjacent nodes. These neighboring nodes then uploaded the collected data to the internet using a GSM module. (Wedashwara *et al.*, 2020, 2021, 2023). The sensors used in the irrigation system include a light sensor (BH1750) with a function to measure light intensity to monitor solar panel energy production. Temperature and Humidity Sensor (DHT11) with a function to measure temperature and humidity. Current sensor (ACS1750) which has the function of measuring the electric current flowing in the solar panel and battery system, voltage sensor (32V DC) which has the function of measuring the voltage on the battery and solar panel to monitor the battery charging status. In this study, data processing will be carried out that has been taken with the drip irrigation system and then data processing will be carried out. After the monitoring process, data management is performed. Effective data management is essential for classifying drip irrigation data, enabling the identification of efficient irrigation patterns.

Classification is a way of grouping based on the characteristics of the classification object. In the process, classification can be done in many ways, both manually and with the help of technology. Several algorithms can be used in classification, such as Logistic Regression, KNN, Decision Tree, Naive Bayes, and SVM. As in previous studies that classified cauliflower leaf diseases using the Convolutional Neural Network (CNN) and Support Vector Machine (SVM) methods, with a precision value for the healthy class of 97.32% and unhealthy of 98.38% (Tanwar *et al.,* 2024). Other studies apply an IoT-based smart farming system, where data collected from sensors is made to be studied using the Support Vector Machine (SVM) classification algorithm, KNN Classifier, and Naive Bayes Classifier, with an accuracy of 87.5 percent using the SVM method, which means SVM has better performance (Bhanu *et al.,* 2020). Previous research has been conducted with the aim of mapping with 3 categories using the SVM method with an accuracy of 84.7% for linear kernels and 74.36% for polynomial kernels (Angellina *et al.,* 2023).

Several studies have been described, which show that the SVM method can produce better accuracy; therefore, this study uses the SVM method, and SVM is also considered suitable for the data that will be used in the study. The SVM method can be used to classify numeric data, image data, or text data. Several studies use SVM to classify many things, such as disease classification (Sudeep *et al.,* 2022), sensing classification (Ibrahim *et al.,* 2019; Putra *et al.,* 2020; Sudeep *et al.,* 2022; Toledo-Pérez *et al.,* 2019), and rice quality classification (Ibrahim *et al.,* 2019), with high classification results in these studies producing high accuracy using the SVM method. The SVM method is used to find the best classification and distinguish between two classes in the training data (Harikiran, 2020), (Guo *et al.,* 2019; Harikiran, 2020).

Based on several studies have been conducted on algorithms such as SVM, K-Means, CNN, and others for various purposes such as health, agriculture, education, and the environment. SVM has proven to be superior in classification accuracy in various cases such as brain tumors, lung cancer, and hyperspectral images. Therefore, this study will conduct IoT-based drip data classification. This study uses the Support Vector Machine algorithm for drip irrigation data classification. This study is expected to be used to optimize the drip irrigation system so that an optimal drip irrigation schedule is created.

2 RELATED RESEARCH

This research proposes a Smart Small-Scale Solar Energy Planning system utilizing IoT, incorporating a Wireless Sensor Network (WSN) for data collection and Evolutionary Fuzzy Association Rule Mining (EFARM) as a Decision Support System (DSS). The system was evaluated over a two-week period using two solar panels equipped with sensors for light intensity, temperature, and humidity to compare environmental conditions with the energy generated. Through this evaluation, EFARM successfully identified patterns in energy generation and consumption, achieving high average values for support (0.247, 0.236), confidence (0.393, 0.219), and score (0.335, 0.127). Additionally, the system demonstrated its capability to support decisions regarding the optimal number of panels and batteries needed, with a relatively low mean squared error (6.094). (Wedashwara *et al.,* 2020).

This study proposes the application of the Internet of Things (IoT) to classify the condition of energy sources in the application of a solar-powered hydroponic system. This study involves the development of an IoT device to collect data on energy generated by solar cells and stored in batteries and environmental conditions, namely light intensity, temperature, and humidity. The simulation results demonstrate that the developed solar-powered hydroponic system can continuously supply energy for 24 hours using three 10Wp solar panels and a 7Ah battery. This setup supports a deep flow hydroponic system consisting of 24 pots on a 1-meter stand, powered by a 20-watt water pump. The integrated IoT system effectively collects environmental data, while the proposed FARM (Fuzzy Assciation Rule-based Model) regression algorithm achieves a low error rate, with a mean absolute error of 0.06 and a mean squared error of 0.24 using just three days of data. Additionally, FARM can extract relevant environmental patterns, with an accuracy value of 0.32. (Wedashwara *et al.,* 2021).

Previous research has explored the use of fuzzy logic in drip irrigation systems to address challenges in crop cultivation, such as unfavorable climatic conditions and limited cultivation intensity. These systems aim to improve water efficiency by delivering controlled irrigation to crops. Their stability can be further enhanced through decision-making methods like fuzzy logic. Such studies typically involve reading temperature and humidity data, which is then processed to estimate optimal watering times. In contrast, the proposed study does not estimate watering time but instead determines the watering status by processing temperature, humidity, and light intensity data using the Support Vector Machine (SVM) method. (Umam *et al.,* 2023).

This study explores the development of a solar-powered Internet of Things (IoT)-based Smart Hydroponic Nutrient Management System using Fuzzy Association Rule Mining (FARM). The system integrates an IoT platform connected to a Total Dissolved Solids (TDS) sensor, a relay module controlling two 5V mini pumps for AB Mix nutrient delivery, and a solenoid valve for water supply. It also includes sensors for temperature, humidity, light intensity, and rainfall to monitor environmental conditions that influence changes in TDS levels. The system is designed to maintain a target TDS of 1200, ideal for leafy vegetables such as lettuce. Test results showed a standard deviation of 2.345 from an average TDS of 1196.17, with a control threshold of 50. Over a one-week evaluation, three rainfall events and four instances of hot weather were found to impact TDS levels, prompting seven relay activations. FARM successfully extracted fuzzy rules with an average support of 0.401 and confidence of 0.826. (Wedashwara *et al.,* n.d.)

Related research has been conducted on IoT-based drip irrigation systems using the Thingspeak platform. This research aims to develop a database system for monitoring water parameters via IoT, utilizing the Thingspeak platform. It focuses on real-time monitoring of water parameters such as temperature, pH, and turbidity. However, this study primarily focused on data collection through sensor-based monitoring. (Hutabarat *et al.,* n.d.)

Several studies in previous research indicate that IoT-based drip irrigation is an efficient method for controlling the water supply to plants, offering a viable alternative to traditional irrigation methods. (Hassan *et al.,* 2022; S *et al.,* 2021).

Several related studies have also explored drip irrigation systems that have successfully addressed irrigation challenges in dry or water-scarce regions. These studies have demonstrated significant success in conserving water and improving irrigation efficiency, which are critical issues in many areas. (Evaluation of Machine Learning Approaches for Precision Farming in Smart Agriculture System: A Comprehensive Review I IEEE Journals & Magazine I IEEE Xplore, t.t.; Halim *et al.,* 2023; Hassan *et al.,* 2022).

Other related research also discusses the success of using drip irrigation systems. This system utilizes cutting-edge technologies such as cloud computing, integrated systems, and IoT for real-time monitoring, control, and data analysis. Using the ThingSpeak cloud, the system enables wireless communication be-tween devices and farmers, who can access environmental data such as humidity, temperature, and air levels from anywhere. This technology supports more efficient irrigation by saving up to 70% of air compared to conventional methods. This solution, developed with the V model, has shown positive impacts on food security and supports sustainable agriculture (Morchid *et al.,* 2024).

The research states that the (Harsha *et al.,* 2022) increasing demand for electricity in India is driving the need for efficient irrigation systems to support home farming and mini gardens. A smart irrigation system based on IoT and Machine Learning with Cloud has been developed to optimize water and energy usage. Using Support Vector Machine (SVM) and Decision Tree algorithms, the system can predict irrigation requirements based on environmental parameters and weather forecasts to aid crop growth with effective air usage and power-saving initiatives. SVM provides weather descriptions with 99% accuracy, while Decision Tree predicts rainfall with 97% accuracy. This approach can help farmers utilize air more efficiently and support energy savings.

Related research has applied the SVM method for the classification and extraction of brain tumors. This study demonstrates that the use of the SVM method in brain tumor classification achieves high accuracy in identifying tumor types. (Sudeep *et al.,* 2022).

Previous studies have conducted detection and classification of grape leaf diseases using SVM. Then clustering was carried out to segment to find areas affected by the disease in the image. This study successfully detected and classified the disease studied with an accuracy of 88.89% (Padol & Yadav, 2016).

In the study entitled Classification of Cauliflower Leaf Diseases was Accomplished with a High Degree of Accuracy using the CNN-SVM Approach, cauliflower leaf disease classification was carried out. This was achieved through the combined use of the Convolutional Neural Network (CNN) and Support Vector Machine (SVM) methods. The results with a precision value for the health class of 97.32% and

98.38% for unhealthy. This model shows a high level of accuracy in recognizing correctly (Tanwar *et al.,* 2024).

In the study entitled Hyperspectral image classification using support vector machines, a study was conducted that aimed to classify hyperspectral images using PCA, BEMD, and SVM. The SVM method with the best accuracy value achieved an accuracy of 96.7% and 97.6% on the Indian Pines and Pavia Univer-sity datasets, respectively. The experimental results on the standard dataset have proven that the proposed model has excellent performance compared to other methods (Harikiran, 2020).

This study (Syahputra & Wibowo, 2023) aims to determine whether negative content can be identified and used to prevent adverse events before they occur, ultimately contributing to a safer and more positive online environment. The research employs Support Vector Machine (SVM) and Random Forest (RF) algorithms to detect negative content on websites. The results showed that SVM out-performed Random Forest, achieving an accuracy of 97%, precision of 90%, and recall of 91%. In comparison, Random Forest achieved an accuracy of 92%, precision of 71%, and recall of 86%. These findings indicate that SVM was more effective in detecting negative content in this study.

In this study (Tanwar *et al.,* 2024), we conducted a classification of cauliflower leaf diseases. The model exhibited excellent precision, recall, and F1 score to distinguish healthy and unhealthy classes and unhealthy leaves. This was achieved through the combined use of Convolutional Neural Network (CNN) and Support Vector Machine (SVM) methods. With a precision value of 97.32% for the healthy class and 98.38% for the unhealthy, the model demonstrated a high level of accuracy in correctly recognizing and showing that the model can accurately represent most of the actual cases of each condition. The model was trained and evaluated using 2000 photos. The model was highly accurate at 97.52%.

3 METHODOLOGY

Figure 1. Research methodology

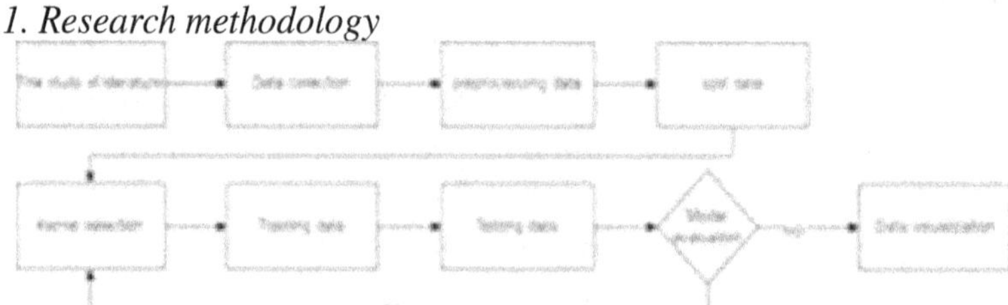

Methodology begins with conducting a literature study that aims to gain an in-depth understanding of research by reading and analyzing literature relevant to the research topic, such as in journals with related topics. Furthermore, data collection is collecting data used in research, such as in this study collecting data on dates, temperature, humidity, light, ambient humidity, ambient temperature. After collecting the data, the data is preprocessed to ensure the data is ready for analysis, in this process the date and index data is deleted and ensures that there is no empty data. After the data is ready to be split data, at this stage the data will be divided into two, namely training data and testing data. The next stage is to select the kernel that will be used to train the model. It then trains the model using training data and testing data. Then check whether the results of the kernel are appropriate or not, if they are appropriate the data will be visualized otherwise the data will be trained again with the kernel.

Data collection using sensors on the drip irrigation system. The drip irrigation system used in Dusun Amor-Amor in previous studies consisted of nodes with Arduino Nano as an ATmega328P microcontroller to process data from sensors and control the irrigation system in real-time, and nRF24L01 as a wireless data communication module between five nodes in the IoT-based smart irrigation system. This irrigation system has a pump that flows water to the drip irrigation network with stable pressure and flow. This irrigation system uses mini solar energy consisting of a 6W 1A solar panel with a function to convert sunlight into electrical energy for the drip irrigation system, a TP4056 charging module with a charging module function to charge battery data from the solar panel so that the microcontroller and sensor can operate without an external power source and ensure the battery is charged safely and prevent damage due to unstable charging, a 18650 battery with a voltage specification of 3.7V functions to store electrical energy generated from the solar panel and supply power to the microcontroller and sensor, provide sufficient voltage for the nRF24L01 and SIM800L communication modules, and maintain the sustainability of the IoT system so that it continues to operate even though there is no sunlight for some time. This irrigation system uses a GSM (Global System for Mobile Communication) module, with the GSM module used being SIM800L, with the function of sending measurement data to the server or monitoring application via the GSM network.

The sensors used in the irrigation system are, current sensor (ACS1750) which has a current measurement range specification of 5A, 20A, 30A and has a function to measure the electric current flowing in the solar panel and battery system, voltage sensor (32V DC) with a maximum voltage specification of 32V and a function to measure the voltage on the battery and solar panel to monitor the battery charging status, light sensor (BH1750) with a function to measure light intensity to monitor the energy production of solar panels, soil moisture and temperature sensors are used to measure soil moisture and soil temperature and finally the air temperature and humidity sensor (DHT11) with a function to measure air temperature and humidity affects the rate of water evaporation from the soil.

Figure 2. Drip irrigation system

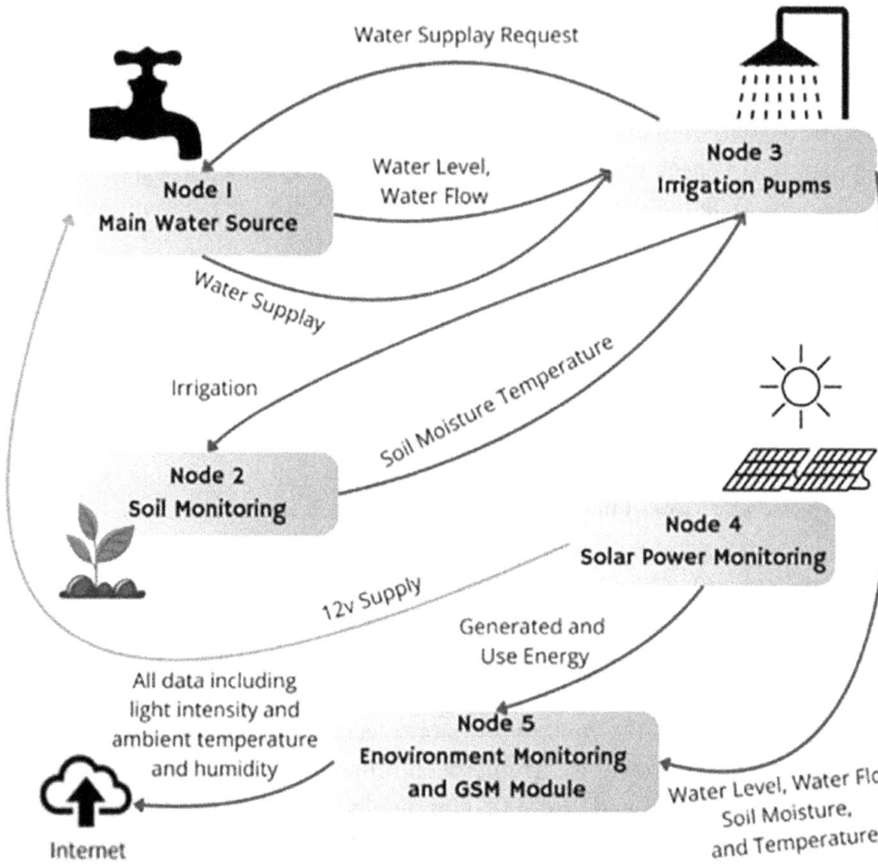

Figure 2 illustrates the interaction of data communication between certain nodes. Node 2 acts as a data collector of soil moisture and temperature that sends watering requests to node 3 as an irrigation pump. Meanwhile, Node 1 acts as the main water provider that receives water requests from node 3 when the water supply in the irrigation reservoir is low. Node 3 also sends data on water flow and supply to node 5 together with node 2. On the other hand, Node 4 is responsible for recording electricity production from solar panels and electricity usage from batteries. The battery operates the main pump and solenoid valve on Node 1. Data from Node 4 is sent directly to Node 5 to be uploaded to the internet. The data summary is divided into two parts, namely Node 3 and 4, due to the limited data package on the device (Wedashwara *et al.,* 2020, 2021, 2023).

The drip irrigation system is used to collect data; the system consists of five nodes with Arduino Nano as a microcontroller and nRf24L01 as a data communication module. nRF24L01 is connected to 3.3v and other digital inputs. The nRF24L01 on both vibration nodes only sends data to the neighboring node, which will upload the data to the internet via the GSM module (Wedashwara *et al.,* 2020, 2021, 2023).

The Solar Powered Drip Irrigation System in this study is very relevant because it focuses on drip irrigation systems and energy control development. This drip irrigation system uses nodes with Arduino Nano as a microcontroller and nRF24L01 as a data communication module. This irrigation system provides an efficient and environmentally friendly irrigation solution, utilizing renewable energy to distribute air precisely and economically directly to plant roots. This system is the main focus of research to measure the impact of environmental influences on the quantity of watering, starting from air temperature, air humidity, soil temperature, soil humidity, and light intensity.

The data used in this study were taken from one of the gardens in Amor-Amor Hamlet, North Lombok Regency. The data used in this study were taken on May 1, 2023, May 3, 2023, and May 4, 2023 with a data collection time span of around 2-3 hours. The characteristics of the data used include air temperature, air humidity, soil temperature, soil humidity, and light intensity. The total data used in this study was 198000.

4 RESULT

Classification of IoT-based drip irrigation data using support vector machine (SVM) with data obtained from the IoT-based drip irrigation system consisting of nodes with Arduino Nano as a microcontroller and using five nodes that communicate via the nRF24101 network and disintegrated with the FRBR algorithm for regression. The five nodes consist of temperature monitoring, air humidity, main irrigation pump, drip irrigation pump, sunlight monitoring, temperature and humidity monitoring, light intensity, and a GSM module to stream data to the internet. Some of the sensors used in the irrigation system include a light sensor (BH1750), which functions to measure light intensity to integrate solar panel energy production. The Temperature and Humidity Sensor (DHT11) functions to measure temperature and humidity. In this study, data processing will be carried out that has been taken with the drip irrigation system and then data processing will be carried out.

In previous studies, this study began with data collection using a drip irrigation system. The data used in this study were taken from one of the gardens in Amor-Amor Hamlet, North Lombok Regency. The data used in this study were taken on May 1 2023, May 3 2023, and May 4 2023, with a data collection period of around 2-3 hours. The characteristics of the data used include air temperature, air humidity, soil temperature, soil humidity, and light intensity.

The total data used in this study was 198000. The data used in this study were date, temperature, humidity, light, environmental humidity, environmental temperature, and status. This study aims to determine soil evaporation and water discharge in drip irrigation involving various types of soil and weather, namely dry cloudy, dry bright, evenly sunny, and evenly cloudy.

In the early stages of this study, data collection was carried out on dry cloudy, evenly sunny, evenly sunny, and evenly cloudy data objects. After data collection using the irrigation system, the data was processed first by deleting the date and index data. After pre-processing the data, the data was divided into training data and test data, with 30% being training data and the rest being test data.

Table 1. Data description

	humidity	**temperature**	**ambient humidity**	**ambient temperature**	**light**
count	19800	19800	19800	19800	19800
mean	66.274	36.658	56.47	33.152	29564.75
std	20.214	2.771	3.280	1.034	14093.898
min	39	30.191	46	29	16615
25%	47	34.425	55	32.99	16668
50%	66	36.984	56	33	16789.5
75%	85	39.268	57	33.994	45000
max	98	40.852	67	34	45066

The data description table displays the results of all the data combined at the preprocessing stage. There is one column of data that is deleted at this stage, namely the date column, the date data is deleted because it is not needed in the next stage. Table 1 displays the total amount of data to be used, which is 19,800.

Table 1 above provides a descriptive overview of the dataset used in the study, with a total of 19,800 observations for each variable. This data includes five main variables: humidity, temperature, ambient humidity, ambient temperature, and light intensity.

In the Humidity column, the humidity value ranges from 39 to 98, with an average value of 66,274 and a standard deviation of 20,214. The quartile values show a fairly wide distribution, where 25% of the data are below 47, and 75% are below 85, reflecting significant variations in humidity.

In the temperature column, the temperature value has an average of 36,658 with a standard deviation of 2,771, and the data ranges from 30,191 to 40,852. This distribution is relatively normal with a small difference between the first quartile (34.425) and the third quartile (39.268), indicating a more moderate temperature compared to humidity.

In the ambient humidity column, the average value of ambient humidity is 56.47 with a standard deviation of 3.280. The data range for this variable is 46 to 67, with the quartile distribution also quite narrow. This variation may indicate the stability of ambient humidity in the environment where the data was collected.

In the ambient temperature column, the ambient temperature value has an average value of 33.152 with a relatively small standard deviation (1.034), indicating the consistency of ambient temperature in the environment. The minimum and maximum values range from 29 to 34, with 50% of the data centered around a temperature of 33.

In the light column, the value of the light intensity variable has an average of 29564.75 and a high standard deviation of 14093.898, indicating a large variation in the data. The range of light intensity is from 16615 to 45066, with a fairly wide distribution in each quartile.

Figure 3. Humidity histogram

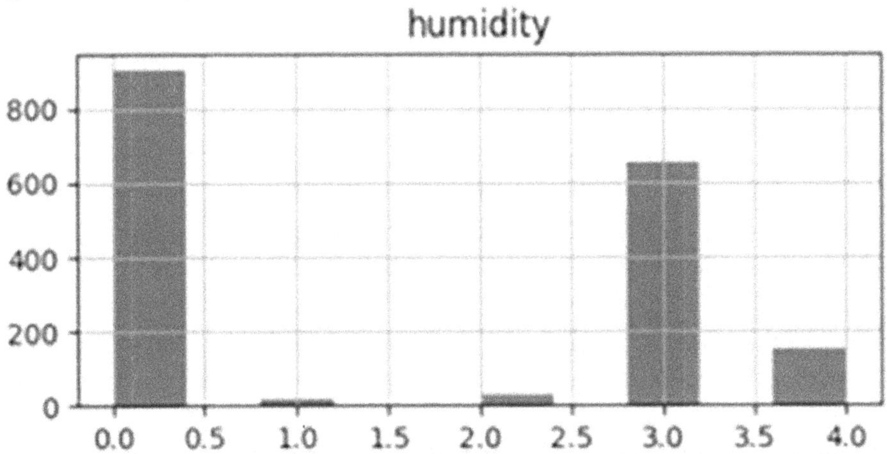

Figure 4. Light histogram

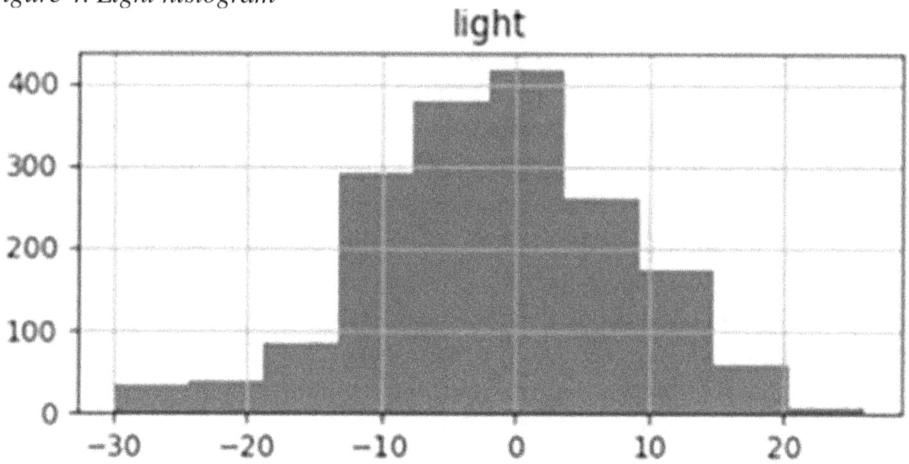

The humidity histogram shows that the most common value is 0, which is due to the sensor's resolution of 1, limiting the rapid fluctuations in humidity values. In contrast, the light histogram demonstrates more dynamic changes in light intensity, with values fluctuating more noticeably.

Table 2. Accuracy data

Data Description	Accuracy
Data before diff	0.8334
Data after diff	0.9127

Table 2 above shows the level of accuracy of the data obtained before and after the differentiation (diff) process is carried out on the data. Based on the table, the original data before going through the diff process has an accuracy level of 0.8334. After the diff process is carried out on the data, the accuracy level increases to 0.9127.

This increase in accuracy shows that the diff process has a positive impact on the accuracy of the model in predicting or classifying data. This can be caused by the ability of the diff process to reduce variability or unwanted heights in the original data, so that the model can capture more relevant and consistent patterns in the data. Diff functions to increase data stability which ultimately improves model performance.

Based on the results in Table 2, it can be concluded that the diff process has an important role in increasing model accuracy. These results form the basis that further data processing using diff can be an effective step in improving the quality and accuracy of data predictions.

Table 3. Report detail

Accuracy: 83.99%	**precision**	**recall**	**f1-score**	**Support**
Drying	0.84	1.00	0.91	1650
Flush	1.00	0.04	0.08	330
Accuracy			0.84	1980
Macro avg	0.92	0.52	0.49	1980
Weighted avg	0.87	0.84	0.77	1980

In this study, classification was performed using Support Vector Machine (SVM), with an accuracy result of 83.99%, indicating that in general 83.99% of the total test data was correctly classified. However, for a deeper understanding, we need to look at other metrics such as precision, recall, and f1-score, especially considering the imbalance between classes.

Table 3 is the result of the classification report using SVM, the following is an explanation of the classification results table. In the "dry" class, the model has a precision of 0.84 and a recall of 1.00, which means that the model is very good at correctly identifying samples of this class and rarely misclassifies samples of other classes as "dry". The F1-score for this class reaches 0.91, indicating a good balance between precision and recall.

However, in the "flush" class, the performance is much lower. Precision reaches 1.00, but recall is very low, which is only 0.04. This shows that although the model does not classify other classes as "flush" (high precision), it fails to accurately recognize "flush" samples, with only a few samples from this class being correctly identified. This results in the "flush" class f1 score being only 0.08.

Overall, the macro average values show that the model is suboptimal in balancing performance across both classes, with an average recall of 0.52 and an average f1 score of 0.49. Meanwhile, a higher weighted average indicates that the model is focusing more on the majority class.

Figure 5. Hyperplane graph

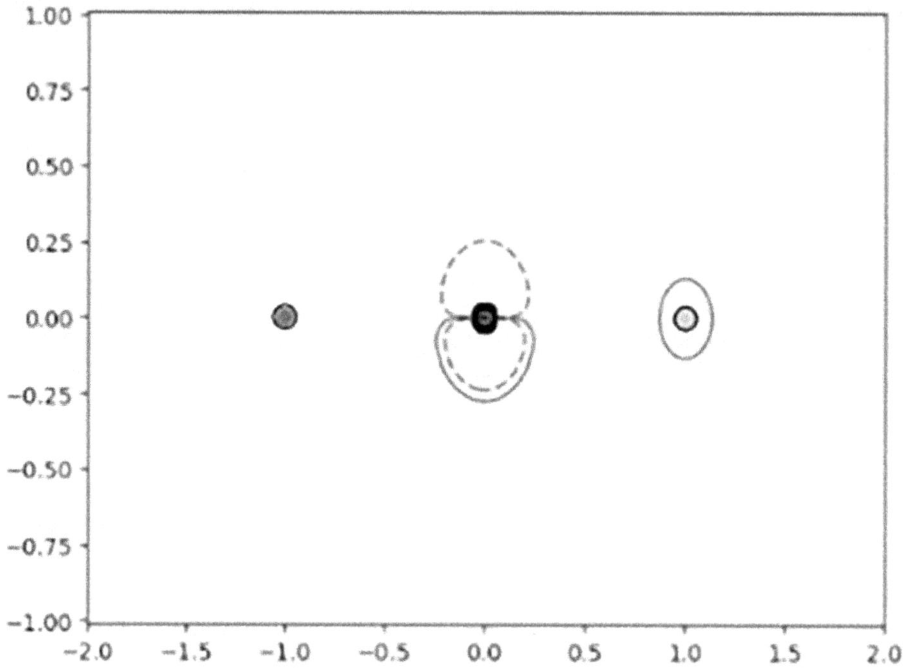

Figure 6. RBF graph

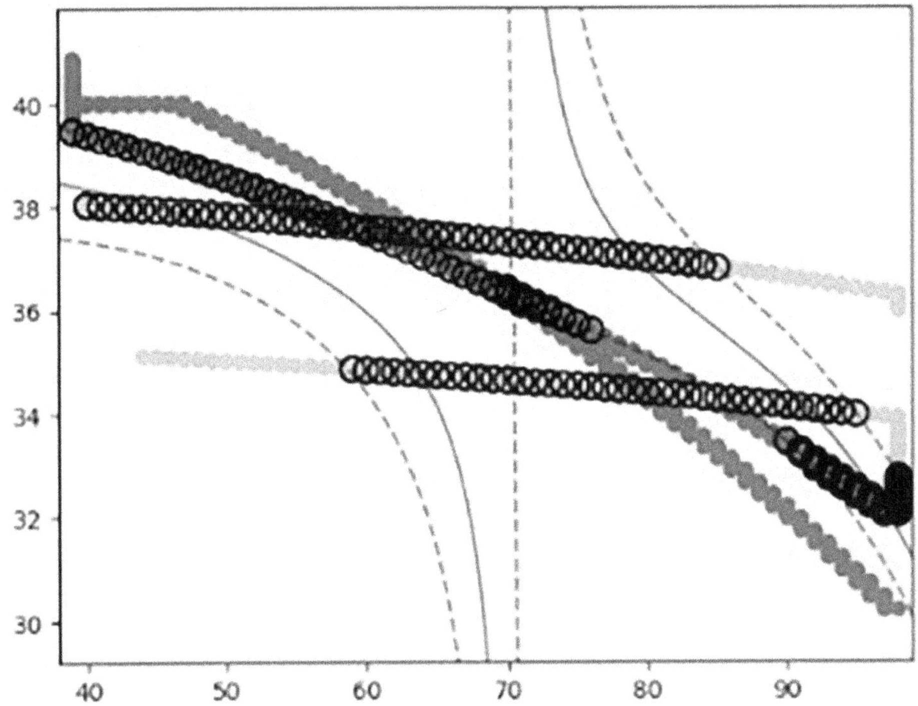

The image above consists of two graphs, namely Figure 5 "Hyperplane" and Figure 6 "RBF Graph" which presents the results of data analysis that has gone through the diff process. Both graphs mainly aim to understand the behavior of soil moisture affected by temperature changes, with different visual interpretations.

In Figure 5, the "Hyperplane" shows changes in data on a limited scale, with this graph focusing on results that have a height around the values -1 and 1. This graph functions as a representation that separates data into several classes based on the division of space (hyperplane) which is classified into two main categories, namely positive and negative data.

Meanwhile, Figure 6 namely the "RBF Graph" or Radial Basis Function Graph, shows a more comprehensive data analysis by considering two main variables on the x and y axes. The x-axis in this graph is soil moisture, while the y-axis describes temperature. This graph also displays two types of points with different colors, namely red points that describe the "watering" status or wet conditions and blue points that indicate the "drying" status or dry conditions. With these two variables displayed, Figure 3 (b) can help us understand the relationship between temperature and soil moisture and how moisture conditions change as temperatures rise and fall.

From the RBF graph, it can be seen that the higher the soil temperature, the higher the watering process required to maintain stable soil moisture. This shows that soil at high temperatures tends to dry out faster and requires more water to maintain its moisture. Conversely, when the temperature drops, the soil will experience a slower drying process.

This graph visualization is useful for analysis that requires an in-depth understanding of how temperature changes directly affect soil moisture conditions and the watering needs required in a land management system.

Figure 7. Flush line graph

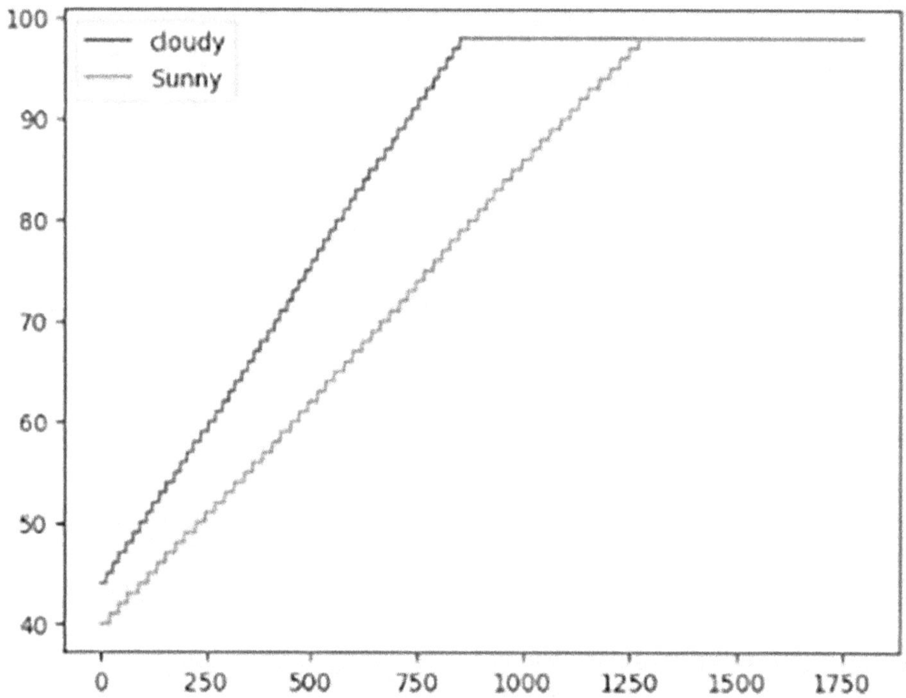

Figure 8. Dry line graph

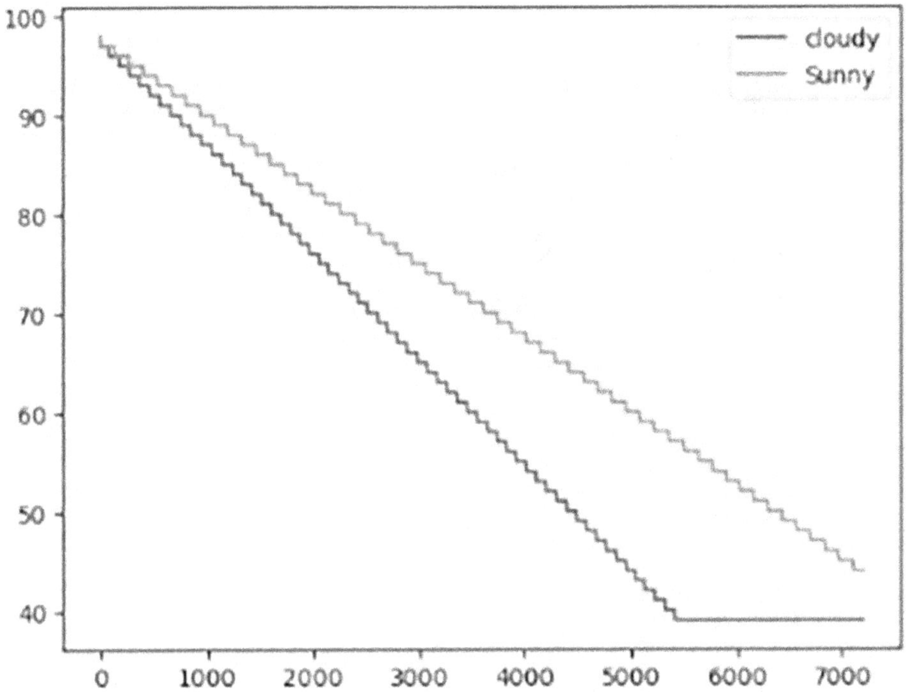

Figure 7 & 8 shows the line graphs of watering and drying results in cloudy and sunny weather. The watering line graph shows that the humidity and temperature values increase, while the drying line graph has decreasing values. In Figure 7, the flush line graph shows that in cloudy weather, watering is faster and in fig. 8 the drying line graph shows that in cloudy weather drying is longer. The watering graph shows that in sunny weather it can reach 1,250 and in cloudy weather it only reaches 750. Meanwhile, the drying graph in cloudy weather reaches 5,500 and in sunny weather reaches 7,000.

Figure 9. Confusion matrix

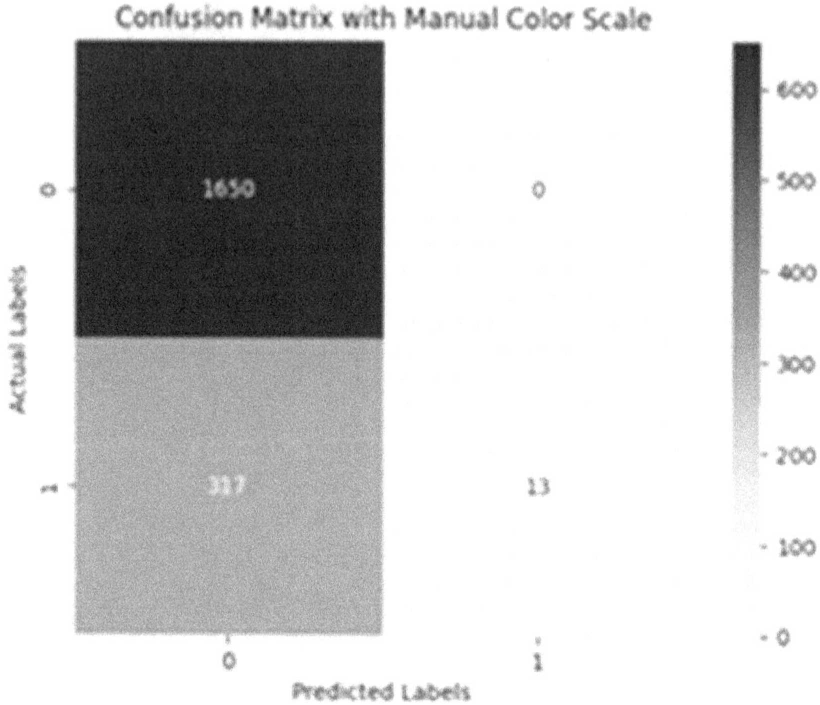

Confusion matrix Figure 9 above shows the results of the model classification with two classes, namely the dry class (label 0) and the flush class (label 1). This matrix shows how the model classifies the test data by comparing the model prediction with the actual label.

True Negative (TN): A total of 1650 dry class samples (0) were successfully classified correctly by the model as the negative class. This shows that the model has a good ability to identify negative class samples.

False Positive (FP): No samples from the negative class were classified as the flush class (1). This is a good result because the model does not produce false positive prediction errors.

False Negative (FN): A total of 317 positive class samples were incorrectly classified as the negative class. This shows the weakness of the model in recognizing some positive samples.

True Positive (TP): Only 13 positive class samples were successfully classified correctly as the positive class. This number is relatively low and shows that the model has difficulty in detecting the positive class.

Based on the results of this confusion matrix, the model shows good performance in identifying the negative class, but is less effective in detecting the positive class. The comparison between a relatively high FN and a low TP indicates possible problems in handling class imbalance or that the model needs to be improved to better recognize the characteristics of the positive class.

5 CONCLUSION

The conclusion of this research is that the classification results show that higher temperatures lead to faster water absorption in the soil, requiring more frequent watering to maintain moisture. Conversely, lower humidity levels also necessitate watering, while higher light intensity accelerates evaporation, thereby increasing the need for watering. The humidity histogram reveals that the most common value is 0, which is due to the sensor's resolution of 1, preventing rapid fluctuations in the humidity values. In contrast, the light histogram shows more dynamic changes in light intensity.

Using the SVM method, data accuracy before the differential process was 0.8334, and after the differential process, the accuracy increased to 0.9127. This indicates that the data after differentiation provides greater accuracy. Cloudy weather results in faster watering and longer drying periods. Additionally, the RBF graph demonstrates that higher temperatures increase the need for watering, leading to higher soil moisture levels, whereas lower temperatures cause the soil to dry out.

REFERENCES

Adibi, M., Sholichin, M., & Prayogo, T. B. (2024). Comparative Planning Of Sprinkler Irrigation and Drip Irrigation for Orange And Avocado Crops in Research Land. *JTRESDA : Jurnal Teknologi dan Rekayasa Sumber Daya Air, 04*(01), 790–803. DOI: 10.21776/ub.jtresda.2024.004.01.067

Angellina, A., Herwindiati, D. E., & Hendryli, J. (2023). Performa Support Vector Machine Pada Klasifikasi Lahan dan Air Tanah. *JURNAL MEDIA INFORMATIKA BUDIDARMA*, *7*(1), 231. DOI: 10.30865/mib.v7i1.5279

Bhanu, K. N., Mahadevaswamy, H. S., & Jasmine, H. J. (2020). IoT based Smart System for Enhanced Irrigation in Agriculture. *2020 International Conference on Electronics and Sustainable Communication Systems (ICESC)*, 760–765. DOI: 10.1109/ICESC48915.2020.9156026

Enjellina, D. (2021). Effect of Duration and Interval of Drip Irrigation System on Growth and Yield of Pakcoy Mustard (Brassica rapa L. Ssp. Chinensis). *Agrifor*, *20*(2), 311. DOI: 10.31293/agrifor.v20i2.5757

Ghosh, S., Sayyed, S., Wani, K., Mhatre, M., & Hyder Ali, H. (2016). *Smart irrigation: A smart drip irrigation system using cloud, android and data mining* (hlm. 239). DOI: 10.1109/ICAECCT.2016.7942589

Glória, A., Dionisio, C., Simões, G., Cardoso, J., & Sebastião, P. (2020). Water Management for Sustainable Irrigation Systems Using Internet-of-Things. *Sensors (Basel)*, *20*(5), 5. Advance online publication. DOI: 10.3390/s20051402 PMID: 32143482

Guo, Y., Yin, X., Zhao, X., Yang, D., & Bai, Y. (2019). Hyperspectral image classification with SVM and guided filter. *EURASIP Journal on Wireless Communications and Networking*, *2019*(1), 56. DOI: 10.1186/s13638-019-1346-z

Halim, A. A. A., Mohamad, R., Rahman, F. Y. A., Harun, H., & Anas, N. M. (2023). IoT based smart irrigation, control, and monitoring system for chilli plants using NodeMCU-ESP8266. *Bulletin of Electrical Engineering and Informatics*, *12*(5), 3053–3060. DOI: 10.11591/eei.v12i5.5266

. Harikiran, J. (2020). Hyperspectral image classification using support vector machines. *IAES International Journal of Artificial Intelligence (IJ-AI)*, *9*, 684. DOI: 10.11591/ijai.v9.i4.pp684-690

. Harsha, D., Mohammed, Z. A., & Arra, R. R. (2022). Cloud Based Irrigation System with IoT and Machine Learning. *International Journal of Innovative Research in Engineering & Management*, 106–111. DOI: 10.55524/ijirem.2022.9.3.16

Hassan, E. E., Chung, L. L., Sulaima, M. F., Bahaman, N., & Kadir, A. F. A. (2022). Smart irrigation system with photovoltaic supply. *Bulletin of Electrical Engineering and Informatics, 11*(1), 29–41. DOI: 10.11591/eei.v11i1.3338

Hutabarat, B. F., Peslinof, M., Afrianto, M. F., & Fendriani, Y. (t.t.). Water Parameter Monitoring Database System Based On Internet of Things (IoT) Using Thingspeak Platform. *Journal Online of Physics, 8*(2), 42–50.

Ibrahim, S., Zulkifli, N. A., Sabri, N., Shari, A. A., & Noordin, M. R. M. (2019). Rice grain classification using multi-class support vector machine (SVM). *IAES International Journal of Artificial Intelligence (IJ-AI), 8*(3), Article 3. DOI: 10.11591/ijai.v8.i3.pp215-220

Jain, R. K. (2023). Experimental performance of smart IoT-enabled drip irrigation system using and controlled through web-based applications. *Smart Agricultural Technology, 4*, 100215. DOI: 10.1016/j.atech.2023.100215

. Kavianand, G., Nivas, V. M., Kiruthika, R., & Lalitha, S. (2016). Smart drip irrigation system for sustainable agriculture. *2016 IEEE Technological Innovations in ICT for Agriculture and Rural Development (TIAR)*, 19–22. DOI: 10.1109/TIAR.2016.7801206

Krishnan, R. S., Julie, E. G., Robinson, Y. H., Raja, S., Kumar, R., Thong, P. H., & Son, L. H. (2020). Fuzzy Logic based Smart Irrigation System using Internet of Things. *Journal of Cleaner Production, 252*, 119902. DOI: 10.1016/j.jclepro.2019.119902

Mohyuddin, G., Khan, M. A., Haseeb, A., Mahpara, S., Waseem, M., & Saleh, A. M. (2024). Evaluation of Machine Learning approaches for precision farming in Smart Agriculture System-A comprehensive Review. *IEEE Access : Practical Innovations, Open Solutions.*

Morchid, A., Muhammad Alblushi, I. G., Khalid, H. M., El Alami, R., Sitaramanan, S. R., & Muyeen, S. M. (2024). High-technology agriculture system to enhance food security: A concept of smart irrigation system using Internet of Things and cloud computing. *Journal of the Saudi Society of Agricultural Sciences.* Advance online publication. DOI: 10.1016/j.jssas.2024.02.001

Ndunagu, J. N., Ukhurebor, K. E., Akaaza, M., & Onyancha, R. B. (2022). Development of a Wireless Sensor Network and IoT-based Smart Irrigation System. *Applied and Environmental Soil Science, 2022*(1), 7678570. DOI: 10.1155/2022/7678570

Padol, P. B., & Yadav, A. A. (2016). SVM classifier based grape leaf disease detection. *2016 Conference on Advances in Signal Processing (CASP)*, 175–179. DOI: 10.1109/CASP.2016.7746160

Parvathi Sangeetha, B., Kumar, N., Ambalgi, A. P., Abdul Haleem, S. L., Thilagam, K., & Vijayakumar, P. (2022). IOT based smart irrigation management system for environmental sustainability in India. *Sustainable Energy Technologies and Assessments*, *52*, 101973. DOI: 10.1016/j.seta.2022.101973

Pereira, G. P., Chaari, M. Z., & Daroge, F. (2023). IoT-enabled smart drip irrigation system using ESP32. *IoT*, *4*(3), 221–243.

Putra, R. E., Tjandrasa, H., & Suciati, N. (2020). Severity classification of non-proliferative diabetic retinopathy using convolutional support vector machine. *International Journal of Intelligent Engineering and Systems*, *13*(4), 156–170. DOI: 10.22266/ijies2020.0831.14

Sudeep, U. S., Naidu, K. N., Girish, P. S., Nikesh, T. N., & Sunanda, C. (2022). Brain Tumor Classification using a Support Vector Machine. *International Journal of Computer Applications*, *184*(28), 15–17. DOI: 10.5120/ijca2022922347

Syahputra, H., & Wibowo, A. (2023). Comparison of Support Vector Machine (SVM) and Random Forest Algorithm for Detection of Negative Content on Websites. *Jurnal Ilmiah Teknik Elektro Komputer Dan Informatika*, *9*(1), 1. Advance online publication. DOI: 10.26555/jiteki.v9i1.25861

Tanwar, V., Anand, V., Upadhyay, D., & Singh, M. (2024). Classification of Cauliflower Leaf Diseases was Accomplished with a High Degree of Accuracy using the CNN-SVM Approach. *2024 4th International Conference on Intelligent Technologies (CONIT)*, 1–5. DOI: 10.1109/CONIT61985.2024.10626016

Toledo-Pérez, D. C., Rodríguez-Reséndiz, J., Gómez-Loenzo, R. A., & Jauregui-Correa, J. C. (2019). Support Vector Machine-Based EMG Signal Classification Techniques: A Review. *Applied Sciences (Basel, Switzerland)*, *9*(20), 20. Advance online publication. DOI: 10.3390/app9204402

. Umam, F., Dafid, A., & Cahyani, A. D. (2023). *Implementation of Fuzzy Logic Control Method on Chilli Cultivation Technology Based Smart Drip Irrigation System*. *9*(1).

Wedashwara, W., Jatmika, A. H., Zubaidi, A., & Arimbawa, I. W. A. (2021). *Smart solar powered hydroponics system using internet of things and fuzzy association rule mining*. IOP Publishing. DOI: 10.1088/1755-1315/712/1/012007

. Wedashwara, W., Jatmika, A. H., Zubaidi, A., & Arimbawa, I. W. A. (t.t.). Solar-powered IoT based smart hydroponic nutrition management system using FARM. *IOP Publishing*. DOI: 10.1088/1755-1315/913/1/012010

Wedashwara, W., Wayan Agus Arimbawa, I., Hidayat Jatmika, A., Zubaidi, A., & Mulyana, T. (2020). IoT based Smart Small Scale Solar Energy Planning using Evolutionary Fuzzy Association Rule Mining. *2020 International Conference on Advancement in Data Science, E-learning and Information Systems (ICADEIS)*, 1–6. DOI: 10.1109/ICADEIS49811.2020.9276905

Wedashwara, W., Yadnya, M. S., Sudiarta, I. W., Arimbawa, I. W. A., & Mulyana, T. (2023). Solar Powered Vibration Propagation Analysis System using nRF24l01 based WSN and FRBR. *JOIV: International Journal on Informatics Visualization*, *7*(1), 15–21. DOI: 10.30630/joiv.7.1.1592

Chapter 7

K-Means Clustering for Profiling Logical-Mathematical Intelligence and Problem-Solving Abilities

Syahroni Hidayat
Universitas Negeri Semarang, Indonesia

Budi Sunarko
Universitas Negeri Semarang, Indonesia

Uswatun Hasanah
Universitas Negeri Semarang, Indonesia

ABSTRACT

Intelligence is the ability to think, experience learning, problem-solving, and adaptation to new situations. Thus, intelligence is an essential foundation in the learning process. Logical-mathematical intelligence (KLM) is a strong indicator for assessing individual intelligence levels and learning achievements. KLM significantly correlates to the ability to solve story problems (KSC). However, exploring these two variables to group students to help educators prepare good learning strategies has not yet been done. Therefore, this study applied the K-means algorithm to reveal it all by utilizing KLM and KSC data. Silhouette index (SI) and Davies-Bouldin (DBI) were used to validate the results of the K-means cluster. The results show that K-means can classify students based on their KLM and KSC data into three groups: high, medium, and low. The SI and DBI values are 0.46 and 0.75, respectively. It indicates

DOI: 10.4018/979-8-3693-9846-3.ch007

the formed clusters are quite good and there is still an opportunity for improvement.

1 INTRODUCTION

In today's educational landscape, understanding students' cognitive abilities is crucial for tailoring instructional strategies that foster better learning outcomes. Among the various forms of intelligence, logical-mathematical intelligence stands out due to its close association with critical thinking, reasoning, and problem-solving skills (Hana et al., 2020). According to Howard Gardner's theory of multiple intelligences, logical-mathematical intelligence is one of the nine types individuals exhibit, including verbal-linguistic, spatial-visual, kinesthetic, and interpersonal intelligence (Gardner, 2011). Logical-mathematical intelligence involves the capacity to handle abstract concepts, recognize patterns, and solve complex problems, making it essential for success in STEM fields. This intelligence type has historically been considered one of the most reliable academic performance indicators, especially in mathematics, physics, and computer science (Wiguna et al., 2018).

Logical-Mathematical intelligence type is still the main benchmark for assessing the level of intelligence of individuals and is one of the strong indicators in learning achievement (Hana et al., 2020; Roth et al., 2015). Many studies have been conducted to prove the relationship and influence of intelligence with learning achievement. It is found that there is and influence between logical - mathematical intelligence with math learning outcomes (Akhmad, 2019; Al-Kinani & Al-Omari, 2021; Milsan & Wewe, 2018; Nurzaelani & Wibowo, 2015; Raehanah & Apriani, 2019; Roikha, 2016), solving story problems (Budiyanto et al., 2018; Muncarno & Yulina, 2017; Puspitasari et al., 2018; Subariyanto et al., 2017)· learning physics (Fauziah et al., 2015; Umaeza & Budhi, 2016), financial accounting (Pehlİvan & Durgut, 2017)· algorithms (Korkmaz, 2012), language skills (Esmaeili et al., 2014; Khodadady & Dastgahian, 2013), and disciplinary attitudes (Fauzi & Monawati, 2018)·

Numerous studies have established the relationship between logical-mathematical intelligence and academic achievement (Balovsyak et al., 2023; Hana et al., 2020). Research has shown that students with higher logical-mathematical intelligence tend to excel in mathematics (Akhmad, 2019; Al-Kinani & Al-Omari, 2021; Milsan & Wewe, 2018; Nurzaelani & Wibowo, 2015; Raehanah & Apriani, 2019; Roikha, 2016), problem-solving (Budiyanto et al., 2018; Muncarno & Yulina, 2017; Puspitasari et al., 2018; Subariyanto et al., 2017), and analytical tasks (Roth et al., 2015). For example, Milsan & Wewe (2018) observed a positive correlation between students' mathematical reasoning abilities and their performance in logical problem-solving tasks (Al-Kinani & Al-Omari, 2021; Imamah et al., 2024; Milsan & Wewe, 2018). However, while these studies highlight the importance of logical-mathematical intelligence in learning, they often fail to capture the individual variations in cognitive abilities among students (Hana et al., 2020; Mohamed Nafuri et al., 2022). Moreover, traditional assessment methods, such as standardized tests, are limited in providing a holistic view of students' problem-solving skills and reasoning capabilities (Tirri et al., 2013).

To address the limitations of traditional assessments, data-driven approaches like clustering have gained traction in educational research (Ashari et al., 2022; Rahman et al., 2024; Yu & Wang, 2022). Clustering is an unsupervised machine learning technique that groups data based on shared characteristics, allowing for the identification of patterns and trends within datasets (Roski et al., 2024). Among the different clustering algorithms, K-Means is one of the most widely used due to its simplicity and efficiency in handling large datasets (Y. Wang, 2024). K-Means clustering works by partitioning data into K distinct clusters, where each data point belongs to the cluster with the nearest mean (Ashari et al., 2022; Idrus et al., 2022; Qomariyah & Siregar, 2022; Sulistiyawati & Supriyanto, 2020; Y. Wang, 2024). This method is particularly effective for educational data mining, where it can be used to group students based on various factors such as academic performance, learning behaviours, and cognitive traits (Araka et al., 2022; Batool et al., 2023; Yağcı, 2022).

The results of these studies have indeed been able to explain well about the level of relationship and influence of students' logical-mathematical intelligence with learning achievement and students' attitudes, but still cannot describe the level of logical-mathematical intelligence and these variables for each individual student simultaneously. Whereas in the learning process it is very important at the beginning to know the level of mathematical-logical intelligence and/or other special abilities that each learner has (Ismaya et al., 2022; Purnamaningsih & Aziz, 2014; Sulistiyawati & Supriyanto, 2020). Because with this initial knowledge, educators can design an appropriate learning to be able to achieve the maximum learning process and results for each individual learner (Jiménez-Macías et al., 2024; Karsenti, 2019). In order to know the level of logical-mathematical intelligence along with other special abilities of each individual learner, it is necessary to do segmentation or grouping (Mohd Talib et al., 2023; Muthmainnah et al., 2020).

The use of K-Means clustering in education has shown promise in uncovering meaningful patterns in student performance. (Jiménez-Macías et al. (2024) applied K-means clustering to group students based on their learning styles and problem-solving strategies (Jiménez-Macías et al., 2024). They found that students could be categorized into distinct clusters, allowing educators to identify those needing specific interventions (Mohd Talib et al., 2023). Despite its success in profiling general learning behaviours, applying K-Means clustering specifically for profiling logical-mathematical intelligence and problem-solving abilities remains underexplored. This gap in literature is significant, as understanding these cognitive traits is critical for improving student outcomes in subjects that rely heavily on logic and reasoning.

There are two types of segmentation methods i.e., hard clustering and soft clustering (López-Oriona et al., 2023). When data is only a member of one cluster, the method is categorized as hard clustering, while if the data can be a member of several clusters based on its proximity to some center value, it is called soft clustering. The method included in the soft clustering category is Fuzzy C Means, while for hard clustering it is the K-Means algorithm (Nazari et al., 2023). Both Fuzzy C Means and K-Means clustering algorithms are unsupervised and have been widely used and easy to implement. Their wide application has reached various fields (Nazari et al., 2023), including industry (Sharma et al., 2023), information technology, agriculture, education (Chen, 2022; Hidayat et al., 2020; Q. Wang et al., 2022), and others. Its application to very large education data (big data) has been proven to provide information mining results that are very useful for the education process (Al-Doulat et al., 2020; Fan & Yan, 2016; Maylawati et al., 2020; Qomariyah & Siregar, 2022; Sya'iyah et al., 2019; Yu & Wang, 2022).

Fuzzy C Means works to group data with the possibility that the data can be combined into several different clusters. So, this condition is not suitable to be applied to academic data analysis (Chen, 2022; Q. Wang et al., 2022). Meanwhile, based on its properties, K Means works by grouping data based on their minimum distance within the same group. It will work on its own without being guided (unsupervised), identifying similarities between objects, which are grouped according to the same characteristics that distinguish them from others, and will stop when there is no more movement of members in the formed group. These characteristics and ways of working of the K-means algorithm are certainly very suitable to be applied to unidentified data such as logical-mathematical intelligence and the special variables mentioned above. Therefore, this research focuses on how to apply the k means algorithm to cluster education data and interpret the results. The data explored is the level of logical-mathematical intelligence and story problem solving ability of students.

2 METHODOLOGY

2.1 Research Implementation

The implementation of this research is shown in Figure 1. The research begins with a literature study related to the object to be studied, the algorithm to be used, and the extent to which similar research has been conducted. After the variables are determined, the preparation of instruments to measure Logical-Mathematical

Intelligence (KLM) and Ability to Solve Story Problems (KSC) is continued. The instruments that have been prepared are then given to experts in their fields to be validated. Criticism and suggestions obtained from experts are then used to revise the instrument. After the instrument was revised, it was distributed to the respondents so they could collect their KLM and KSC data. The data obtained then analyzed again to ensure the data used in the subsequent process was valid.

After data validation, the following process determines the best number of clusters used in the clustering stage. In this process, the Elbow method is applied. The results of the Elbow method are then used to calculate the number of clusters in the clustering process using the K Means algorithm. The K-means algorithm will work until the process converges. The results of the K Means clustering are then revalidated to ensure that the groups formed can represent the respondents. Each group formed is then analyzed to obtain information/knowledge about each respondent's KLM and KSC levels.

Figure 1. Research activities

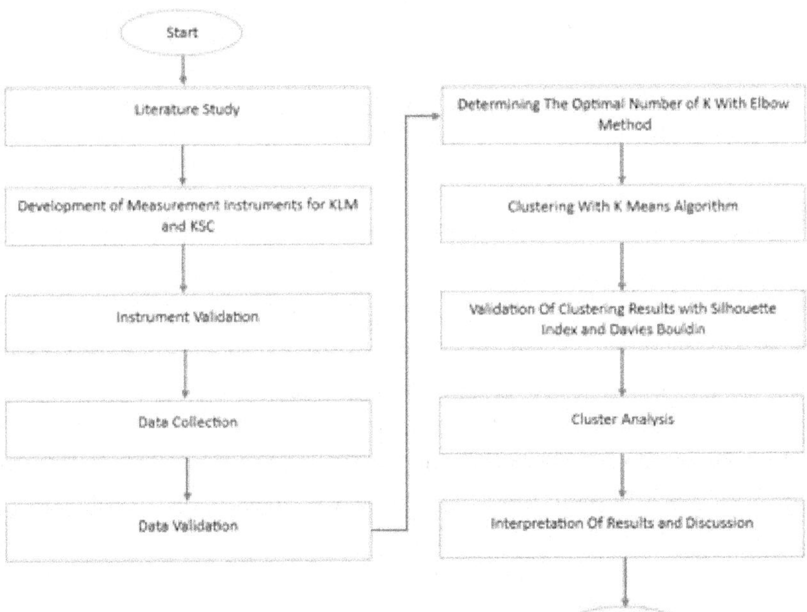

2.2 Data Collection

The data used in this study are data on logical-mathematical intelligence (KLM) and the ability to solve story problems (KSC) obtained from a total of 157 samples of grade VI elementary school students who are members of cluster 5 Gunungsari sub-district, West Lombok district, namely, SDN 1 Gunungsari, SDN 3 Gunungsari, SDN 1 Tamansari, SDN 2 Tamansari, SDI Al-Aziziyah, and SDI Ad-Dinul Qayyim. Determination of the number of student samples using the proportionate stratified random sampling method. KLM is obtained using an instrument in the form of a multiple-choice test that measures indicators of understanding the rhythm of numbers and sorting numbers or patterns (IK1), grouping numbers or patterns (IK2), identifying numbers or patterns (IK3), and causal relationships (IK4). The total number of multiple-choice questions is 15 items with correct answers having a score of one (1) and wrong answers worth zero (0). The KSC was obtained using an instrument in the form of a description test of mathematical story problems with an assessment rubric in the form of identifying what is known and asked, making mathematical models, performing calculations, and interpreting answers to the form of questions. The number of story problems used was 10 questions with the lowest KSC score of 1 and the highest score of 4. The scores obtained were then converted into scores using equation (1). In equation (1) S states the score and Ns states the total number of questions.

$$Score = \frac{S}{N_s} x100 \tag{1}$$

The collected KLM and KSC data are tested for validity and reliability of the instrument. Two methods carried out the instrument validity test: construct validity test and empirical validity test. The construct validity test was carried out before data collection, namely by giving the test instrument to experts to be validated and then making improvements based on the expert's suggestions. While the empirical validity test is carried out after data collection, namely the Cronbach's Alpha method which compares the Pearson product-moment correlation index at the 5% significance level. This validation aims to test the feasibility of answering each item to be used in the next process.

2.3 Elbow Method

The elbow method is a heuristic technique to determine the optimal number of clusters (K) in K-Means clustering. The main idea behind the elbow method is to find where the decrease in the within-cluster sum of squares (WCSS) becomes less significant as the number of clusters increases. WCSS measures the squared distance between each data point and its assigned cluster centroid, and a lower WCSS indicates better-defined clusters (Febrianti et al., 2018). The elbow method involves plotting the WCSS against the number of clusters. As the number of clusters increases, WCSS decreases, but there is a point where the rate of decrease sharply slows down, creating an "elbow" shape in the plot. This point is considered the optimal number of clusters because adding more clusters beyond this point does not significantly improve the clustering solution. The elbow method is a valuable tool for K-Means clustering, particularly when profiling logical-mathematical intelligence and problem-solving abilities, as it helps ensure that the clusters generated represent meaningful groupings of students based on their cognitive traits. However, the elbow method is not always definitive, as the "elbow" can sometimes be ambiguous or difficult to identify. In such cases, other evaluation metrics like the silhouette index or Davies-Bouldin index may provide additional insights into the quality of the clustering solution (Ashari et al., 2023). In cluster analysis, the elbow method is a heuristic method used to determine the appropriate number of clusters k in a data set and improve model performance (Dewi & Pramita, 2019; Febrianti et al., 2018; Hartanti, 2020). The procedure consists of plotting the explained variation as a function of the number of clusters and selecting the elbow of the curve as the number of clusters to be used from the results of calculating the sum of squared points and calculating the average distance or WCSS (Within-Cluster Sum of Square) (Cárcamo et al., 2024).

2.4 K Means Algorithm

K-means is one of the most widely used clustering algorithms that finds the minimum distance value within the same cluster (Maylawati et al., 2020). K-means is a simple algorithm with fast processing time and produces optimal clusters. The K-means algorithm is as follows:

1. Determine the number of clusters to be formed.
2. Initialize the initial centroid value for each cluster ($\mu_1, \mu_2, ..., \mu_k \in \mathbb{R}$) randomly.
3. Repeat the calculation process of equation (2) and equation (3) until it converges.
4. For each i, calculate:

$$c^{(i)} \ : = \text{arg}min_j \left\|x^i - \mu_j\right\|^2 \tag{2}$$

5. For each j, calculate:

$$\mu_j := \frac{\sum_{i=1}^{m} 1\{c^{(i)} = j\}x^{(i)}}{\sum_{i=1}^{m} 1\{c^{(i)} = j\}} \quad (3)$$

2.5 Silhouette Index

The silhouette index is a metric used to evaluate the cohesion and separation of clusters in a clustering solution. It measures how similar each data point is to its assigned cluster (cohesion) compared to other clusters (separation) (Petrovic, 2006; Starczewski & Krzyżak, 2015). The silhouette score ranges from -1 to 1, with higher values indicating better clustering. A silhouette score close to 1 suggests that data points are well clustered, as they are close to other points in their cluster and far from points in different clusters. A score near 0 means that data points lie on the boundary between two clusters, while negative values indicate that data points are likely assigned to the wrong cluster (Dewi & Pramita, 2019; Maylawati et al., 2020). In the context of profiling logical-mathematical intelligence and problem-solving abilities, the silhouette index can provide valuable insights into the quality of K-means clusters. A high silhouette score indicates that students within each cluster share similar cognitive traits, while those in different clusters exhibit distinct profiles. This information can help educators identify students with similar learning needs and design appropriate interventions (Hartanti, 2020).

The silhouette method is analyzed by looking at the value of s. Testing is done by finding the value of a(i), which is the average distance of the core point with all points in the same group, and the value of b(i), which is the average distance of the core point with all points in different groups using the Euclidean distance. After obtaining the values of a(i) and b(i), the silhouette index value is then sought and expressed in the equation (4) (Dewi & Pramita, 2019; Maylawati et al., 2020).

$$S(i) = \frac{(b(i) - a(i))}{\max(a(i), b(i))} \quad (4)$$

2.6 Davies-Bouldin Index (DBI)

The Davies-Bouldin index is another metric used to assess the quality of a clustering solution. It measures the average similarity between each cluster and its most similar cluster, with lower values indicating better clustering (Wijaya et al., 2021). The Davies-Bouldin index evaluates the intra-cluster similarity (how compact each cluster is) and the inter-cluster separation (how far apart the clusters are). A lower Davies-Bouldin index indicates that the clusters are well-separated and compact, while a higher index suggests that the clusters are overlapping or poorly defined (Idrus et al., 2022).

For profiling logical-mathematical intelligence and problem-solving abilities, the Davies-Bouldin index can help determine the effectiveness of the K-Means clustering solution. A lower index indicates that the clusters represent distinct groups of students with different cognitive profiles, allowing educators better to understand the diversity of learning abilities in their classrooms. This, in turn, can inform more personalized and targeted teaching strategies (Umagapi & Umaternate, 2023). The Davies-Bouldin index (DBI) computational process is complex (Petrovic, 2006). Validation is done on the clustering results by maximizing the inter-cluster distance and minimizing the intra-cluster distance (Mughnyanti et al., 2020; Widiarina & Wahono, 2015). The DBI value getting smaller or closer to 0 shows that the resulting cluster is getting better and optimal (Dewi & Pramita, 2019). DBI is determined by the equation (5):

$$DBI = \frac{1}{K}\sum_{i=1}^{k} max_{i \neq j}\left(R_{i,j}\right) \quad (5)$$

With k being the number of clusters used. $\left(R_{i,j}\right)$ is a ratio measure used to compare the i-th and j-th clusters. The ratio $\left(R_{i,j}\right)$ is determined by the equation (6):

$$R_{i,j} = \frac{SSW_i + SSW_j}{SSB_{i,j}} \quad (6)$$

SSW is the inter-cluster value and SSB is the intra-cluster value. Both are expressed as equation (7) and equation (8) respectively:

$$SSW_i = \frac{1}{m_i}\sum_{j=1}^{m_i} d\left(x_j, c_i\right) \quad (7)$$

$$SSB_{i,j} = d\left(c_i, c_j\right) \quad (8)$$

Where m_i is the number of data in the i-th cluster, c_i is the centroid of the i-th cluster $d\left(x_j, c_i\right)$ is the euclidean distance of each data to the centroid, and $d\left(c_i, c_j\right)$ is the distance between centroids.

2.7 Validity and Reliability of Data in Profiling

When creating profile systems for educational research, validity and reliability are crucial factors to consider because they guarantee that the information gathered appropriately represents students' cognitive capacities and learning styles. Reliability is the consistency and stability of the measurement across time, whereas validity is the degree to which an evaluation or measurement tool measures what it is supposed to measure (Cohen et al., 2017; Taherdoost, 2016). To guarantee that the clusters produced by K-means appropriately reflect students' cognitive characteristics, data validity is crucial for profiling logical-mathematical intelligence and problem-solving skills. For instance, profiling can miss other essential facets of logical-mathematical intelligence, such as the capacity for reasoning or problem-solving, if the data used for grouping exclusively depends on students' performance on standardized tests. Thus, the validity of the profile system can be improved by utilizing a range of data sources, such as evaluations of reasoning, critical thinking, and abstract problem-solving (Ahvan & Pour, 2016; Scholar & Zamir, 2018; Taherdoost, 2016).

Reliability is also crucial in K-Means clustering, ensuring the profiling results are consistent and reproducible. If the clustering results vary significantly when different datasets are used, the reliability of the profiling system may be called into question. In educational research, ensuring reliability often involves collecting data from multiple sources or administering the same assessment numerous times to evaluate consistency (Heale & Twycross, 2015).

2.8 Cronbach's Alpha

Cronbach's alpha is a commonly used statistic for measuring the internal consistency or reliability of a set of items in a survey or assessment. It estimates how closely related the items are as a group, with values ranging from 0 to 1. A higher Cronbach's alpha indicates greater internal consistency, with values above 0.7 generally considered acceptable for research purposes (Vaske et al., 2017). In the context of profiling logical-mathematical intelligence and problem-solving abilities using K-means clustering, Cronbach's alpha can be used to assess the reliability of the data collected for clustering. For example, if the profiling system includes multiple assessments or measures of cognitive traits, calculating Cronbach's alpha can help determine whether these measures consistently evaluate the same underlying construct. A high Cronbach's alpha suggests that the assessments are reliable and can be used to generate accurate and consistent clusters of students based on their cognitive profiles (Bonett & Wright, 2015) (DeVellis, 2017). Furthermore, Cronbach's alpha can be used to evaluate the clustering results' internal consistency. If the same clustering solution is generated across different iterations or datasets, this would indicate high reliability in the clustering process. Conversely, if the clustering results vary significantly between different datasets, this may suggest that the profiling system is unreliable and requires further refinement (Taber, 2018).

In educational profiling, Cronbach's alpha can be indirectly linked to K-Means clustering, even though it is typically employed to evaluate the internal consistency of survey or assessment data. For the clustering method to be successful, the data utilized for clustering must be reliable because faulty data can result in poorly defined groups that do not fairly represent the cognitive characteristics of the students (Vaske et al., 2017). The clustering solution produced by K-Means may not be trustworthy if, for instance, the data used to profile logical-mathematical intelligence and problem-solving skills has a low Cronbach's alpha, indicating that the tests or measures used to gather the data are inconsistent. On the other hand, a high Cronbach's alpha suggests that the data is trustworthy and that the K-Means clusters are more likely to correctly represent the cognitive profiles of the pupils (Bujang et al., 2018). Researchers and educators can increase the accuracy and dependability of the profiling system by ensuring the data used for clustering has a high Cronbach's alpha. As a result, it is possible to identify children with comparable learning requirements more successfully and create focused interventions to help them succeed academically (Taber, 2018).

3 DISCUSSION

3.1 Data Collection Results

A total of 157 KLM and KSC data were obtained from cluster 5 of Gunungsari sub-district. More details from SDN 1 Gunungsari with 33 samples with 12 boys and 21 girls, SDN 3 Gunungsari with 31 samples with 13 boys and 18 girls, SDN 1 Tamansari with 47 samples with 20 boys and 27 girls, SDN 2 Tamansari with 20 samples with 7 boys and 13 girls, SDI Al-Aziziyah with 16 samples with 10 boys and 6 girls, and SDI Ad-Dinul Qoyim with 10 samples with 5 boys and 5 girls. The total number of participants based on gender was 67 males and 90 females. The data contains the final scores of indicators IK1 - IK4, KLM, and KSC as shown in Table 1.

Table 1. Research data collection

No	IK1	IK2	IK3	IK4	KLM	KSC
1	100.00	33.33	75.00	80.00	73.30	50.00
2	66.67	33.33	50.00	100.00	66.70	55.00
3	66.67	66.67	100.00	80.00	80.00	60.00
4	66.67	100.00	50.00	100.00	80.00	60.00
5	100.00	66.67	50.00	100.00	80.00	60.00
6	100.00	33.33	50.00	80.00	66.70	55.00
7	100.00	66.67	50.00	80.00	73.30	47.50
8	100.00	66.67	50.00	80.00	73.30	52.50
9	66.67	33.33	50.00	80.00	60.00	50.00
...	...	...	...	...	...	...
155	66.67	0.00	50.00	80.00	53.30	25.00
156	100.00	0.00	50.00	80.00	60.00	25.00
157	0.00	66.67	50.00	40.00	40.00	25.00

The data collected was then tested for the validity of each item of KLM and KSC questions. From the statistical processing results obtained that all questions used in this study are valid as shown in Table 2. After the validity test is carried out, it is continued by testing its reliability. The results show that at the 5% significance level, the reliability coefficient for both the KLM and KSC instruments is > 0.6 as shown in Table 3. Both results indicate that the data is suitable for use in the clustering process with the K Means algorithm.

Table 2. Validity test results

Variables	Question Item	rcount	rtable	Description
KLM	1	0,247	0,157	Valid
	2	0,451	0,157	Valid
	3	0,398	0,157	Valid
	4	0,417	0,157	Valid
	5	0,305	0,157	Valid
	6	0,170	0,157	Valid
	7	0,505	0,157	Valid
	8	0,576	0,157	Valid
	9	0,507	0,157	Valid
	10	0,407	0,157	Valid
	11	0,501	0,157	Valid
	12	0,536	0,157	Valid
	13	0,562	0,157	Valid
	14	0,313	0,157	Valid
	15	0,357	0,157	Valid
KSC	1	0,792	0,157	Valid
	2	0,849	0,157	Valid
	3	0,779	0,157	Valid
	4	0,904	0,157	Valid
	5	0,731	0,157	Valid
	6	0,771	0,157	Valid
	7	0,826	0,157	Valid
	8	0,674	0,157	Valid
	9	0,802	0,157	Valid
	10	0,783	0,157	Valid

Table 3. Reliability test results

Instrument	*Cronbach Alpha* (α)	r table	Description
KLM	0,650	0,60	Reliable
KSC	0,932	0,60	Reliable

3.2 Clustering Result

Before entering the clustering stage, it is necessary to determine the best number of clusters using the Elbow method. This is done so that the clustering results form an optimal group. The results of applying the elbow method in Figure 2 show that for KLM and KSC data grouping, the best number of clusters is 3 clusters.

Figure 2. Elbow method results

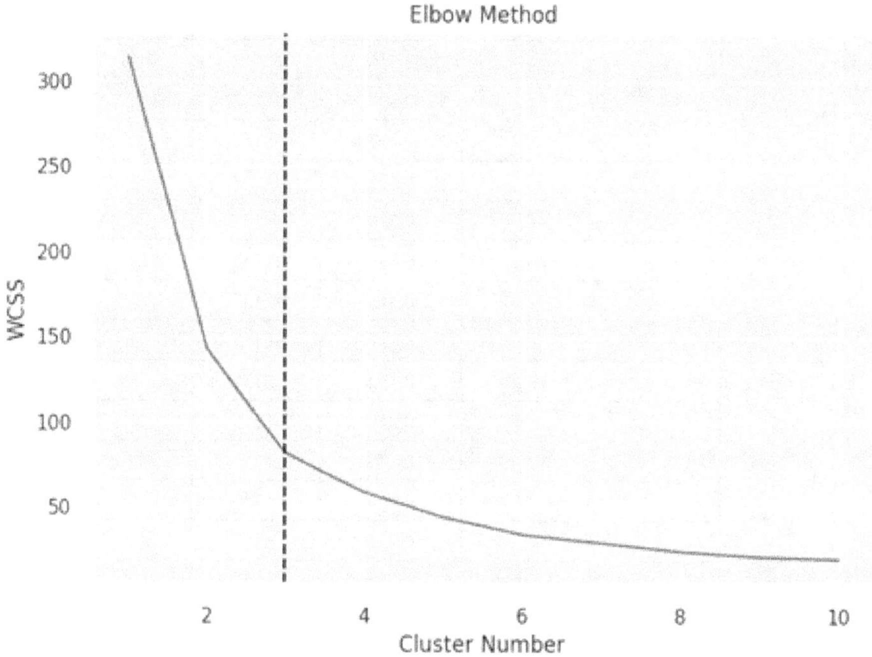

The number of clusters obtained from the elbow method is used as the initial number of clusters in the k-means algorithm. Determination of the initial centroid value is done randomly. The maximum number of iterations is 300. After the implementation of k-means clustering algorithm, the clusters are obtained as shown in Figure 3. Green dots represent the 0th cluster, red dots represent the 1st cluster, and blue dots represent the 2nd cluster. The 0th cluster has a midpoint value of KLM 48.92 and KSC 28.23, the 1st cluster has a midpoint value of KLM 86.67 and KSC 84.35, and the 2nd cluster has a midpoint value of KLM 71.01 and KSC 48.11. The cluster center value for each cluster is shown in Table 4. The clusters formed were then validated with SI and DBI. It was found that the SI and DBI values for the 3 clusters were 0.46 and 0.75, respectively.

Figure 3. Distribution of respondents whose cluster values have been determined

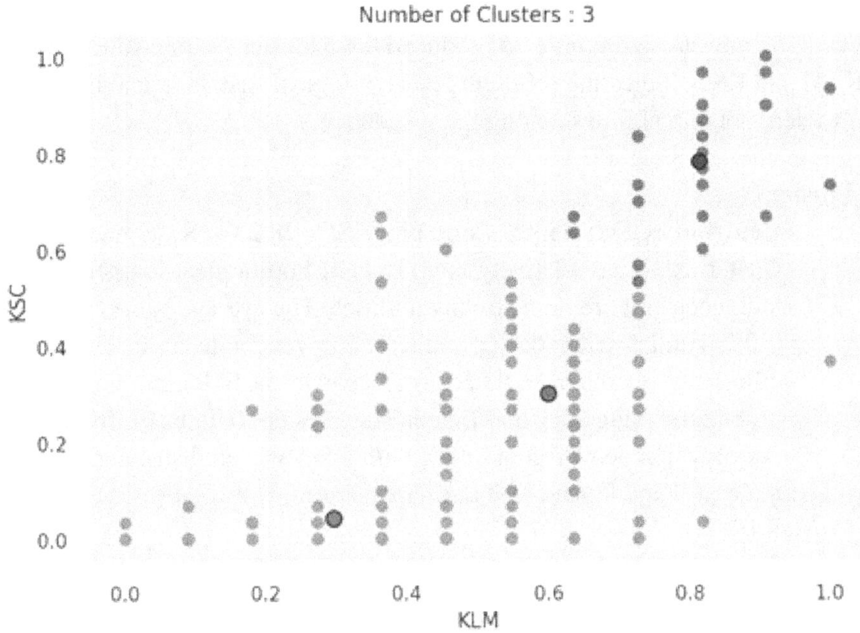

Table 4. Midpoint value of KLM and KSC after clustering process

Cluter	Centroid KLM	Centroid KSC	Av.IK1	Av.IK2	Av.IK3	Av.IK4	Total Respondents	nL	nP
0	48.92	28.23	70.26	25.64	31.54	64.00	65	31	34
1	86.67	84.35	97.10	72.46	84.78	90.43	23	9	14
2	71.01	48.11	95.17	43.00	61.23	81.16	69	26	42
SI	0.46							56	90
DBI	≈ 0.75								

The clustering analysis aimed to profile logical-mathematical intelligence (KLM) and problem-solving abilities (KSC) among respondents based on a set of attributes, including Av.IK1, Av.IK2, Av.IK3, and Av.IK4, which represent different dimensions of cognitive skills. Table 4 presents the midpoint values of KLM and KSC after the K-Means clustering process, with a breakdown of the number of respondents per cluster, including gender (nL for male, nP for female). The results also include evaluation metrics such as the silhouette index (SI) and Davies-Bouldin index (DBI), which assess the quality of the clustering solution.

Cluster Characteristics

The K-Means clustering process produced three clusters with distinct centroids for KLM and KSC, indicating different levels of logical-mathematical intelligence and problem-solving abilities among the respondents.

1. **Cluster 0**:
   - **Centroid KLM: 48.92**, **Centroid KSC: 28.23** – Respondents in this cluster exhibited relatively low levels of both logical-mathematical intelligence and problem-solving abilities. The average values for Av.IK1 (70.26) and Av.IK4 (64.00) suggest some moderate competence in specific areas, but the overall performance in terms of logical reasoning and problem-solving appears limited. This cluster contains the largest number of respondents (65 in total), with a fairly even distribution between male (31) and female (34) participants.
2. **Cluster 1**:
   - **Centroid KLM: 86.67**, **Centroid KSC: 84.35** – Cluster 1 represents the highest-performing group, with significantly higher levels of logical-mathematical intelligence and problem-solving abilities compared to the other clusters. The average values of Av.IK1 (97.10) and Av.IK3 (84.78) further demonstrate the advanced cognitive abilities of the respondents in this cluster. However, this cluster contains the fewest respondents (23 in total), with a notable gender imbalance—more males (9) than females (14) are represented.
3. **Cluster 2**:
   - **Centroid KLM: 71.01**, **Centroid KSC: 48.11** – Respondents in this cluster exhibited moderate levels of logical-mathematical intelligence, but their problem-solving abilities were lower than their KLM scores. The average values of Av.IK1 (95.17) and Av.IK4 (81.16) suggests that this group has potential but may struggle with specific problem-solving tasks. This cluster contains 69 respondents, again showing a gender imbalance, with more females (42) than males (26).

Cluster Evaluation

The silhouette index (SI) of 0.46 indicates moderate cohesion and separation among the clusters. A silhouette score closer to 1 would indicate more distinct clusters, while a score closer to 0 implies that many respondents are near the decision boundary between clusters. The SI of 0.46 suggests that while the clusters are somewhat well-defined, there is room for improvement, especially in refining the separation between groups.

The Davies-Bouldin index (DBI) is approximately 0.75, which indicates that the clusters are moderately compact and well-separated. A lower DBI value typically reflects better clustering quality, suggesting that the clusters are more distinct. The DBI value of 0.75 supports the conclusion that the clustering solution is reasonably good but imperfect, with some overlap between clusters.

Discussion

The clustering analysis successfully profiled respondents based on their logical-mathematical intelligence and problem-solving abilities, revealing distinct groups with varying cognitive strengths. The relatively high performance of Cluster 1 suggests that a subset of students excels in both domains, indicating a potential group of learners who may benefit from more advanced educational interventions. In contrast, Cluster 0, with its lower centroids, may represent students who struggle with both logical reasoning and problem-solving, highlighting the need for additional support and instructional strategies tailored to their needs.

The moderate SI and DBI values indicate that while the clusters are meaningful, there may be some overlap between the respondents in different clusters, particularly between Clusters 0 and 2, where logical-mathematical intelligence scores are similar, but problem-solving abilities differ. Further refinement of the clustering process, such as incorporating additional features or using alternative clustering techniques, could enhance the accuracy of the profiling system.

3.3 Cluster Analysis

The results of the application of the K-means algorithm on KLM and KSC data divide them into 3 categories of levels, namely high, medium, low levels. For students with KLM and KSC, the high-level category is gathered in cluster 1, which has 23 members, with 9 male respondents and 14 female respondents. Their KLM centroid value is 86.67 and their KSC centroid value is 84.35. When considering the KLM value per category, respondents who are members of the high level have an average ability of IK1 97.10, IK2 72.46, IK3 84.78, and IK4 90. Respondents who are members of the medium level category are grouped in cluster 2, having 69 members with 26 male members and 42 female members. Their KLM and KSC centroid values are 71.01 and 48.11, respectively. The average KLM scores per category are IK1 95.17, IK2 43, IK3 61.23, and IK4 81.16. The remaining respondents in cluster 0 belong to the low-level KLM and KSC categories. Their KLM centroid value is 48.92 and KSC 28.23. Meanwhile, the average KLM value per category is IK1 70.26, IK2 25.64, IK3 31.54, and IK4 64.

Figure 4. Distribution of silhouette score for KMeans clusters

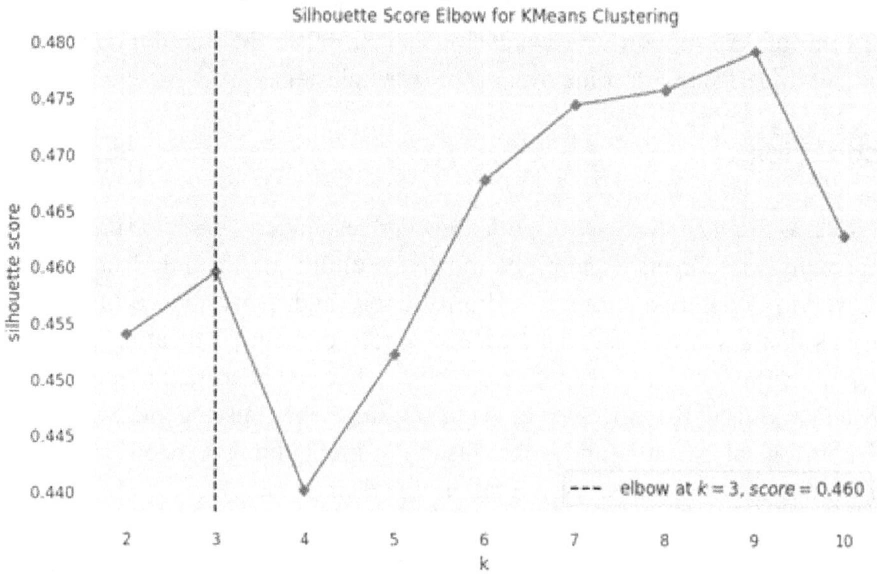

Based on the average value per KLM category, IK1 is always the highest for all three clusters, and IK2 is the lowest. This means that all respondents have a good understanding of number rhythm and sequencing of numbers or patterns but are less able to classify patterns. Then, the ability to understand causal relationships and identify numbers and patterns is still in the medium category. Each level category has its advantages and disadvantages. Groups that are members of the high level have an even ability for each KLM indicator. The group in the medium category is dominant in IK1 and IK4, but a little weak in IK2 and IK3. Meanwhile, the group in the low-level category also has a similar pattern of ability to the medium category but with a lower average value with the highest difference almost reaching 2 times, especially in IK3.

Figure 5. Distribution of DBI score for KMeans clusters

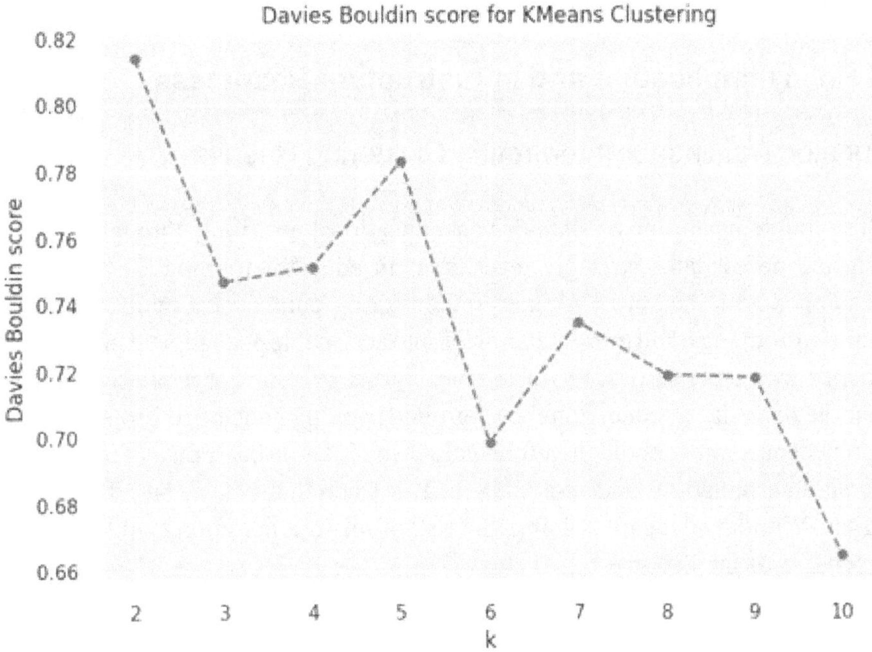

When viewed from the number of respondents, the most cluster members, 69 people, are occupied by clusters with medium-level KLM and KSC abilities (cluster 2), followed by low-level ones (cluster 0) as many as 65 people, and the last is a high level of 23 people. As for the gender of respondents, each cluster is always dominated by female P respondents with the highest number in cluster 2 and the lowest in cluster 1. Based on the SI and DBI values, the validity level of the clusters formed reaches 0.46 and 0.75. This value indicates that there is still an opportunity for improvement of the clustering method used in profiling students based on KLM and KSC. Even though the SI and DBI scores obtained are relatively low, these are the best values for clusters formed with the K-Means algorithm using the KLS and KSC datasets. Based on the SI and DBI values, it can be said that cluster number 3 is the best number for profiling logical-mathematical intelligence (KLM) and the ability to answer story questions (KSC).

This can be strengthened by the condition that the SI and DBI values are always the same when the clustering simulation is repeated. Even though in Figure 4 and Figure 5 the SI value increases and DBI decreases as the number of clusters increases, this result does not persist when repeated training with random initial values is carried out. The best SI and DBI values will change along with changes in the number of clusters. This is because an increasing number of clusters can reduce the strength of the degree of connection between categories and cluster labels (Dubes, 1987).

3.4 Policy Implication and Infrastructure Readiness

Educator Training for Interpreting Clustering Results

The implementation of data-driven educational profiling through K-Means clustering necessitates strategic investment in educator training. Teachers must be equipped with the ability to interpret clustering outputs accurately, understand cognitive groupings, and translate findings into actionable pedagogical interventions (Jiménez-Macías et al., 2024). In Indonesia, this becomes particularly important given the diversity in educational backgrounds among educators. Professional development programs should therefore include modules on basic data literacy, cluster analysis interpretation, and individualized instructional design based on student profiles. Without adequate training, the risk of misinterpretation could undermine the benefits of intelligence-based clustering.

Internet Access Disparities and Their Impact

Another critical challenge within the Indonesian context is the persistent digital divide, particularly in rural and remote areas. Limited access to reliable internet infrastructure impedes the broader application of data-driven learning and clustering analysis at the national scale (Batool et al., 2023). Schools in disadvantaged regions may struggle to adopt digital assessment tools necessary for gathering and processing intelligence data. This digital gap exacerbates educational inequality, as students in underserved areas are less likely to benefit from precision educational strategies informed by clustering results. National initiatives to expand equitable internet access are therefore foundational to the success of educational data mining efforts (Roski et al., 2024).

Proposal for Data-Driven Education Policies

To institutionalize the use of clustering-based intelligence profiling, Indonesia's educational policy should promote data-driven decision-making frameworks at both school and district levels (Rahman et al., 2024). Policies could mandate the integration of cognitive profiling at key transition points in education (e.g., late primary to early secondary). Furthermore, standardized, ethical data collection protocols must be established to ensure student privacy and responsible data use. Strategic partnerships with technology providers and research institutions could accelerate the development of secure platforms capable of supporting clustering and other educational data mining applications (Araka et al., 2022).

Real-World Applications: Forming Effective Learning Groups

The real-world implication of clustering results is profound: educators can utilize cognitive profiles to form homogenous or heterogeneous learning groups tailored to specific objectives (Mohd Talib et al., 2023). For instance, high logical-mathematical students could be grouped to engage in complex problem-solving projects, while mixed-ability groups could foster peer learning opportunities. By aligning group composition with students' strengths and weaknesses as identified through clustering, teachers can create more dynamic, inclusive, and effective classroom environments.

Societal Benefits: Teaching Efficiency and Administrative Relief

Adopting clustering-based student profiling holds several social advantages. First, it enhances teaching efficiency by allowing educators to design instruction that resonates with distinct learner profiles, thereby reducing trial-and-error approaches in pedagogy (Yağcı, 2022). Second, it minimizes administrative burden by automating aspects of student needs analysis, traditionally a time-intensive task. Third, it improves the accuracy of identifying high-potential students, thereby enabling more effective talent development and targeted academic interventions (Batool et al., 2023).

Economic Potentials: Supporting Affirmative Education Policies

Profiling students' cognitive traits also presents substantial economic implications. By enabling precise identification of students requiring additional support or those who can benefit from acceleration programs, resources can be allocated more effectively, optimizing educational investments (Roski et al., 2024). This data-driven approach can reinforce affirmative education policies, particularly for students from marginalized communities, ultimately contributing to broader goals of social mobility and economic equity in Indonesia.

Influence of Curriculum Changes and Digitalization

The ongoing transformation of Indonesia's national curriculum, particularly the shift toward “Merdeka Belajar” (Freedom to Learn) and the rapid digitalization of education, amplifies the importance of individualized intelligence mapping (Ashari et al., 2023). Flexible curricula and technology integration necessitate deeper insights into each student's cognitive profile to truly personalize learning pathways. As educational environments become increasingly digitized, the ability to dynamically map and respond to students' logical-mathematical abilities will be pivotal in maintaining curriculum relevance and ensuring no student is left behind.

4 CONCLUSION

The application of the K-Means clustering algorithm in profiling students' logical-mathematical intelligence (KLM) and story problem-solving abilities (KSC) successfully segmented students into three distinct groups: high, medium, and low levels. This segmentation not only provides deeper insights into students' cognitive profiles but also offers a strategic foundation for personalized learning interventions. However, the moderate SI and DBI scores (0.46 and 0.75, respectively) suggest opportunities for refining the clustering approach, including exploring alternative algorithms or enhancing initial data preprocessing.

Beyond methodological considerations, the findings underscore critical policy and infrastructure readiness needs in Indonesia. To maximize the benefits of intelligence-based student profiling, systemic support is required in the form of educator training in data interpretation, expansion of equitable internet access, and the development of ethical, data-driven educational policies. Real-world applications such as forming targeted learning groups, improving instructional efficiency, reducing administrative burdens, and supporting affirmative education policies highlight the transformative potential of integrating clustering results into everyday teaching practices. Furthermore, the shift towards flexible national curricula and educational digitalization strengthens the urgency of embedding cognitive profiling into Indonesia's educational framework to ensure inclusivity and responsiveness to individual learner needs. In sum, the strategic integration of intelligence clustering into educational practice and policy represents a pivotal step toward achieving more equitable, efficient, and personalized education outcomes in Indonesia.

ACKNOWLEDGMENT

The Authors would like to thank the financial support from the Informatics and Computer Engineering Education (ICEE) Program, Engineering Faculty, Universitas Negeri Semarang.

REFERENCES

Ahvan, Y. R., & Pour, H. Z. (2016). The correlation of multiple intelligences for the achievements of secondary students. *Educational Research Review*, *11*(4), 141–145. DOI: 10.5897/ERR2015.2532

Akhmad, M. W. (2019). *Pengaruh Kecerdasan Logis-Matematis dan Kecerdasan Spasial-Visual terhadap Kemampuan Pemecahan Masalah Geometri Siswa Kelas IV di MI Darul Ulum Kota Batu*. UIN Maulana Malik Ibrahim.

Al-Doulat, A., Nur, N., Karduni, A., Benedict, A., Al-Hossami, E., & Maher, M. Lou, Dou, W., Dorodchi, M., & Niu, X. (2020). Making Sense of Student Success and Risk Through Unsupervised Machine Learning and Interactive Storytelling. *Artificial Intelligence in Education*, 9. https://doi.org/https://doi.org/10.1007/978-3-030-52237-7_1

Al-Kinani, H. K. R., & Al-Omari, A. H. M. (2021). Analytical thinking and its relationship to logical intelligence for scientific fifth-grade students in mathematics. *Turkish Journal of Computer and Mathematics Education*, *12*(10), 3148–3161. https://www.proquest.com/scholarly-journals/analytical-thinking-relationship-logical/docview/2628339997/se-2?accountid=62707

Araka, E., Oboko, R., Maina, E., & Gitonga, R. (2022). Using educational data mining techniques to identify profiles in self-regulated learning: An empirical evaluation. *International Review of Research in Open and Distance Learning*, *23*(1), 131–162. DOI: 10.19173/irrodl.v22i4.5401

Ashari, I. F., Banjarnahor, R., Farida, D. R., Aisyah, S. P., Dewi, A. P., & Humaya, N. (2022). Application of Data Mining with the K-Means Clustering Method and Davies Bouldin Index for Grouping IMDB Movies. *Journal of Applied Informatics and Computing, 6*(1), 07–15. DOI: 10.30871/jaic.v6i1.3485

Ashari, I. F., Dwi Nugroho, E., Baraku, R., Novri Yanda, I., & Liwardana, R. (2023). Analysis of Elbow, Silhouette, Davies-Bouldin, Calinski-Harabasz, and Rand-Index Evaluation on K-Means Algorithm for Classifying Flood-Affected Areas in Jakarta. *Journal of Applied Informatics and Computing*, *7*(1), 89–97. DOI: 10.30871/jaic.v7i1.4947

Balovsyak, S., Derevyanchuk, O., Kravchenko, H., Ushenko, Y., & Hu, Z. (2023). Clustering students according to their academic achievement using Fuzzy Logic. *International Journal of Modern Education and Computer Science*, *15*(6), 31–43. DOI: 10.5815/ijmecs.2023.06.03

Batool, S., Rashid, J., Nisar, M. W., Kim, J., Kwon, H.-Y., & Hussain, A. (2023). Educational data mining to predict students' academic performance: A survey study. *Education and Information Technologies*, *28*(1), 905–971. DOI: 10.1007/s10639-022-11152-y

Bonett, D. G., & Wright, T. A. (2015). Cronbach's alpha reliability: Interval estimation, hypothesis testing, and sample size planning. *Journal of Organizational Behavior*, *36*(1), 3–15. DOI: 10.1002/job.1960

Budiyanto, R., Yohanie, D. D., & Widodo, S. (2018). Pengaruh Kecerdasan Logis-Matematis terhadap Kemampuan Pemecahan Masalah Matematis Siswa pada Penyelesaian Soal Cerita Materi SPLDV. *Simki-Techsain*, *2*(10), 1–8.

Bujang, M. A., Omar, E. D., & Baharum, N. A. (2018). A review on sample size determination for Cronbach's alpha test: A simple guide for researchers. *The Malaysian Journal of Medical Sciences : MJMS*, *25*(6), 85–99. DOI: 10.21315/mjms2018.25.6.9 PMID: 30914882

Cárcamo, J., Cuevas, A., & Rodríguez, L. A. (2024). On uniqueness of the set of k-means. arXiv preprint arXiv:2410.13495.

Chen, S. (2022). Improved Fuzzy Algorithm for College Students' Academic Early Warning. *Mathematical Problems in Engineering*, *2022*, 1–9. Advance online publication. DOI: 10.1155/2022/5764800

Cohen, L., Manion, L., & Morrison, K. (2017). Validity and reliability. In *Research methods in education* (pp. 245–284). Routledge. DOI: 10.4324/9781315456539-14

Dewi, D. A. I. C., & Pramita, D. A. K. (2019). Analisis Perbandingan Metode Elbow dan Silhouette pada Algoritma Clustering K-Medoids dalam Pengelompokan Produksi Kerajinan Bali. *Matrix : Jurnal Manajemen Teknologi Dan Informatika*, *9*(3), 102–109. DOI: 10.31940/matrix.v9i3.1662

Dubes, R. C. (1987). How many clusters are best? - An experiment. *Pattern Recognition, 20*(6), 645–663. DOI: 10.1016/0031-3203(87)90034-3

Esmaeili, F., Behnam, B., & Esmaeili, K. (2014, September). A Study of Relationship between Multiple Intelligences and Writing Ability of Iranian Female and Male Students. *Mediterranean Journal of Social Sciences.* Advance online publication. DOI: 10.5901/mjss.2014.v5n20p2663

Fan, Z., & Yan, S. (2016). Clustering of College Students Based on Improved K-means Algorithm. *International Computer Symposium.* DOI: 10.1109/ICS.2016.0139

Fauzi, F., & Monawati, M. (2018). Hubungan Antara Kecerdasan Logis Matematika dan Kedisiplinan Belajar Siswa Kelas V SD Negeri 1 Pagar Air Kabupaten Aceh Besar. Pesona Dasar: Jurnal Pendidikan Dasar dan Humaniora, 6(1).

Fauziah, K. R., Nurhayati, N., & Arsyad, M. (2015). Analisis Hubungan Antara Kecerdasan Logis-Matematis dengan Hasil Belajar Fisika Peserta Didik Kelas XI IPA SMA Negeri Di Kabupaten Jeneponto. Jurnal Sains dan Pendidikan Fisika, 11(3), 239-244.

Febrianti, A. F., Cabral, A. H., & Anuraga, G. (2018). K-Means Clustering Dengan Metode Elbow Untuk Pengelompokan Kabupaten Dan Kota Di Jawa Timur. *Seminar Nasional Hasil Riset Dan Pengabdian -SNHRP*, 863–870. http://karyailmiah .unipasby.ac.id/wp-content/uploads/2019/04/K-Means-Artikel.pdf

Gardner, H. E. (2011). *Frames of mind: The theory of multiple intelligences.* Basic books.

Hana, Z., Nujumuddin, N., & Sucilestari, R. (2020). Memahami dan Mengukur Kecerdasan Logis Matematis: Sebuah Telaah Pustaka. *DIKDAS MATAPPA: Jurnal Ilmu Pendidikan Dasar, 3*(2), 197–206.

Hartanti, N. T. (2020). Metode Elbow dan K-Means Guna Mengukur Kesiapan Siswa SMK Dalam Ujian Nasional. *Jurnal Nasional Teknologi Dan Sistem Informasi, 6*(2), 82–89. DOI: 10.25077/TEKNOSI.v6i2.2020.82-89

Heale, R., & Twycross, A. (2015). Validity and reliability in quantitative studies. *Evidence-Based Nursing, 18*(3), 66–67. DOI: 10.1136/eb-2015-102129 PMID: 25979629

Hidayat, S., Rismayati, R., Tajuddin, M., & Merawati, N. L. P. (2020). Segmentation of university customers loyalty based on RFM analysis using fuzzy c-means clustering. *Jurnal Teknologi Dan Sistem Komputer, 8*(2), 133–139. DOI: 10.14710/ jtsiskom.8.2.2020.133-139

Idrus, A., Tarihoran, N., Supriatna, U., Tohir, A., Suwarni, S., & Rahim, R. (2022). Distance Analysis Measuring for Clustering using K-Means and Davies Bouldin Index Algorithm. *TEM Journal, 11*(4), 1871–1876. https://doi.org/DOI: 10.18421/TEM114-55

Imamah, Y., U. L., Djunaidy, A., Pardede, E., & Purnomo, M. H. (2024). Clustering Approach for Modeling Course Difficulty Level in Adaptive Learning. *2024 IEEE International Conference on Computational Intelligence and Virtual Environments for Measurement Systems and Applications (CIVEMSA)*, 1–6. DOI: 10.1109/CIVEMSA58715.2024.10639956

Ismaya, N., & Zyen, A. K. (2022). Penentuan siswa berprestasi menggunakan metode k-means clustering di smp takhassus al qur'an sadamiyyah. Jtinfo. *Jurnal Teknik Informatika*, *1*(1), 64–48.

Jiménez-Macías, A., Muñoz-Merino, P. J., & Kloos, C. D. (2024). Student Clustering Through KMeans to Enhance Teaching. *2024 XVI Congreso de Tecnología, Aprendizaje y Enseñanza de La Electrónica (TAEE)*, 1–5. DOI: 10.1109/TAEE59541.2024.10604956

Karsenti, T. (2019). *Artificial intelligence in education : The urgent need to prepare teachers for tomorrow ' s schools. 27*(1), 105–111. DOI: 10.18162/fp.2019.a166

Khodadady, E., & Dastgahian, B. S. (2013). Logical-Mathematical Intelligence and Its Relationship with English Language Proficiency. *American Journal of Scientific Research*, *89*, 57–68.

Korkmaz, O. (2012). The Impact of Critical Thinking and Logico-Mathematical Intelligence on Algorithmic Design Skills *. *Journal of Educational Computing Research*, *46*(2), 173–193. https://doi.org/http://dx.doi.org/10.2190/EC.46.2.d. DOI: 10.2190/EC.46.2.d

López-Oriona, Á., Vilar, J. A., & D'Urso, P. (2023). Hard and soft clustering of categorical time series based on two novel distances with an application to biological sequences. *Information Sciences*, *624*, 467–492. DOI: 10.1016/j.ins.2022.12.065

Maylawati, D. S., Priatna, T., Sugilar, H., & Ramdhani, M. A. (2020). *Data science for digital culture improvement in higher education using K-means clustering and text analytics. 10*(5), 4569–4580. DOI: 10.11591/ijece.v10i5.pp4569-4580

Milsan, A. L., & Wewe, M. (2018). Hubungan Antara Kecerdasan Logis Matematis dengan Hasil Belajar Matematika. *Journal of Education Technology*, *2*(2), 65–69. DOI: 10.23887/jet.v2i2.16183

Mohamed Nafuri, A. F., Sani, N. S., Zainudin, N. F. A., Rahman, A. H. A., & Aliff, M. (2022). Clustering Analysis for Classifying Student Academic Performance in Higher Education. *Applied Sciences (Basel, Switzerland), 12*(19), 9467. Advance online publication. DOI: 10.3390/app12199467

Mohd Talib, N. I., Abd Majid, N. A., & Sahran, S. (2023). Identification of student behavioral patterns in higher education using k-means clustering and support vector machine. *Applied Sciences (Basel, Switzerland), 13*(5), 3267. DOI: 10.3390/app13053267

Mughnyanti, M., Efendi, S., & Zarlis, M. (2020). Analysis of determining centroid clustering x-means algorithm with davies-bouldin index evaluation. *IOP Conference Series. Materials Science and Engineering, 725*(1), 012128. Advance online publication. DOI: 10.1088/1757-899X/725/1/012128

Muthmainnah, R. N., Eminita, V., & Ramadhan, A. I. (2020). CLUSTER ANALYSIS OF STUDENTS ' MULTIPLE INTELLIGENCES IN FIRST GRADE AT MADRASAH IBTIDAIYAH AL-INAYAH IN INDONESIA. *Journal of Critical Reviews, 7*(13), 929–933.

Nazari, M., Hussain, A., & Musilek, P. (2023). Applications of Clustering Methods for Different Aspects of Electric Vehicles. *Electronics (Basel), 12*(4), 790. Advance online publication. DOI: 10.3390/electronics12040790

Nurzaelani, M. M., & Wibowo, S. (2015). Hubungan Antara Kecerdasan Logis-Matematis dan Komunikasi Interpersonal dengan Hasil Belajar Mata Pelajaran Matematika. *Jurnal Teknologi Pendidikan, 4*(1), 56–72.

Pehlivan, A., & Durgut, M. (2017). The effect of logical-mathematical intelligence on financial accounting achievement according to multiple intelligence theory. *Journal of Education & Social Policy, 4*(3), 132–139.

Petrovic, S. (2006). A comparison between the silhouette index and the davies-bouldin index in labelling ids clusters. *Proceedings of the 11th Nordic Workshop of Secure IT Systems,* 2006, 53–64.

Purnamaningsih, C., & Aziz, A. (2014). Pemanfaatan Metode K-Means Clustering dalam Penentuan Penjurusan Siswa SMA. *JURNAL ITSMART, 3*(1), 27–33. DOI: 10.20961/its.v3i1.644

Puspitasari, A., Rapani, & Ambarita, A. (2018). Hubungan Kecerdasan Verbal dan Logika-Matematika dengan Meyelesaikan Soal Cerita Matematika. *Jurnal Pedagogi, 7*(2), 430–439.

Raehanah, R., & Apriani, R. (2019). Pengaruh kecerdasan logis matematis terhadap hasil belajar matematika dasar. *Jurnal Pijar MIPA, 14*(3), 112–117.

Rahman, S. I., Ahmed, S., Fariha, T. A., Mohammad, A., Haque, M. N. M., Chellappan, S., & Noor, J. (2024). Unsupervised machine learning approach for tailoring educational content to individual student weaknesses. *High-Confidence Computing, 4*(4), 100228. DOI: 10.1016/j.hcc.2024.100228

Roikha, M. (2016). *Pengaruh Kecerdasan Mtematis-Logis dan Kecerdasan Visual-Spasial terhadap Prestasi Belajar Matematika Materi Bangun Ruang Kelas V di SD Tamansiswa Turen*. UIN Maulana Malik Ibrahim, Malang.

Roski, M., Sebastian, R., Ewerth, R., Hoppe, A., & Nehring, A. (2024). Learning analytics and the Universal Design for Learning (UDL): A clustering approach. *Computers \& Education, 214*, 105028.

Roth, B., Becker, N., Romeyke, S., Schäfer, S., Domnick, F., & Spinath, F. M. (2015). Intelligence and school grades : A meta-analysis. *Intelligence, 53*, 118–137. DOI: 10.1016/j.intell.2015.09.002

Scholar, Q., & Zamir, S. (2018). A Comparative Study of Multiple Intelligence of Science and Arts Group Students. *Journal of Research in Social Sciences, 6*(2), 2305–6533.

Sharma, A., Singh, S. K., Badwal, E., Kumar, S., Gupta, B. B., Arya, V., Chui, K. T., & Santaniello, D. (2023). Fuzzy Based Clustering of Consumers' Big Data in Industrial Applications. *2023 IEEE International Conference on Consumer Electronics (ICCE)*, 1–3. DOI: 10.1109/ICCE56470.2023.10043451

Siregar, M. U. Comparative Study of K-Means Clustering Algorithm and K-Medoids Clustering in Student Data Clustering. JISKA (Jurnal Informatika Sunan Kalijaga), 7(2), 91-99.

Starczewski, A., & Krzyżak, A. (2015). Performance evaluation of the silhouette index. *International Conference on Artificial Intelligence and Soft Computing*, 49–58.

Subariyanto, M. I., Muncarno, & Supriyadi. (2017). Hubungan Kecerdasan Verbal dan Kecerdasan Logika-Matematika dengan Kemampuan Menyelesaikan Soal Cerita Matematika. *Jurnal P, 6*(7), 1–13.

Sulistiyawati, A., & Supriyanto, E. (2020). Implementasi Algoritma K-means Clustring dalam Penetuan Siswa Kelas Unggulan. *Jurnal Tekno Kompak, 15*(2), 25–36. DOI: 10.33365/jtk.v15i2.1162

Sya'iyah, K., Yuliansyah, H., & Arfiani, I. (2019). Clustering Student Data Based On K - Means Algorithms. *INTERNATIONAL JOURNAL OF SCIENTIFIC & TECHNOLOGY RESEARCH, 8*(08), 1014–1018.

Taber, K. S. (2018). The use of Cronbach's alpha when developing and reporting research instruments in science education. *Research in Science Education, 48*(6), 1273–1296. DOI: 10.1007/s11165-016-9602-2

Taherdoost, H. (2016). Validity and reliability of the research instrument; how to test the validation of a questionnaire/survey in a research. *International Journal of Academic Research in Management (IJARM), 5.*

Tirri, K., Nokelainen, P., & Komulainen, E. (2013). Multiple intelligences : Can they be measured? *Psychological Test and Assessment Modeling, 55*(4), 438–461.

Umaeza, A., & Budhi, W. (2016). Hubungan Antara Kemampuan Numerik, Verbal dan Menyelesaikan Soal Cerita Fisika Dengan Prestasi Belajar Fisika. *Jurnal Ilmiah Pendidikan Fisika, 3*(1), 45–46.

Umagapi, I. T., & Umaternate, B. (2023). Uji Kinerja K-Means Clustering Menggunakan Davies-Bouldin Index Pada Pengelompokan Data Prestasi Siswa. *Prosiding SISFOTEK*, 303–308. https://seminar.iaii.or.id/index.php/SISFOTEK/article/view/411%0A

Vaske, J. J., Beaman, J., & Sponarski, C. C. (2017). Rethinking internal consistency in Cronbach's alpha. *Leisure Sciences, 39*(2), 163–173. DOI: 10.1080/01490400.2015.1127189

Wang, Q., Liu, L., Liu, Z., Qiao, W., & Liu, F. (2022). Application of Fuzzy Clustering in Higher Education General Management Based on Internet Environment. *Mathematical Problems in Engineering, 2022*, 1–11. Advance online publication. DOI: 10.1155/2022/3438666

Wang, Y. (2024). Exploring the Application of Early Childhood Psychological Education Based on K-means Cluster Analysis. *Journal of Electrical Systems, 20*(9s), 64–71.

Widiarina, W., & Wahono, R. S. (2015). Algoritma cluster dinamik untuk optimasi cluster pada algoritma k-means dalam pemetaan nasabah potensial. *Journal of Intelligent Systems, 1*(1), 33–36.

Wiguna, B. J. P. K., Suwarma, I. R., & Liliawati, W. (2018). STEM-based science learning implementation to identify student's personal intelligences profiles. *Journal of Physics: Conference Series, 1013*(1), 012082. Advance online publication. DOI: 10.1088/1742-6596/1013/1/012082

Wijaya, Y. A., Kurniady, D. A., Setyanto, E., Tarihoran, W. S., Rusmana, D., & Rahim, R. (2021). Davies bouldin index algorithm for optimizing clustering case studies mapping school facilities. *TEM J, 10*(3), 1099–1103. DOI: 10.18421/TEM103-13

Yağcı, M. (2022). Educational data mining: Prediction of students' academic performance using machine learning algorithms. *Smart Learning Environments, 9*(1), 11. DOI: 10.1186/s40561-022-00192-z

Yu, C., & Wang, Y. (2022). College Student Management System Based on K-means Clustering Algorithm. *International Journal of New Developments in Education, 4*(2), 28–33. DOI: 10.25236/IJNDE.2022.040206

Chapter 8
Machine Learning Techniques for Network Security:
A Comprehensive Review

Shyamalendu Paul
https://orcid.org/0009-0008-1028-4867
Department of Computer Science and Engineering, Brainware University, West Bengal, India

Nobhonil Roy Choudhury
https://orcid.org/0009-0009-1046-7492
Brainware University, West Bengal, India

Ayan Bandyopadhyay
https://orcid.org/0009-0005-5927-1715
Brainware University, West Bengal, India

ABSTRACT

Machine learning methods are now recognized as mighty instruments for bolstering network security in our progressively interconnected world. This article presents an extensive overview of machine learning's utilization in network security, encompassing critical areas like intrusion detection, malware detection, phishing detection, anomaly detection, and beyond. The paper starts with a summary of machine learning fundamentals, it is included supervised learning, unsupervised learning, reinforcement learning, and evaluation metrics. It explores how machine learning is applied across different domains of network security, highlighting the character of machine learning techniques in bolstering consecutive security measures and addressing emerging threats. The paper identifies challenges and limitations in

DOI: 10.4018/979-8-3693-9846-3.ch008

applying machine learning to network security, encompassing concerns about data quality and quantity, feature selection, model interpretability, adversarial attacks, scalability and real time processing constraints.

1. INTRODUCTION

In recent years, the convergence of machine learning and network security has gained noticeable attention, spurred by the increasing frequency and complexity of cyber threats. This review paper seeks to investigate and unite the current state of machine learning techniques in network security, emphasizing their capabilities, challenges and future directions.

1.1 Background and Motivation

Network security is a crucial element of information technology infrastructure, safeguarding data integrity, confidentiality and availability from malicious attacks. Foundational to this protection are traditional security measures like firewalls, antivirus software and intrusion detection systems (IDS). However, the ever changing landscape of cyber threats requires more development and flexible strategies. The emergence of machine learning (ML) provides promising methods to developed network security by utilizing patterns and irregularities within large datasets, allowing for protecting threat detection and mitigation.

The drive to incorporate ML into network security arises from its ability to automate analysis and response tasks, thereby minimizing human error and enhancing response times to incidents. Moreover, ML algorithms have the capacity to continually learn and adjust to new threats, rendering them highly efficient in dynamic and complex environments.

1.2 Importance of Network security

The significance of network security in today's digital era cannot be emphasized enough. As businesses increasingly depend on interconnected systems and the Internet for their functions, the dangers linked to cyber threats have surged dramatically. Cyberattacks can cause substantial financial losses, harm to reputational, legal repercussions and interruptions in operations. For example, the worldwide cost of cybercrime is expected to hit $10.5 trillion per year by 2025, up from $3 trillion in 2015.

Effective network security measures are crucial for protecting sensitive data, maintaining customer trust and ensuring business continuity. The increase in remote work, cloud computing and the Internet of Things (IoT) has extended the attack surface, making strong security measures more vital than ever. Additionally, regulatory frameworks like the General Data Protection Regulation (GDPR) and the Health Insurance Portability and Accountability Act (HIPAA) require stringent security practices, underscoring the necessity for advanced security solutions.

1.3 Role of Machine learning in network security

Machine learning significantly transforms network security by providing tools and techniques to analyse large amount of data for detecting patterns and anomalies indicative of cyber threats. Unlike traditional rule-based systems, ML algorithms can uncover previously unknown threats by learning from data. This ability is especially crucial for identifying zero-day attacks and advanced malware that can bypass standard detection methods.

Various machine learning techniques have been employed in different areas of network security, such as supervised learning for classification tasks, unsupervised learning for anomaly detection and reinforcement learning for adaptive defence mechanisms. For example, deep learning (LeCun et al., 2015), a subset of machine learning, has been successfully utilized to identify and classify malware by examining code behaviour and network traffic patterns (Chandola et al., 2009).

Moreover, ML boosts the effectiveness of intrusion detection systems (IDS) by enhancing accuracy and minimizing false positives. By constantly learning from new data, ML-based systems can adjust to the changing threat landscape, offering a dynamic and responsive defence mechanism.

1.4 Objective of the Review

This review paper aims to achieve several key objectives:

a) **Comprehensive Overview:** Give a elaborate summary of the current situation of machine learning applications in network security, unite a various techniques and their execution.
b) **Evaluation of Techniques:** Evaluate the effectiveness of various ML techniques in addressing specific network security challenges, emphasizing their strengths and weaknesses.

c) **Identification of Challenges:** Identify the primary challenges and weaknesses linked with the application of ML in network security, containing issues related to data quality, interpretability and adversarial attacks.
d) **Future Directions:** Explore potential future research and development directions in this field, focusing on emerging trends and technologies that could improve the effectiveness of ML-based security solutions.
e) **Case Studies and Practical Insights:** Offer case studies and practical insights from real-world applications of ML in network security, providing a practical perspective on the theoretical concepts discussed.

2. OVERVIEW OF NETWORK SECURITY

Network security is a crucial component of contemporary information technology, safeguarding data and resources from various cyber threats. This section offers a thorough overview of network security, including its definitions and concepts, the different types of network threats, traditional security measures, and the limitations associated with these conventional approaches.

2.1 Definition and Concepts

Network security involves policies, practices, and technologies aimed to guard the integrity, confidentiality and availability of networked systems and data. Key concepts in network security include:

a) **Confidentiality:** Ensuring that sensitive information is accessible only to authorized users. Techniques like encryption and access controls are employed to preserve confidentiality.
b) **Integrity:** Protecting data from unauthorized modifications to ensure its accuracy and reliability. Techniques such as hashing and digital signatures are used for this purpose.
c) **Availability:** Ensuring that network services remain accessible and functional when needed. Measures such as redundancy, failover mechanisms and protection against denial-of-service attacks.
d) **Authentication:** Verifying the identity of users and devices before allowing access to network resources. This is commonly accomplished through passwords, biometrics, or multi-factor authentication.
e) **Authorization:** Defining permissions for authenticated users based on their roles and responsibilities. Access control lists and role-based access control are frequently used authorization mechanisms for this purpose.

f) **Non-repudiation:** Ensuring the verifiability of actions taken within a network, thereby preventing repudiation by involved parties. This is commonly achieved through digital signatures and logging mechanisms.

2.2 Types of Network Threats

Network threats are diverse and continuously evolving. They can be broadly categorized into several types:

a) **Malware:** Malicious software intended to disrupt, damage, or gain unauthorized access to computer systems. Examples include viruses, worms, Trojans, ransomware and spyware. Malware can propagate through different methods, such as email attachments, infected websites and removable media (Saxe & Berlin, 2015).
b) **Phishing:** Phishing is a form of social engineering attack where perpetrators deceive users into revealing sensitive information such as usernames, passwords or credit card details, under the guise of a trustworthy entity. These attacks typically occur through email, fraudulent websites or SMS.
c) **Denial-of-Service (DoS) and Distributed Denial-of-Service (DDoS) attacks:** Attacks are aimed at overwhelming a network, service or website with an excessive amount of traffic, making it inaccessible to legitimate users. These DDoS attacks utilize numerous compromised devices to launch the attack simultaneously.
d) **Man-in-the-Middle (MitM) attack:** Attacks involve an attacker intercepting and potentially alter communication between two parties without their awareness. This can be done using methods such as session hijacking, packet sniffing, or DNS spoofing.
e) **SQL Injection:** An attack that involves inserting malicious SQL code into an input field to manipulate the backend database and gain unauthorized access to sensitive data. This type of attack exploits vulnerabilities in web applications.
f) **Cross-site Scripting (XSS):** An attack that involves injecting malicious scripts into web pages that are viewed by other users. XSS can be used to steal cookies, session tokens, or other sensitive information, or to execute actions on behalf of the victim.
g) **Advanced Persistent Threats (APTs):** Long-term targeted attacks where an adversary gains unauthorized access to a network and remains undetected for an extended period. APTs typically involve sophisticated techniques and aim to steal sensitive information or disrupt operations.

h) **Insider Threats:** Threats come from within the organization, either from malicious insiders who intentionally cause harm or from negligent insiders who inadvertently compromise security. Insider threats can be particularly difficult to detect and mitigate.

The types of Network Threats are shown in Figure 1.

Figure 1. Types of network threats

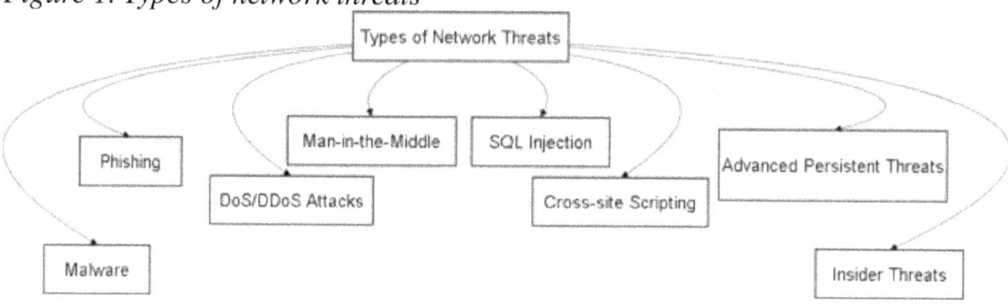

2.3 Traditional Security Measures

Conventional network security protocols have been the foundation of organizational defences for many years. These measures include:

a) **Firewalls:** Hardware and software-based system that watch and manage incoming and outgoing network traffic based on predetermined security laws. Firewalls act as a defences between trusted and untrusted networks, preventing unauthorized access.
b) **Antivirus and anti-Malware Software:** Programs are planned to detect, prevent and delete malicious software. These tools based on signature-based detection, heuristic analysis and behavioural observation to identify and neutralize threats (Saxe & Berlin, 2015).
c) **Intrusion Detection System (IDS) and Intrusion Prevention System (IPS):** IDS observe network traffic for any symptoms of malicious activity and aware administrators, while IPS take active measures to stop or mitigate detected threats. IDS can be signature-based or anomaly-based.
d) **Virtual Private Networks (VPNs):** Technologies that make secure, encrypted connections in the internet, give access to remote users to securely access the organization's network as if they were physically present. VPNs protect information in transit from eavesdropping and interception.

e) **Access Control Lists (ACLs):** Mechanism that means and enforce what users and devices can control in the network. ACLs are used to resist control to sensitive resources and assured that only eligible entities can do the specific actions.
f) **Encryption:** The process to convert data in a coded format to block not secure access. Encrypting in used for protect data at rest and in transit, ensuring confidentiality and integrity.
g) **Patch Management:** The process of daily updating software and system to fix vulnerabilities and make advanced security. Timely applications of patches are hard to guard against known exploits and emerging threats.
h) **Security Policies and Training:** Formalized rules and procedures that govern how an organization's information and systems are protected. Security policies, coupled with regular training and awareness programs, help ensure that employees understand and adhere to security best practices.

2.4 Limitations of Traditional Approaches

While traditional security measures are essential, they have several limitations that necessitate the adoption of more advanced techniques, such as machine learning:

a) **Static Nature:** Many conventional security methods, like signature-based antivirus programs and firewalls, depend on established rules and signatures. This renders them ineffective against new, unidentified threats and zero-day exploits that lack existing signatures.
b) **High False Positive and Fals Negative Rates:** Detection systems based on signatures can result in numerous false positives (legitimate activities identified as malicious) and false negatives (malicious activities that go undetected). This can inundate security teams and enable threats to evade defences.
c) **Limited Adaptability**: Traditional systems often lack the ability to adapt to evolving threats in real-time. Attackers constantly develop new techniques to evade detection and static defences struggle to keep up with these changes.
d) **Resource Intensive**: Maintaining and updating traditional security measures requires significant resources, including time, personnel and financial investment. This can be challenging for organizations with limited resources.
e) **Insider Threats**: Traditional measures primarily focus on external threats and may not effectively address insider threats. Detecting and mitigating malicious or negligent actions by insiders requires advanced monitoring and analysis capabilities.

f) **Complexity and Scalability**: As networks grow in size and complexity, traditional security measures can become cumbersome to manage and scale. Large organizations with diverse and distributed networks need more flexible and scalable security solutions.
g) **Lack of Contextual Awareness:** Traditional security systems frequently function in isolation, lacking the contextual awareness required to make informed decisions.

3. MACHINE LEARNING BASICS

Machine learning (ML) has transformed numerous fields, including network security, by offering best techniques to analyse large datasets and uncover patterns that traditional methods may overlook. This section delves into the basics of machine learning, containing supervised, unsupervised and reinforcement learning methods and explores the evaluation metrics used to assess ML models (Sommer & Paxson, 2010).

3.1 Introduction to Machine Learning

Machine learning is subset of AI that take part in development of algorithms and statistical models that allowing computers to do tasks without detailed information. Instead, these algorithms learns from data by observing patterns, making predictions and enhance over time through experience.

Key Concepts in Machine Learning

a) **Training Data:** The datasets used to train the ML model. It contain input output pairs where the input data are features and the output is the target variable.
b) **Features:** The input variable used to create predictions. Features are typically shown as a vector in a feature space.
c) **Model:** The mathematical representation of the relationship between input features and output variables. The model's parameters are adjusted during training to minimize prediction errors.
d) **Training**: The process of adjusting the model's parameters using the training data to minimize the error in predictions.
e) **Validation**: The process of tuning the model's hyperparameters using a separate dataset, known as the validation set, to avoid overfitting and ensure generalization.
f) **Testing:** Assessing the performance of the trained model on a distinct dataset, referred to as the test set, to evaluate its ability to generalize.

Types of Machine Learning

a) **Supervised Learning:** The algorithm learns from labelled data, with each training example associated with an output label. The objective is to learn a mapping from inputs to corresponding outputs.
b) **Unsupervised Learning**: The algorithm learns from unlabelled data, aiming to identify underlying patterns or structures without predefined labels.
c) **Reinforcement Learning:** The algorithm learns by interacting with an environment, receiving feedback through rewards or penalties, and adjusting its actions to maximize cumulative rewards.

3.2 Supervised Learning Techniques

Supervised learning is one of the most thoroughly used types of machine learning, where the aim is to learn a operation that maps input features to output labels. This phase of paper covers some of the key supervised learning techniques used in network security.

Classification Techniques

a) **Decision Trees**: A decision tree is a tree-structured model in which each internal node signifies a decision based on a feature, each branch indicates an outcome, and each leaf node represents a class label.
b) **Support Vector Machines (SVM)**: SVMs are robust classifiers that determine the optimal hyperplane to separate data into distinct classes. They excel in high-dimensional spaces and are especially valuable for binary classification tasks.
c) **Neural Networks and Deep Learning**: Neural networks consist of layers of interconnected nodes (neurons) that can learn complex representations of the data. Deep learning, a subset of neural networks, involves multiple hidden layers that enable the model to learn hierarchical features. Deep learning has shown remarkable success in tasks such as image and speech recognition and its applications in network security are growing (LeCun et al., 2015b) (Goodfellow et al., 2016).
d) **Random Forest**: Random forests are ensemble learning methods that combine multiple decision trees to improve prediction accuracy and reduce overfitting. Each tree is trained on a random subset of the data and features, and the final prediction is made by averaging the predictions of all trees.
e) **Naïve Bayes:** Naive Bayes classifiers leverage Bayes' theorem and assume that features are conditionally independent given the class label. Despite this simplification, Naive Bayes classifiers are highly efficient and demonstrate strong performance across various text classification tasks.

Regression Techniques

a) **Linear Regression:** Linear regression models establish the connection between input features and a continuous output variable by fitting a linear equation to the data. It's straightforward to implement and interpret, yet it may not capture intricate relationships within the data.
b) **Logistic Regression**: Logistic regression is employed for binary classification tasks. It models the probability of a binary outcome using a logistic function, making it appropriate for problems where the output variable is categorical.

3.3 Unsupervised Learning Techniques

Unsupervised learning entails analysing data without predefined labelled output, with the goal of discovering secret patterns or structures. This part addresses several common unsupervised learning methods used in network security.

Clustering Techniques

a) **K-means:** K-means is a famous clustering algorithm that partitions the data in K-clusters by shorting the addition of the squared distances between data points and their respective cluster centroids. It is simple and enough but it needed specifying the number of clusters in advance.
b) **DBSCAN (Density-Based Spatial Clustering of Application with Noise:** DBSCAN unite data spot that are tightly packed together, create points in low-density levels as outliers. It does no needed to require specifying the number of cluster and can handle cluster of arbitrary shapes.
c) **Hierarchical Clustering:** Hierarchical clustering makes a tree-like skeleton (dendrogram) to represent nested clusters. It can be agglomerative (bottom-up) or divisive (top-down) and does not needed specifying the number of clusters in advanced.

Dimensionality Reduction Techniques

a) **Principal Components Analysis (PCA):** PCA is a methods utilized to decrease the dimensionality of the data by showing it onto a set of orthogonal axes (Principal Components) that caught the maximum variance. PCA has massive used for data visualization and noise reduction.
b) **t-Distributed Stochastic Neighbour Embedding (t-SNE):** t-SNE is a non-linear dimensionality reduction methods that store the local structure of the data, making it eligible for visualizing high-dimensional data in two or three dimensions.

Anomaly Detection Techniques

a) **Autoencoders**: Autoencoders are neural networks designed to learn efficient representations of the data by encoding the input into a lower-dimensional space and then decoding it back. They can be used for anomaly detection by measuring the reconstruction error (Chandola et al., 2009).
b) **Isolation Forest**: Isolation Forest is an ensemble method specifically designed for anomaly detection. It isolates anomalies by randomly partitioning the data and measuring how many splits are required to isolate a point.

3.4 Reinforcement Learning

Reinforcement learning (RL) is a branch of machine learning where an agent learns to create judgment by interacting with an environment and receiving remarks in the form of rewards or penalties. The aim is to know a policy that maximizes cumulative rewards over time (Sutton & Barto, 2018).

Key Concepts in Reinforcement Learning

a) **Agent**: The learner or decision-maker that interacts with the environment.
b) **Environment**: The system or domain with which the agent interacts.
c) **State**: A representation of the current situation or configuration of the environment.
d) **Action**: A decision or move made by the agent that affects the state of the environment.
e) **Reward**: Feedback received by the agent based on its actions, indicating the immediate benefit or cost.
f) **Policy**: A strategy or mapping from states to actions that the agent follows.
g) **Value Function**: A function that estimates the expected cumulative reward for a given state or state-action pair.

Applications in Network Security

Reinforcement Learning is used in many types of network security tasks, including:

a) **Intrusion Detection:** RL can be used in advancement of adaptive intrusion detection systems that learn to locate new types of attacks by interacting with the network environment.
b) **Network Defence:** RL agent can adapt optimal defence strategies by imitate different attack scenarios and learn their response to minimize damage.
c) **Resource Allocation:** RL can optimize the allocation of security resources like bandwidth and computational power to improve the overall security posture.

3.5 Evaluation Metrics for Machine learning Models

Evaluating machine learning models is important to increase their effectiveness and reliability. Different metrics are used to rely on the type of task (classification, regression, clustering etc) (Pedregosa et al., 2012).

Classification Metrics

a) **Accuracy:** The ratio of correctly classified instances to the total number of instances. Although simple to understand, accuracy can be deceptive in imbalanced datasets.
b) **Precision:** The ratio of true positive predictions to all positive predictions. It evaluates the precision of positive predictions.
c) **Recall (Sensitivity)**: The proportion of true positive predictions out of all actual positive instances. It measures the ability to identify positive instances.
d) **F1-Score**: The harmonic mean of precision and recall, providing a single metric that balances both. It is useful when the class distribution is imbalanced.
e) **Receiver Operating Characteristic (ROC) Curve**: A graphical representation of the true positive rate (recall) against the false positive rate at various threshold settings.
f) **Area under the ROC Curve (AUC-ROC):** A single value that encapsulates the ROC curve, indicating the likelihood that the model assigns a higher rank to a randomly selected positive instance compared to a randomly selected negative instance.

Regression Metrics

a) **Mean Absolute Error (MAE):** The average of the absolute difference between predicted and actual values. It gives a simple measure of prediction accuracy.
b) **Mean Squared Error (MSE):** The average of the square difference between predicted and actual values. MSE punished big errors more than MAE.
c) **Root Mean Square Error (RMSE):** The square root of MSE, give a measure of prediction accuracy in the same units as the target variables.
d) **R-Squared (R^2):** The proportion of variance in the aimed variable that is elaborate by the model. R^2 values range from 0 to 1, with large values indicating better fit.

Clustering metrics

a) **Silhouette Score**: Measures the cohesion and separation of clusters, with values ranging from -1 to 1. Higher values means well-defined clusters.

b) **Davies-Bouldin Index**: Measures the average similarity ratio of each cluster with its most similar cluster, with lower values indicating better clustering.
c) **Adjusted Rand Index (ARI)**: Measures the similarity between the clustering result and the ground truth, adjusted for chance. Values range from -1 to 1, with higher values indicating better agreement.

Anomaly Detection Metrics

a) **Precision-Recall AUC**: Assesses the area under the precision-recall curve, providing valuable insights for evaluating models on imbalanced datasets with infrequent anomalies (Chandola et al., 2009).
b) **F1-Score**: Utilized to evaluate the trade-off between precision and recall in anomaly detection tasks.

4. APPLICATION OF MACHINE LEARNING IN NETWORK SECURITY

Machine learning (ML) has emerged as a vital element in improving network security by allowing systems to learn from data, recognize patterns, and make intelligent decisions. This section delves into multiple applications of machine learning in network security, such as intrusion detection systems (IDS), malware detection, phishing detection, spam detection, botnet detection, and network traffic analysis (Pedregosa et al., 2012) (Mohri et al., 2012).

4.1 Intrusion Detection System (IDS)

Intrusion Detection system are planed to observe network traffic for suspicious activities and potential threats. Machine learning methods have significantly improve the functionality of IDS by allowing the detection of new and advanced threats. IDS can be extensively categorized into three types: anomaly-based, signature-based and hybrid IDS.

4.1.1 Anomaly-based IDS

Anomaly-based IDS employ machine learning algorithms to create a baseline of typical network behaviour and detect deviations from this baseline as possible threats. This method is effective in identifying unknown and novel attacks.

Techniques and Models

a) **Clustering Algorithm:** Algorithms like K-means and DBSCAN are utilized to cluster similar network behaviours and identify outliers as anomalies. These techniques can uncover unusual patterns that suggest intrusions (Ambusaidi et al., 2014).
b) **Statistical Models**: Techniques like Gaussian Mixture Models (GMM) and Principal Component Analysis (PCA) analyse the statistical properties of network traffic to identify deviations from normal behaviour.
c) **Neural Networks:** Deep learning models, particularly autoencoders and recurrent neural networks (RNNs), employed to learn hard representations of normal traffic and detect anomalies based on rebuild errors or temporal patterns (Goodfellow et al., 2016).

4.1.2 Signature-based IDS

Signature-based IDS depend on pre-defined patterns and signatures of known threats. Machine learning enhances these systems by automating the generation and updating of signatures, thereby improving their capability to detect new variations of known attacks.

Techniques and Models

a) **Pattern Matching:** Traditional methods use pattern matching to identify known attack signatures. Machine learning can automate and optimize the creation of these patterns.
b) **Decision Tress and Random Forests:** These models can be trained on labelled datasets containing known attack signatures, enabling the system to classify network traffic based on learned signatures.
c) **Support Vector Machines (SVM):** SVMs are effective in distinguishing between normal and malicious traffic by leveraging learned signatures, especially in high-dimensional feature spaces (Cristianini & Shawe-Taylor, 2000).

4.1.3 Hybrid IDS

Hybrid IDS combine the advantages of both anomaly-based and signature-based approaches to offer thorough protection. These systems employ machine learning to marge various detection methods and enhance the overall accuracy.

Techniques and Models

a) **Ensemble Learning**: Methods such as bagging and boosting integrate various models to enhance detection rates and minimize false positives. Random forests and gradient boosting are frequently employed ensemble techniques (Rokach, 2009).
b) **Neural Network Ensembles**: Integrating various types of neural networks like, convolutional neural networks (CNNs) and RNNs, improves the system's ability to identify a large range of threats.
c) **Feature Fusion**: Hybrid systems employ machine learning to combine features from both anomaly-based and signature-based detectors, creating a stronger detection mechanism.

4.2 Malware Detection

Malware detection entails identifying and countering harmful software that can damage network systems. Machine learning methods are used in both static and dynamic analysis to enhance malware detection effectiveness (Saxe & Berlin, 2015).

4.2.1 Static Analysis

Static analysis inspects the code and structure of software without running it, detecting potential malware based on its attributes.

Techniques and Models

a) **Feature Extraction**: Machine learning models examine features like byte sequences, API calls, and control flow graphs. Techniques such as TF-IDF (Term Frequency-Inverse Document Frequency) are utilized to represent these features numerically.
b) **Classification Models**: Algorithms like decision trees, SVM, and neural networks are trained on labelled datasets of benign and malicious samples to classify new software.
c) **Signature Generation**: Machine learning automates the generation of signatures by detecting unique patterns in malware samples. These signatures are employed to identify known malware variants.

4.2.2 Dynamic Analysis

Dynamic analysis involves running the software in a controlled environment to observe its behaviour and identify malicious activities.

Techniques and Models

a) **Behavioural Analysis**: Machine learning models monitor runtime behaviors such as file system changes, network activity, and process creation. Anomalies in these behaviors can indicate malware.
b) **Sandboxing**: Malware is run in isolated environments (sandboxes) and machine learning algorithms examine the final behaviours to detect malicious patterns.
c) **Sequence Analysis**: Techniques such as RNNs and Long Short-Term Memory (LSTM) networks are employed to examine sequences of system calls and identify deviations indicative of malware.

4.3 Phishing Detection

Phishing attacks has goal to trick users by giving sensitive information by masquerading as legitimate entities. Machine learning models are employed to identify phishing attempts rely on various features.

Techniques and Models

a) **URL Analysis**: Machine learning algorithms examine features of URLs like, domain age, length, and the presence of suspicious keywords. Methods like logistic regression and random forests are effective in detecting phishing URLs.
b) **Email Content Analysis**: Models examine the content of emails, including text, links and metadata to identify phishing attempts. Natural language processing (NLP) methods and neural networks like, LSTM are commonly employed.
c) **Website Features**: Machine learning models assess website attributes like, SSL certificates, favicon and webpage structure to detect phishing sites.

4.4 Spam Detection

Spam detection engaged to detecting and filtering unwanted and potentially harmful emails. Machine learning techniques has large used to enhance the accuracy of spam filters.

Techniques and Models

a) **Bayesian Filtering**: Naive Bayes classifiers are compute the probability of an email being spam based on the occurrence of words and phrases.
b) **Text Classification**: Machine learning models such as SVM and neural networks, examine email content and metadata to classify emails as spam or legitimate.
c) **Ensemble Methods**: Integrating multiple classifiers like random forests and gradient boosting, enhance spam detection accuracy and minimize false positives (Rokach, 2009).

4.5 Botnet Detection

Botnets are networks of compromised devices manipulated by attackers to carry out malicious activities. Machine learning methods are employed to identify botnet traffic and activities.

Techniques and Models

a) **Traffic Analysis**: Machine learning models examine network traffic patterns to detect anomalies indicative of botnet activity. Methods like clustering and anomaly detection are frequently employed.
b) **Behavioural Analysis**: Models monitor the behaviour of devices like unusual communication patterns and resource usage to identify botnets. Sequence analysis methods integrating HMM and LSTM are effective.
c) **Flow-based Detection**: Machine learning algorithms examine network flow data like IP addresses, port numbers and flow duration to detect botnet traffic.

4.6 Network Traffic Analysis

Network traffic examine involves monitoring and examining data packets traveling across the network to detect potential security threats.

Techniques and Models

a) **Deep Packet Inspection (DPI)**: Machine learning models examine the content of data packets to identify malicious patterns and activities. Methods like CNNs and RNNs are employed for deep packet inspection.

b) **Flow Analysis**: Models examine network flow data to identify anomalies and potential threats. Clustering and classification algorithms are frequently employed for this task.
c) **Time-series Analysis**: Machine learning methods integrating ARIMA and LSTM networks, examine the temporal patterns of network traffic to detect irregularities indicative of security threats.

5. MACHINE LEARNING TECHNIQUES USED IN NETWORK SECURITY

Machine learning methods have become important in network security for identifying and mitigating several types of cyber threats. This part elaborates the key machine learning methods employed in network security integrating classification, clustering, anomaly detection and ensemble methods (Paricherla et al., 2023).

5.1 Classification Techniques

Classification techniques are employed to categorize data into predefined classes. In network security, classification algorithms help to detect whether network traffic or activities are normal or malicious.

5.1.1 Applications in Network Security

a) **Intrusion Detection Systems (IDS)**: Decision trees can categorised network traffic as normal or malicious based on several features like packet size, duration, and protocol (Ambusaidi et al., 2014).
b) **Malware Detection**: They are employed to identify malware by examine features extracted from software binaries or network behaviour.
c) **Phishing Detection**: Decision trees help to categorised emails or URLs as phishing or legitimate based on content and metadata.

5.1.2 Support Vector Machines (SVM)

Support Vector Machines (SVM) are powerful classifiers that discover the optimal hyperplane to separate data into various classes. They are particularly effective for binary classification tasks (Cristianini & Shawe-Taylor, 2000).

Applications in Network Security

a) **Anomaly Detection**: SVMs are employed to identify anomalies in network traffic by detecting patterns that deviate from normal behaviour (Bhuyan et al., 2013).
b) **Spam Detection**: They categorized emails as spam or legitimate based on features like word frequency and sender information.
c) **Botnet Detection**: SVMs help to detect botnet traffic by examining network flow attributes (Mahoney & Chan, 2003).

5.1.3 Neural Networks and Deep Learning

Neural networks, especially deep learning models, have transformed network security with their ability to learn complex representations from large datasets.

Applications in Network Security

a) **Intrusion Detection**: Deep learning models such as Convolutional Neural Networks (CNNs) and Recurrent Neural Networks (RNNs) can identify sophisticated attacks by learning patterns in network traffic.
b) **Malware Detection**: Neural networks examine features from software binaries or runtime behaviours to identify malware.
c) **Phishing Detection**: Deep learning models employ natural language processing (NLP) methods to examine email content and detect phishing attempts.

5.1.4 Random Forest

Random Forest is an ensemble learning technique that integrate multiple decision trees to enhance classification accuracy and robustness.

Applications in Network Security

a) **Intrusion Detection**: Random forests categorise network traffic by integrating predictions from multiple decision trees, improving detection accuracy.
b) **Malware Detection**: They are employed to identify malware by examine features from software binaries and network traffic.
c) **Spam Detection**: Random forests categorised emails as spam or legitimate by aggregating predictions from multiple trees.

5.2 Clustering Techniques

Clustering techniques group same data points together and are employed in network security to detect patterns and anomalies in network traffic.

5.2.1 K-Means

K-Means is a famous clustering algorithm that partitions data into K clusters by reducing the addition of squared distances between data points and their respective cluster centroids.

Applications in Network Security

a) **Anomaly Detection**: K-Means clusters normal network traffic with outliers indicating potential anomalies or attacks.
b) **Botnet Detection**: It helps to detect botnet activity by clustering network traffic patterns.
c) **Malware Detection**: K-Means clusters software binaries based on extracted features, detecting outliers as potential malware.

5.2.3 Hierarchical Clustering

Hierarchical clustering makes a tree-like skeleton (dendrogram) representing nested clusters, which can be agglomerative (bottom-up) or divisive (top-down).

Applications in Network Security

a) **Anomaly Detection**: Hierarchical clustering detects anomalies by examine the hierarchical skeleton of network traffic.
b) **Malware Detection**: It helps to identify malware by clustering software binaries based on feature similarity.
c) **Phishing Detection**: Hierarchical clustering groups similar phishing emails, aiding in the identification of phishing campaigns.

5.3 Anomaly Detection Techniques

Anomaly detection methods detect deviations from normal behaviour, making them important for identifying unknown and novel attacks in network security.

5.3.1 Principal Component Analysis (PCA)

PCA is a dimensionality reduction method that shows data onto a set of orthogonal axes (principal components) capturing the maximum variance.

Applications in Network Security

a) **Intrusion Detection**: PCA minimize the dimensionality of network traffic data, highlighting anomalies that deviate from normal patterns.
b) **Malware Detection**: It helps to identify malware by examine deviations in software features from normal behaviour.
c) **Botnet Detection**: PCA detects botnet traffic by projecting network flow data onto principal components.

5.3.2 Autoencoders

Autoencoders are neural networks planned to learn efficient representations of data by encoding input into a lower-dimensional space and decoding it back.

Applications in Network Security

a) **Anomaly Detection**: Autoencoders identify anomalies by measuring reconstruction errors, which are higher for anomalous data.
b) **Malware Detection**: They detect malware by learning normal software behaviour and identifying deviations.
c) **Intrusion Detection**: Autoencoders identify intrusions by examine reconstruction errors in network traffic data.

5.3.3 One-Class SVM

One-Class SVM is an alternative of SVM employed for anomaly detection by learning a decision boundary around normal data points.

Applications in Network Security

a) **Anomaly Detection**: One-Class SVM identify anomalies in network traffic by detecting points outside the learned decision border.
b) **Malware Detection**: It helps to detect malware by learning the border of normal software behaviour and identifying outliers.

c) **Botnet Detection**: One-Class SVM identify botnet traffic by detecting deviations from normal network behaviour.

5.4 Ensemble Methods

Ensemble methods integrate various models to enhance prediction accuracy and robustness. They are widely used in network security to improve identification rates and reduce false positives.

5.4.1 Bagging

Bagging (Bootstrap Aggregating) participate training various models on different subsets of the data and aggregating their predictions.

Applications in Network Security

a) **Intrusion Detection**: Bagging enhance identification accuracy by integrating predictions from various models trained on different subsets of network traffic.
b) **Malware Detection**: It improve malware identification by aggregating predictions from multiple categorised.
c) **Spam Detection**: Bagging minimizes false positives in spam identification by integrating predictions from different models.

5.4.2 Boosting

Boosting improves overall prediction accuracy by training models sequentially, with each model addressing the errors of its predecessors.

Applications in Network Security

a) **Intrusion Detection**: Boosting enhance identification rates by integrating weak learners into a strong classifier.
b) **Malware Detection**: It improves malware identification by sequentially training models to correct previous errors.
c) **Phishing Detection**: Boosting minimizes false positives in phishing identification by integrating predictions from various models.

5.4.3 Stacking

In stacking, multiple models are trained, followed by a meta-model that integrates their predictions to enhance accuracy Applications in Network Security.

Applications in Network Security

a) **Intrusion Detection**: Stacking improves identification accuracy by integrating predictions from different models with a meta-model.
b) **Malware Detection**: It enhance malware identification by integrating the strengths of various classifiers.
c) **Spam Detection**: Stacking minimizes false positives in spam identification by aggregating predictions from various models with a meta-classifier.

6. CHALLENGES AND LIMITATIONS IN APPLYING MACHINE LEARNING TO NETWORK SECURITY

While machine learning (ML) methods have significantly improved network security, several challenges and weaknesses still hinder their effectiveness. This section explores these challenges, integrating data quality and quantity, feature selection, model interpretability, adversarial attacks, scalability and performance and real-time processing (Breiman, 2001).

6.1 Data Quality and Quantity

The effectiveness of machine learning models heavily rely on the quality and quantity of the data employed for training and evaluation.

Data Quality

High-quality data is essential for training robust machine learning models. However, in network security, data can often be noisy, incomplete, or imbalanced, leading to several issues:

a) **Noisy Data**: Network traffic data may integrate a significant amount of noise like irrelevant or redundant features, which can degrade the execution of machine learning models (Zhang et al., 2020).
b) **Incomplete Data**: Traceless values and not done records are common in network logs and can lead to biased or inaccurate model predictions (Wang et al., 2021).

c) **Imbalanced Data**: In network security, there is often a class imbalance where the number of malicious events is much smaller than normal events. This imbalance can cause models to be biased towards the majority class, resulting in poor identification rates for rare but critical attacks (Haider et al., 2016).

Data Quantity

The volume of data available for training machine learning models is another critical factor.

a) **Large Datasets**: While large datasets can enhance model execution, they also need significant computational resources and storage, which can be challenging to manage (Kumar et al., 2021).
b) **Small Datasets**: On the other hand, limited datasets might fail to reflect the diversity and intricacy of network traffic, resulting in models that overtrain on the given data and struggle with new data (Gao et al., 2019).

Addressing Data Issues

To address these issues, various methods can be used:

a) **Data Cleaning**: Methods like outlier identification and removal, imputation of traceless values, and normalization can help to enhance data quality.
b) **Data Augmentation**: Generating synthetic data through methods such as SMOTE (Synthetic Minority Over-sampling Technique) can help balance datasets (Chawla et al., 2002).
c) **Big Data Tools**: Utilizing massive data frameworks such as Apache Hadoop and Spark can help manage and process massive amounts of data efficiently (Chen et al., 2014).

6.2 Feature Selection

Feature selection participate to detecting the most relevant features for training machine learning models. This process is important for enhancing model performance and interpretability (Cai et al., 2018).

Challenges in Feature Selection

a) **High Dimensionality**: Network security data often have high dimensionality, making it challenging to identify the most relevant features without significant computational overhead (Song et al., 2018).
b) **Redundant Features**: Many features may be redundant or highly correlated, which can negatively impact model performance and increase training time (Shah et al., 2020).
c) **Irrelevant Features**: Including irrelevant features can introduce noise into the model, reducing its accuracy and generalization ability (Fayyad et al., 2020).

Feature Selection Techniques

Several techniques can be used to address these challenges:

a) **Filter Methods**: Network security data typically possess high dimensionality, making it difficult to pinpoint the most pertinent features without considerable computational effort (Song et al., 2018).
b) **Wrapper Methods**: Methods such as recursive feature elimination (RFE) and genetic algorithms assess feature subsets based on model execution, giving more accurate feature selection at the cost of higher computational requirements (Guyon et al., 2002).
c) **Embedded Methods**: Algorithms such as Lasso (Least Absolute Shrinkage and Selection Operator) and decision tree-based techniques perform feature selection during the model training process, balancing accuracy and computational efficiency (Tibshirani, 1996).

6.3 Model Interpretability

Interpretability is important in network security to understand and believe the decisions made by machine learning models.

Challenges in Model Interpretability

a) **Complex Models**: Many effective machine learning models like deep learning networks, are inherently hard and control as "black boxes," making it difficult to interpret their decisions (Rudin, 2019) (Rudin, 2019).

b) **Regulatory Requirements**: Regulations such as GDPR (General Data Protection Regulation) need explanations for automated decisions, necessitating interpretable models (Goodman & Flaxman, 2017).
c) **Trust and Adoption**: Security professionals required to believe and understand model decisions to accept them in real-world scenarios. Lack of interpretability can hinder this believe and slow down acceptation (Doshi-Velez & Kim, 2017) (Doshi-Velez & Kim, 2017).

Techniques for Improving Interpretability

Several techniques can be used to improve model interpretability:

a) **Simpler Models**: employing simpler models like decision trees or linear models can enhance interpretability at the cost of potentially lower accuracy (Molnar, 2019).
b) **Model-Agnostic Methods**: Methods like LIME (Local Interpretable Model-agnostic Explanations) and SHAP (SHapley Additive exPlanations) provide post-hoc explanations for hard model decisions (Ribeiro et al., 2016; Lundberg & Lee, 2017).
c) **Visualization Tools**: Tools and frameworks that visualize feature importance, decision borders, and model predictions can help security professionals understand and interpret model decisions (Zhou & Hooker, 2016).

6.4 Adversarial Attacks

Adversarial attacks participate manipulating input data to deceive machine learning models, posing a significant threat to their reliability and security.

Types of Adversarial Attacks

a) **Evasion Attacks**: Attackers upgrades inputs during the inference phase to evade identification, like crafting malicious network packets that appear benign (Papernot et al., 2016).
b) **Poisoning Attacks**: Attackers place malicious data into the training set to corrupt the model, causing it to make incorrect predictions (Biggio et al., 2012).
c) **Model Extraction Attacks**: Attackers try to mimic the functionality of a machine learning model by querying it and leveraging the responses to train an alternative model (Tramèr et al., 2016).

Mitigating Adversarial Attacks

Several techniques can be employed to guard against adversarial attacks:

a) **Adversarial Training**: Incorporating adversarial examples into the training process to enhance model robustness against such attacks (Goodfellow et al., 2014).
b) **Defense Mechanisms**: Methods such as input preprocessing, anomaly identification, and model hardening can help to guard against evasion and poisoning attacks (Papernot et al., 2016).
c) **Security Audits**: Daily security audits and penetration examining can help detect vulnerabilities and enhance the resilience of machine learning models (Chio & Freeman, 2018).

6.5 Scalability and Performance

Scalability and execution are critical factors in deploying machine learning models in real-world network security environments.

Scalability Challenges

a) **Large Datasets**: Network security enhance examining large volumes of data in real-time, which can strain computational resources and slow down model training and inference (Chen et al., 2013).
b) **Distributed Environments**: Placing models in distributed environments like cloud-based systems, introduces additional challenges related to data synchronization, communication overhead and fault tolerance (Grolinger et al., 2014).

Performance Optimization

Several path can optimize scalability and performance:

a) **Parallel and Distributed Computing**: Utilizing parallel and distributed computing frameworks such as Apache Spark and Hadoop can assist in handling large-scale data processing and model training (Dean & Ghemawat, 2008).
b) **Model Compression**: Methods like model pruning, quantization and knowledge distillation minimize model size and computational needs, enhancing execution without significantly sacrificing accuracy (Cheng et al., 2017).

c) **Hardware Acceleration**: Employing specialized hardware, like GPUs and TPUs, can speed up model training and inference, especially for deep learning models (Jouppi et al., 2017).

6.6 Real-time Processing

Real-time executing is important for timely identification and response to network security threats.

Real-time Processing Challenges

a) **Latency**: Machine learning models need to process data with minimal delay to identify and respond to threats in real-time. High latency can make models ineffective in rapidly evolving attack situations (Chandola et al., 2009).
b) **Throughput**: Assuring high throughput is necessary to handle the amount of data generated by network traffic without bottlenecks (Bhuyan et al., 2014).
c) **Resource Constraints**: Real-time executing needs significant computational resources, which can be challenging to manage in resource-constrained environments (Roshtkhari & Zulkernine, 2018).

Techniques for Real-time Processing

Several methods can be used to allow real-time processing:

a) **Stream Processing Frameworks**: Employing stream processing frameworks such as Apache Kafka, Apache Storm, and Apache Flink allows for real-time data processing and analysis (Kreps et al., 2011).
b) **Edge Computing**: Placing machine learning models at the edge of the network, closer to data sources, minimizes latency and bandwidth usage, improving real-time capabilities (Shi et al., 2016).
c) **Incremental Learning**: Incremental learning algorithms continuously update models with new data, enabling real-time adaptation to changing threats (Losing et al., 2018).

The Challenges and Limitations in Applying Machine Learning to Network Security are shown in Figure 2.

Figure 2. Challenges and limitations in applying machine learning to network security

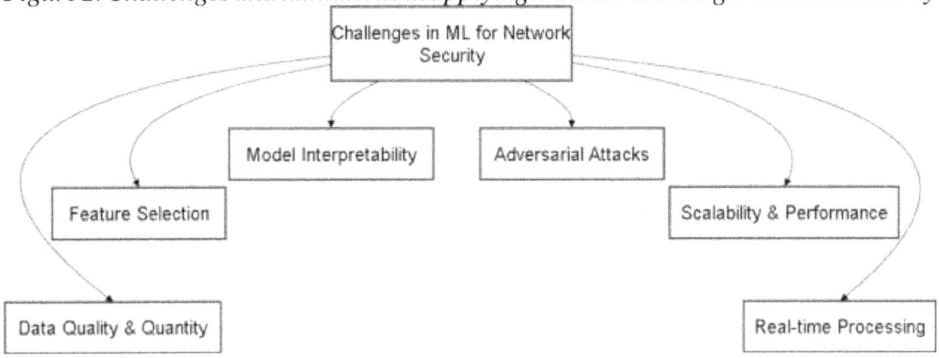

7. FUTURE DIRECTIONS IN MACHINE LEARNING FOR NETWORK SECURITY

As machine learning continues to develops, its integration into network security promises to shape the future of cybersecurity. This part elaborate several promising path for research and development in this field, integrating the integration with emerging technologies, the adoption of explainable AI, transfer learning and domain adaptation, federated learning for privacy preservation, advancements in real-time identification and response and the exploration of interdisciplinary approaches.

7.1 Integration with Emerging Technologies (e.g., IoT, 5G)

The proliferation of emerging technologies like the Internet of Things (IoT) and 5G networks shows both opportunities and challenges for network security. Including machine learning with these technologies can improve security measures and guard against evolving threats.

IoT Security

a) **Device Authentication**: Machine learning algorithms can examine device behaviour patterns to authenticate IoT devices and identify anomalies indicative of unauthorized access (Al-Fuqaha et al., 2015).
b) **Anomaly Detection**: ML models can watch IoT networks for unusual behaviour like sudden spikes in traffic or unexpected data transfers, signalling potential security breaches (Neshenko et al., 2020).

5G Security

a) **Network Slicing Security**: Machine learning methods can examine network traffic and user behaviour to identify anomalous activities within specific network slices, improving security in 5G environments (Giordani et al., 2020).
b) **Edge Computing Security**: ML algorithms placed at the network edge can give real-time threat identification and response, mitigating security risks in distributed 5G infrastructures (Li et al., 2020).

7.2 Explainable AI in Network Security

The need for transparency and accountability in machine learning models has led to the appearance of explainable AI methods. In network security, explainable AI can enhance faith and understanding of model decisions, facilitating better-informed cybersecurity techniques.

Model Interpretability

a) **Feature Importance**: Explainable AI techniques can detect the most influential features in network security models, giving insights into the factors driving security decisions (Guidotti et al., 2018).
b) **Decision Explanation**: Methods like LIME and SHAP allow the explanation of individual model predictions, helping security analysts understand why certain actions are flagged as threats (Lundberg & Lee, 2017).

Regulatory Compliance

a) **GDPR Compliance**: Explainable AI can assist organizations in complying with regulations such as GDPR by offering clear explanations for automated security decisions, thereby ensuring accountability and fairness (Goodman & Flaxman, 2017).
b) **Auditing and Certification**: Explainable AI techniques facilitate auditing and certification processes by giving verifiable justifications for security measures and policies (Doshi-Velez & Kim, 2017).

7.3 Transfer Learning and Domain Adaptation

Transfer learning and domain adaptation methods allow the reuse of knowledge learned from one domain to enhance execution in a related but different domain. In network security, these techniques can locate data scarcity and distribution shifts across various environments.

Cross-Domain Knowledge Transfer

a) **Threat Intelligence Sharing**: Transfer learning allows the transfer of threat intelligence from one organization or sector to another, improving cybersecurity preparedness and response (Cheng et al., 2018).
b) **Zero-Shot Learning**: Methods such as zero-shot learning allow network security models to generalize to new and unseen threats by leveraging knowledge from related domains (Xian et al., 2018).

Domain Adaptation

a) **Adversarial Robustness**: Domain adaptation methods enhance model consistency against adversarial attacks by accepting to changes in network traffic patterns and attacker techniques (Zhang et al., 2021).
b) **Data Augmentation**: Domain acceptation techniques leverage synthetic data generation to augment training datasets and enhance model generalization across various network environments (Zhao et al., 2020).

7.4 Federated Learning for Privacy Preservation

Federated learning allows collaborative model training across decentralized devices or data sources while maintaining data privacy. In network security, federated learning can support collective threat detection without compromising sensitive information.

Collaborative Threat Detection

a) **Privacy-Preserving Analysis**: Federated learning enables organizations to jointly train machine learning models using aggregated network telemetry data without the need to share raw or sensitive information (Bonawitz et al., 2019).

b) **Decentralized Défense**: Distributing model training across network nodes or edge devices through federated learning strengthens resilience against centralized attack vectors and data breaches (Kairouz et al., 2019).

Secure Aggregation

a) **Secure Multiparty Computation (SMC)**: Federated learning uses cryptographic methods such as SMC to securely aggregate model updates from involving nodes without expressing individual contributions (Mohassel & Zhang, 2017).
b) **Differential Privacy**: Federated learning frameworks integrate differential privacy mechanisms to safeguard individual data samples' anonymity during model aggregation, thereby preserving user privacy (McMahan et al., 2017).

7.5 Enhancements in Real-time Detection and Response

Real-time identification and response capabilities are important for effectively combating cyber threats. Upgrades in machine learning can allow faster and more accurate threat detection and mitigation in network security.

Streaming Analytics

a) **Stream Processing Engines**: Machine learning models including with stream processing engines allows real-time examines of network traffic data, facilitating rapid threat identification and response (Shi et al., 2016) (Doshi-Velez & Kim, 2017b).
b) **Online Learning:** Algorithms for online learning continually adjust model parameters based on incoming data streams, ensuring adaptive and responsive security measures (Crammer et al., 2006).

Automated Response

a) **Autonomous Decision-making**: Machine learning algorithms can automate threat response actions like blocking malicious IP addresses or isolating compromised devices, according to predefined security policies (Collier et al., 2020).

b) **Dynamic Policy Adaptation**: Machine learning-powered security systems can adaptively modify access control policies and network configurations in response to changing threat landscapes, thereby reducing the need for manual intervention and enhancing response times (Vilalta & Ma, 2002).

7.6 Interdisciplinary Approaches

Cooperation across various disciplines, such as cybersecurity, data science, psychology, and sociology, can generate creative solutions to intricate network security challenges.

Behavioural Analytics

a) **User Behaviour Modelling**: Including insights from psychology and sociology, machine learning models can examine user attitude patterns to identify insider threats and social engineering attacks (Farnham et al., 2014).
b) **Cognitive Biases**: Knowledge of human cognitive biases aids in developing ML-based security systems that consider human factors in decision-making processes, thereby enhancing overall resilience to cyber threats (Whitaker et al., 2020).

Human-Centric Security

a) **Usable Security Interfaces**: Collaborations between data scientists and user experience designers can improve the uses of security interfaces, enhancing user engagement and compliance with security policies (Herley & Van Oorschot, 2017).
b) **Security Awareness Training**: Machine learning can tailor security awareness training programs to individual learning styles and behavioural preferences, thereby optimizing their effectiveness in mitigating security risks associated with human behaviour (Vance et al., 2018).

8. CASE STUDIES AND PRACTICAL IMPLEMENTATIONS

In this part, we delve into multiple case studies and practical implementations of machine learning methods in network security. We research successful implementations, conduct a comparative analysis of methods and derive lessons learned from real-world applications.

8.1 Successful Implementations

8.1.1 Deep Learning-Based Intrusion Detection Systems

A significant achievement in network security is the adoption of deep learning-based intrusion detection systems (IDS). Traditional IDS, which typically rely on manually crafted rules or signatures, often face challenges in detecting complex and evolving cyber threats. Deep learning models, such as convolutional neural networks (CNNs) and recurrent neural networks (RNNs), provide a more effective approach by autonomously learning intricate patterns and behaviours from network traffic data.

In a recent study conducted by Sgandurra et al. (2021), deep learning models were employed to identify advanced persistent threats (APTs) within enterprise networks. Researchers gathered network traffic data from a sizable corporate network and utilized a deep neural network architecture to differentiate between normal and malicious activities.

8.1.2 Anomaly Detection in Industrial Control Systems

Industrial control systems (ICS) are essential infrastructure components in sectors like energy, manufacturing, and transportation. Safeguarding these systems from cyber threats is crucial for ensuring operational safety and continuity. Anomaly detection techniques, driven by machine learning, have emerged as effective solutions for identifying abnormal behaviours that may indicate cyber-attacks in ICS environments.

In a case study conducted by Goh et al. (2020), machine learning algorithms were utilized to oversee and protect the control systems of an oil and gas pipeline. By scrutinizing sensor data and network traffic patterns, the system could identify anomalies such as unauthorized access attempts and data exfiltration.

8.2 Comparative Analysis of Techniques

8.2.1 Decision Trees vs. Neural Networks for Malware Detection

Decision trees and neural networks are two famous machine learning methods used in malware identification. Decision trees delivers transparency and interpretability, making them appropriate for detecting specific malware characteristics. On the other hand, neural networks excel at capturing hard nonlinear relationships in data, allowing them to identify previously unseen malware variants.

In a comparative analysis conducted by Li et al. (2022), decision trees and neural networks were assessed for their effectiveness in malware detection using real-world datasets. The study revealed that decision trees demonstrated superior accuracy in classifying known malware samples, whereas neural networks excelled in detecting zero-day malware and polymorphic variants.

8.2.2 Support Vector Machines vs. Random Forest for Intrusion Detection

Support vector machines (SVMs) and random forest have large used in intrusion identification systems (IDS) for categorising network traffic as normal or malicious. SVMs delivers robustness against noisy data and are useful in high-dimensional feature spaces. Random forest, on the other hand, is popular for its scalability and ability to handle imbalanced datasets (Cristianini & Shawe-Taylor, 2000).

In a comparative study conducted by Kumar et al. (2021), SVMs and random forest were compared for their efficacy in detecting network intrusions using a benchmark dataset. The findings indicated that SVMs achieved higher accuracy in identifying known attack types, whereas random forest demonstrated superior performance in detecting novel and previously unseen attacks.

8.3 Lessons Learned from Real-world Applications

8.3.1 Importance of Data Quality and Feature Engineering

One recurring lesson from real-world applications of machine learning in network security is the paramount importance of data quality and feature engineering. High-quality, accurately labelled datasets are indispensable for training models that are both accurate and dependable.

In a case study conducted by Garcia et al. (2020), researchers noted that the effectiveness of a machine learning-based IDS was closely tied to the quality and relevance of the features used during training. They observed that meticulous selection and preprocessing of features, such as packet headers, payload content, and temporal behaviour, markedly improved the IDS's ability to detect threats.

8.3.2 Adaptability and Scalability of Models

Another crucial lesson learned is the adaptability and scalability of machine learning models in dynamic network environments. Cyber threats are evolved continuously, needs models to accept and learn from new data to hold effectiveness.

In a practical implementation described by Zhang et al. (2021), researchers devised a scalable machine learning framework for anomaly detection in cloud-based networks. This framework harnessed distributed computing resources and parallel processing techniques to manage large-scale data streams in real-time.

8.3.3 Collaboration and Knowledge Sharing

Collaboration and knowledge sharing among stakeholders are crucial for the effective deployment of machine learning in network security. By exchanging threat intelligence, best practices, and lessons learned, organizations can proactively address emerging threats and collectively enhance their cybersecurity defences.

In a case study detailed by Cimpanu (2020), multiple financial institutions collaborated to create a shared threat intelligence platform driven by machine learning. Through collective pooling of resources and expertise, these institutions enhanced their capability to detect and respond to cyber threats. The platform facilitated real-time exchange of threat indicators and predictive analytics, empowering participating organizations to take proactive measures against shared adversaries and vulnerabilities.

9. CONCLUSION

In this final part, we summarize the key findings, notice the contributions to the field of network security from machine learning methods and give final and recommendations for future research and implementation.

9.1 Summary of Findings

Throughout this review, we analyse the application of machine learning methods in network security, covering a large range of content integrating intrusion identification, malware identification, phishing identification, anomaly identification and more. Here are the key findings:

- **Machine Learning Basics**: We give an over all view of machine learning methods, integrating supervised learning, unsupervised learning, reinforcement learning and evaluation metrics.
- **Application of Machine Learning in Network Security**: We analysed how machine learning is use in multiple areas of network security like intrusion detection systems (IDS), malware identification, phishing identification, spam identification, botnet identification and network traffic analysis.
- **Machine Learning Techniques Used in Network Security**: We talk about classification methods (e.g., decision trees, support vector machines, neural networks), clustering methods (e.g., K-means, DBSCAN), anomaly identification methods (e.g., PCA, autoencoders) and ensemble techniques (e.g., bagging, boosting, stacking).
- **Challenges and Limitations**: We detect challenges and weaknesses in applying machine learning to network security, integrating data quality and quantity issues, feature selection, model interpretability, adversarial attacks, scalability and real-time processing constraints.
- **Future Directions**: We research in future paths in machine learning for network security like integration with emerging technologies (IoT, 5G), explainable AI, transfer learning, federated learning, enhancements in real-time identification and response and interdisciplinary approaches.
- **Case Studies and Practical Implementations**: We check successful implementations of machine learning methods in network security, conducted comparative analyses of different methods and derived lessons learned from real-world applications.

9.2 Contributions to the Field

The observation has made multiple contributions to the field of network security and machine learning:

a) **Comprehensive Overview**: We give a comprehensive over all view of machine learning methods and their application in multiple aspects of network security, offering insights into both foundational concepts and advanced methodologies.
b) **Integration of Recent Research**: By integrating current research discoveries and case analyses, we made certain that the review represents the newest developments and practical applications in the industry.
c) **Identification of Challenges and Future Directions**: We detected key challenges handling the integration of machine learning into network security and outlined future paths for examine and upgradation, guiding researchers and practitioners in addressing emerging issues.

d) **Comparative Analysis and Lessons Learned**: Through relative research of various machine learning methods and lessons learned from practical implementations, we give worthy insights into the strengths, weaknesses and best practices in applying machine learning to network security.

9.3 Final Thoughts and Recommendations

As we conclude this research, it is evident that machine learning holds huge potential for improving network security capabilities. However, realizing this potential needs locate various challenges and embracing future paths in research and implementation.

- **Data Quality and Collaboration**: Organizations ought to prioritize ensuring data quality and encourage collaboration to share threat intelligence and best practices, thereby enabling more robust and efficient machine learning models.
- **Interpretability and Explainability**: Developing interpretable and explainable machine learning models is essential to enhance trust, transparency, and accountability in network security decision-making processes.
- **Adaptability and Scalability**: Future research should aim to create adaptive and scalable machine learning frameworks that can manage dynamic network environments, large-scale data streams, and evolving cyber threats.
- **Integration with Emerging Technologies**: As new technologies like IoT and 5G emerge, researchers should investigate innovative methods to integrate machine learning with these advancements to enhance network security defences.
- **Interdisciplinary Collaboration**: Collaboration among disciplines such as cybersecurity, data science, psychology, and sociology can encourage interdisciplinary methods to tackle complex network security challenges.

REFERENCES

Ambusaidi, M. A., Tan, Z., He, X., Nanda, P., Lu, L. F., & Jamdagni, A. (2014). Intrusion detection method based on nonlinear correlation measure. *International Journal of Internet Protocol Technology*, *8*(2/3), 77. DOI: 10.1504/IJIPT.2014.066377

Bhuyan, M. H., Bhattacharyya, D. K., & Kalita, J. K. (2014). Network Anomaly Detection: Methods, Systems and Tools. *IEEE Communications Surveys and Tutorials. IEEE Communications Surveys and Tutorials*, *16*(1), 303–336. DOI: 10.1109/SURV.2013.052213.00046

Breiman, L. (2001). No Title. *Machine Learning*, *45*(1), 5–32. DOI: 10.1023/A:1010933404324

Cai, J., Luo, J., Wang, S., & Yang, S. (2018). Feature selection in machine learning: A new perspective. *Neurocomputing*, *300*, 70–79. DOI: 10.1016/j.neucom.2017.11.077

Chandola, V., Banerjee, A., & Kumar, V. (2009). Anomaly detection. *ACM Computing Surveys*, *41*(3), 1–58. DOI: 10.1145/1541880.1541882

Chen, J., Chen, Y., Du, X., Li, C., Lu, J., Zhao, S., & Zhou, X. (2013). Big data challenge: A data management perspective. *Frontiers of Computer Science*, *7*(2), 157–164. DOI: 10.1007/s11704-013-3903-7

Cheng, Y., Wang, D., Zhou, P., & Zhang, T. (2017). A Survey of Model Compression and Acceleration for Deep Neural Networks. *arXiv (Cornell University)*. https://doi.org//arxiv.1710.09282DOI: 10.48550

Cristianini, N., & Shawe-Taylor, J. (2000). *An Introduction to Support Vector Machines and Other Kernel-based Learning Methods*. DOI: 10.1017/CBO9780511801389

Dean, J., & Ghemawat, S. (2008). MapReduce. *Communications of the ACM*, *51*(1), 107–113. DOI: 10.1145/1327452.1327492

Doshi-Velez, F., & Kim, B. (2017). Towards A Rigorous Science of Interpretable Machine Learning. *arXiv (Cornell University)*. https://doi.org//arxiv.1702.08608DOI: 10.48550

Goodfellow, I., Bengio, Y., & Courville, A. (2016). *Deep Learning*. MIT Press.

Goodman, B., & Flaxman, S. R. (2017). European Union Regulations on Algorithmic Decision Making and a "Right to Explanation." *˜the œAI Magazine. AI Magazine*, *38*(3), 50–57. DOI: 10.1609/aimag.v38i3.2741

. Grolinger, K., Hayes, M., Higashino, W. A., L'Heureux, A., Allison, D. S., & Capretz, M. A. (2014). *Challenges for MapReduce in Big Data*. DOI: 10.1109/SERVICES.2014.41

. Jouppi, N. P., Young, C., Patil, N., Patterson, D., Agrawal, G., Bajwa, R., Bates, S., Bhatia, S., Boden, N., Borchers, A., Boyle, R., Cantin, P. L., Chao, C., Clark, C., Coriell, J., Daley, M., Dau, M., Dean, J., Gelb, B., . . . Yoon, D. H. (2017). *In-Datacenter Performance Analysis of a Tensor Processing Unit*. https://doi.org/DOI: 10.1145/3079856.3080246

LeCun, Y., Bengio, Y., & Hinton, G. (2015). Deep learning. *Nature*, *521*(7553), 436–444. DOI: 10.1038/nature14539 PMID: 26017442

Mahoney, M. V., & Chan, P. K. (2003). An Analysis of the 1999 DARPA/Lincoln Laboratory Evaluation Data for Network Anomaly Detection. In *Lecture notes in computer science* (pp. 220–237). DOI: 10.1007/978-3-540-45248-5_13

Mohri, M., Rostamizadeh, A., & Talwalkar, A. (2018). *Foundations of Machine Learning* (2nd ed.). MIT Press.

Paricherla, M., Ritonga, M., Shinde, S. R., Chaudhari, S. M., Linur, R., & Raghuvanshi, A. (2023). Machine learning techniques for accurate classification and detection of intrusions in computer network. *Bulletin of Electrical Engineering and Informatics*, *12*(4), 2340–2347. DOI: 10.11591/eei.v12i4.4708

Pedregosa, F., Varoquaux, G., Gramfort, A., Michel, V., Thirion, B., Grisel, O., Blondel, M., Prettenhofer, P., Weiss, R., Dubourg, V., Vanderplas, J., Passos, A., Cournapeau, D., Brucher, M., Perrot, M., & Duchesnay, D. (2012). Scikit-learn: Machine Learning in Python. *arXiv (Cornell University)*. https://doi.org//arxiv.1201.0490DOI: 10.48550

Rokach, L. (2009). Ensemble-based classifiers. *Artificial Intelligence Review*, *33*(1–2), 1–39. DOI: 10.1007/s10462-009-9124-7

Rudin, C. (2019). Stop explaining black box machine learning models for high stakes decisions and use interpretable models instead. *Nature Machine Intelligence*, *1*(5), 206–215. DOI: 10.1038/s42256-019-0048-x PMID: 35603010

Saxe, J., & Berlin, K. (2015). Deep Neural Network Based Malware Detection Using Two Dimensional Binary Program Features. *arXiv (Cornell University)*. https://doi.org//arxiv.1508.03096DOI: 10.1109/MALWARE.2015.7413680

Shi, W., Cao, J., Zhang, Q., Li, Y., & Xu, L. (2016). Edge Computing: Vision and Challenges. *IEEE Internet of Things Journal*, *3*(5), 637–646. DOI: 10.1109/JIOT.2016.2579198

Sommer, R., & Paxson, V. (2010). *Outside the Closed World: On Using Machine Learning for Network Intrusion Detection*. https://doi.org/DOI: 10.1109/SP.2010.25

Sutton, R. S., & Barto, A. G. (2018). *Reinforcement Learning* (2nd ed.). MIT Press.

Tramèr, F., Zhang, F., Juels, A., Reiter, M. K., & Ristenpart, T. (2016). Stealing Machine Learning Models via Prediction APIs. *arXiv (Cornell University)*. https://doi.org//arxiv.1609.02943DOI: 10.48550

Chapter 9
Security Analysis in an Internet of Things Research Perspective Using Bibliometric Analysis

Agni Isador Harsapranata
https://orcid.org/0009-0006-2677-7944
Bina Sarana Informatika University, Indonesia

Eko Sediyono
Satya Wacana Christian University, Indonesia

Hindriyanto Dwi Purnomo
Satya Wacana Christian University, Indonesia

ABSTRACT

Currently, the development of Internet of things technology has become so rapid, it can be seen from every aspect of life, the Internet of things is used to help hu-man life. With rapid development, Security Internet of things is necessary to en-sure that there is no interference in the use of this technology. This research con-ducted a research gap analysis of 2069 journal articles indexed by Scopus from 2017 to 2023. Research analysis using VOSviewer and Biblioshiny, the analysis produced a visual depiction of the relationship between journal articles, global re-search trends, the most cited journal articles, resulting in 3 information The jour-nal with the largest H index is the "IEEE Internet of Things" journal at 94, the "Sensor" Journal at 57, and the "Internet of Things" journal at 30. Understanding this bibliometric

DOI: 10.4018/979-8-3693-9846-3.ch009

analysis can provide direction for global research trends related to security analysis on the Internet of Things.

1 INTRODUCTION

Bibliometric analysis is a method used to measure and analyze scientific publications and patterns of information dissemination in a scientific discipline. This method relies on quantitative data from publications, such as the number of articles, citations, and collaborations between authors or institutions. Bibliometric analysis aims to understand research trends, determine the influence of a scientific work, and identify collaborative networks in research. In this context, bibliometric analysis not only functions as an evaluation tool, but also as a strategic guideline in research development. In recent years, the use of bibliometric analysis has increased along with the growth of the number of scientific publications(Virgil P Diodato, 1994)+. This growth creates new challenges for researchers and institutions to manage and evaluate the existing information. Therefore, bibliometric analysis is important to provide deeper insights into the development and direction of research. There are several commonly used methodologies in conducting bibliometric analysis. These methodologies vary depending on the purpose of the analysis and the type of data available. One approach that is often used is citation analysis, where researchers collect data on the number of citations received by a particular publication. Citation analysis can provide insight into the influence and relevance of work in the scientific community. Another popular methodology is co-authorship analysis, which studies collaboration between authors. This research shows that collaboration between authors can increase the productivity and impact of research. By mapping collaboration networks, researchers can identify groups of researchers who are active in a particular field and explore potential future collaborations. Bibliometric analysis can also be done using specialized software designed to process and analyze publication data. For example, VOSviewer and bibliometric are widely used tools for visualizing citation and collaboration networks. Using these tools, researchers can produce visual maps that illustrate the relationships between authors, institutions, or countries in a particular research context. In conducting bibliometric analysis, it is important to choose the right database. Some commonly used databases include Scopus, Web of Science, and Google Scholar. Each database has its own advantages and disadvantages in terms of coverage, publication type, and accessibility. Therefore, the selection of a database must be adjusted to the purpose of the analysis and the desired research focus. Although bibliometric analysis has many advantages, there are a number of challenges that need to be addressed. One of the main challenges is the quality and consistency of the data used in the analysis. Different databases have different criteria for indexing publications, which can lead to differences in the number and types of publications analyzed. Another challenge is the complexity of data interpretation. While bibliometric analysis produces valuable quantitative data, interpreting the results often requires a deep understanding of the research context. For example, an increase in the number of publications does not necessarily reflect an increase in the quality of research. In addition, bibliometric analysis is often limited to publications indexed in a particular database. This means that non-indexed publications, such as technical reports or policy documents, may not be accounted for in the analysis. This can result in an incomplete picture of research trends in a field. Another challenge is the limitations in measuring the societal impact of research. While bibliometric analysis can provide information about citations and collaborations, it is difficult to measure the broader societal impact of research. For example, research that produces solutions to social problems may not always be highly citations, but its impact can be felt widely in society. Therefore, it is important to develop a more holistic methodology for evaluating research impact. Finally, challenges in bibliometric analysis also include changes in the way scientific publication is conducted. With the advent of open access publishing and new formats such as preprints, the way people access and disseminate research continues to evolve. Therefore, bibliometric analysis must continue to adapt to these changes to remain relevant and provide accurate insights. Looking ahead, bibliometric analysis is expected to continue to evolve as technology advances and the way research is conducted changes. One emerging trend is the use of artificial intelligence (AI) and machine learning to analyze bibliometric data. These technologies can help identify patterns and relationships that may not be apparent in traditional analysis. Additionally, as more data becomes available online, bibliometric analysis will increasingly be integrated with big data. By combining data from multiple sources, researchers can gain a more comprehensive picture of research dynamics. The future of bibliometric analysis will also be influenced by changes in scholarly publication policies. With increasing attention to open access, bibliometric analysis will likely focus more on freely accessible publications. In addition, bibliometric analysis must continue to adapt to new developments in interdisciplinary research. Research that combines multiple disciplines is becoming more common, and bibliometric analysis needs to consider collaboration across disciplines to gain a more complete picture of research trends. Finally, the future of bibliometric analysis will also involve increased transparency and accountability in research evaluation. With increasing attention to research integrity, bibliometric analysis must ensure that the data used in the analysis is accurate and accountable. The Internet of Things (IoT) has become one of the most significant technological innovations in recent years. With more and more devices connected to the internet, from household appliances to smart city infrastructure, IoT offers unprecedented convenience and efficiency. The internet of things is a connection of devices that use computer technology to exchange data using a computer network (Lombardi et al., 2021). According to Cisco, the use of Internet of Things technology was born in the 2008-2009 period. Internet of Things (IoT) technology has been used and its use is so widespread, there are several fields that utilize IoT technology, multimedia(Gupta et al., 2021), with multimedia IoT users are pampered with various kinds of entertainment, the agricultural sector also utilizes IoT technology to increase its production results (Mousavi & Ghaffari, 2021)(Kathole et al., 2023)(Patel & Shrimali, 2023), In the health sector, IoT is used to help monitor patients' conditions using sensors, thereby helping medical staff monitor patients continuously (Lee & Chen, 2021), IoT is also widely used in smart buildings (A. Kumar et al., 2021) as seen in Fig 1, Seeing this does not rule out the possibility of disruption occurring, in overcoming disruptions in IoT, here are several areas being developed in research, blockchain (Brotsis et al., 2021)(Eddine et al., 2021)(R. Kumar & Sharma, 2022)(Khashan & Khafajah, 2023)(Paul et al., 2021)(N et al., 2023), machine learning algorithms (Liu et al., 2021)(Ge et al., 2021)(Sahu et al., 2021)(Rahman et al., 2021)(M. Wu et al., 2021), or even both research are used to secure IoT (Unal et al., 2021)(Y. Wu et al., 2021). However, with this convenience comes great challenges in terms of security. Therefore, security analysis in the context of IoT becomes very important to protect user data and privacy(KOMBATE et al., 2022). In this context, attacks on IoT devices can occur in various forms, from data hacking to DDoS (Distributed Denial of Service) attacks. These incidents show how vulnerable IoT devices are and the importance of implementing strong security measures. In addition, many IoT devices are designed without considering security aspects, which makes them easier to exploit. This shows the need for a more proactive approach in designing and developing IoT devices, including more stringent security testing and regular software updates. With the increasing attention to IoT security, various organizations and government agencies have begun to issue regulations and guidelines to improve device security. This shows that IoT security is not only the responsibility of device developers but also involves collaboration between various stakeholders. The security challenges in IoT are complex and diverse. One major challenge is the lack of consistent security standards across the industry. Different IoT device manufacturers often have different approaches to implementing security, creating loopholes that can be exploited by attackers. This means that many devices are vulnerable to attacks, especially when the data being transmitted is not protected. Many IoT devices have limited computing and storage power, making it difficult to implement more sophisticated security solutions. For example, devices such as temperature sensors or surveillance cameras are often not designed to run complex encryption algorithms. Another issue is the lack of user awareness of the security risks associated with IoT devices. Many users are unaware that the devices they use can be a gateway for attackers. This lack of understanding often causes users to ignore basic security measures, such as changing default passwords or updating software. In addition, cyberattacks targeting IoT devices are often difficult to detect. Many devices lack the ability to detect or report suspicious activity, allowing attackers to operate undetected for long periods of time. Finally, the integration of IoT devices with larger systems, such as corporate networks or critical infrastructure, can create additional risks(Paul et al., 2021). If an infected IoT device manages to make its way into a larger network, the impact can be devastating. Therefore, it is important to understand and address these security challenges thoroughly. In the face of existing security challenges, various techniques and solutions have been developed to protect IoT devices. One of the most common approaches is the use of encryption to protect data transmitted between devices. Encryption can help ensure that data remains secure even if it is successfully accessed by unauthorized parties. In addition to encryption, strong authentication is also very important in IoT security. The use of multi-factor authentication (MFA) methods can add an additional layer of security, making it more difficult for attackers to gain access to the device. Implementing regular software updates is also an important technique in maintaining the security of IoT devices. Many device manufacturers often release updates to fix vulnerabilities that are found. However, users often ignore these updates. Therefore, user awareness and education about the importance of software updates are essential. In addition, the use of blockchain technology is also starting to be introduced as a solution to improve IoT security. Blockchain can provide a secure and transparent way to manage data and transactions between devices. The development of artificial intelligence (AI)-based security solutions also shows great potential in protecting IoT devices. AI can be used to analyze behavioral patterns and detect suspicious activities in real time(Adamova et al., 2024).In this research, the author describes existing research gaps and then proposes research that can be carried out based on a thematic map, namely conducting research in the field of artificial intelligence combined with IoT security.

Figure 1. Smart building (Kumar et al., 2021)

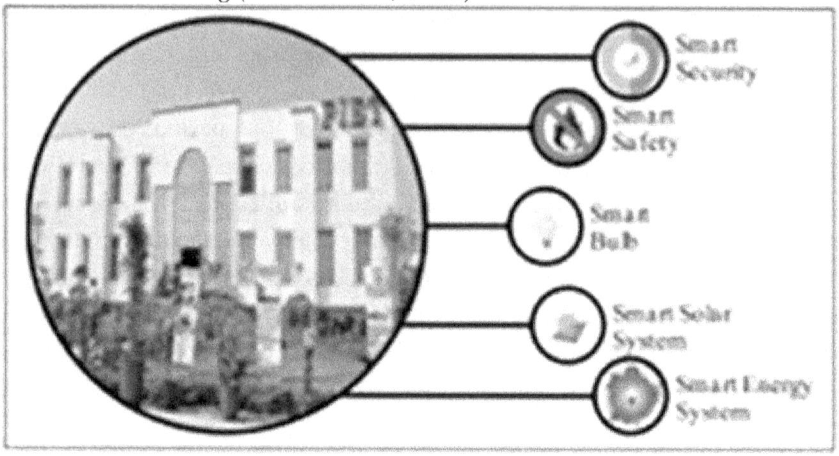

2 RELATED WORKS

The increasing integration of the Internet of Things (IoT) into various sectors has sparked a surge of research aimed at understanding its implications and applications. A notable approach to this exploration is bibliometric analysis, which employs quantitative methods to assess the volume, patterns, and trends of published literature. This method not only illuminates the existing landscape of IoT research but also identifies gaps and future directions. A comprehensive review of recent studies reveals several key areas where bibliometric research has made significant contributions, particularly in irrigation management, communication devices for smart meters, operational management of IoT systems, and the intersection of IoT with agriculture. One of the prominent studies in this domain (Violino et al., 2023), which focuses on utilizing bibliometric analysis to enhance irrigation practices through IoT technologies. The research highlights how IoT can revolutionize traditional irrigation methods by enabling real-time monitoring of soil moisture levels, weather conditions, and crop health. For instance, smart sensors can collect data that inform farmers when and how much to irrigate, ultimately leading to more efficient water usage and improved crop yields. This study not only underscores the importance of data-driven decision-making in agriculture but also illustrates the potential of IoT to address pressing global challenges such as water scarcity and food security. In another significant contribution (Hassan et al., 2022) explore the communication devices employed in smart meters through bibliometric analysis. Smart meters are crucial components of modern energy management systems, allowing for real-time monitoring of energy consumption. The research discusses the various communication technologies that facilitate data transfer between smart meters and utility providers, such as Zigbee, LoRaWAN, and cellular networks. By analyzing the literature surrounding these technologies, the authors provide insights into the evolution of smart meter communication, highlighting the advantages and limitations of each approach. This analysis is vital for stakeholders looking to optimize energy distribution and enhance consumer engagement through more transparent energy usage data.

Rezaee (Rejeb et al., 2022) take a broader perspective by conducting bibliometric research on the operational management of IoT systems. Their findings reveal the complexities involved in managing IoT devices, which often operate in diverse environments and require robust security measures. The study emphasizes the need for effective management strategies that encompass device interoperability, data integrity, and user privacy. For example, the integration of machine learning algorithms can enhance the predictive maintenance of IoT devices, thereby reducing downtime and operational costs. This research not only contributes to the literature on IoT management but also serves as a guide for organizations aiming to harness the full potential of IoT technologies.

Furthermore, the intersection of IoT and agriculture is examined by Rejeb et al. (2022), who employ bibliometric analysis to assess the synergies between these fields. The study highlights various applications of IoT in agriculture, such as precision farming, livestock monitoring, and supply chain optimization. By analyzing the existing literature, the authors identify key trends and emerging technologies that are shaping the future of agricultural practices. For instance, the use of drones equipped with IoT sensors can provide farmers with detailed aerial imagery of their fields, enabling them to make informed decisions about crop management. This research not only illustrates the transformative potential of IoT in agriculture but also emphasizes the importance of interdisciplinary collaboration in driving innovation.

What sets the author's research apart from these existing studies is the focus on bibliometric analysis to uncover trends and future research directions specifically in the realm of IoT and security. As IoT devices proliferate, the security of these systems becomes paramount. The author's work aims to bridge the gap between IoT advancements and security challenges, providing a comprehensive analysis of the current state of research in this critical area. By identifying trends in the literature, the author seeks to highlight the urgent need for robust security frameworks that can protect IoT ecosystems from emerging threats.

In conclusion, the application of bibliometric analysis in the context of IoT research provides a valuable framework for understanding the complexities and interconnections within this rapidly evolving field. The studies reviewed highlight the diverse applications of IoT, from enhancing irrigation practices and optimizing energy management to transforming agricultural methodologies. However, the increasing reliance on IoT also necessitates a focused examination of security challenges, which the author's research aims to address. By synthesizing insights from existing literature and identifying future research trajectories, this work contributes to the ongoing discourse on the responsible and secure deployment of IoT technologies, ultimately paving the way for a more connected and resilient future.

3 METHODS

The initial phase of conducting research in the field of bibliometrics is critical, as it sets the foundation for the entire study. One of the most pivotal decisions researchers faces is selecting the appropriate article database. In this particular research endeavor, the author opted for the Dimension database, a choice that reflects a strategic approach to sourcing relevant literature. Dimension offers a comprehensive repository of scholarly articles, which is essential for ensuring the breadth and depth of the research. Furthermore, the selection of articles fully indexed in Scopus adds an additional layer of quality assurance. Scopus is renowned for its rigorous indexing criteria, which means that the articles it includes have undergone thorough peer review and meet high academic standards. This decision is crucial, as it enhances the credibility of the research findings and allows for a more reliable analysis of trends within the selected literature.

The importance of utilizing high-quality sources cannot be overstated. For instance, in the period from 2017 to 2023, the research identified a substantial total of 2,069 articles. This volume not only reflects the growing interest in the subject matter but also provides a rich dataset for analysis. Each of these articles contributes unique insights and perspectives, making them invaluable for understanding the evolving landscape of research in the field. The researcher's decision to download this data in CSV format demonstrates a methodical approach to data management, facilitating easier manipulation and analysis later in the research process.

Following the data acquisition, the researcher employed the Vosviewer application and R Bibliometrix for data processing. Vosviewer is a powerful tool for visualizing bibliometric networks, allowing researchers to create maps based on citation data, co-authorship, and keyword co-occurrence. This visualization aids in identifying key trends, influential authors, and collaborative networks within the field. For example, a co-authorship network map generated using Vosviewer can reveal which researchers are frequently collaborating, thereby highlighting the most active research groups and potential knowledge hubs. Similarly, R Bibliometrix provides a comprehensive suite of tools for quantitative research in bibliometrics, enabling the researcher to perform detailed statistical analyses and generate insightful reports.

The methodological framework outlined in Figure 2 illustrates the systematic approach taken in this research. The first step, data preparation, involves cleaning and organizing the data to ensure its integrity. This is a crucial phase, as any errors or inconsistencies in the dataset can significantly impact the quality of the bibliometric analysis. Following this, the bibliometric analysis itself is conducted, which involves applying various metrics to assess the impact and relevance of the articles. Metrics such as citation counts, h-index, and journal impact factors are employed to gauge the significance of the research outputs. Finally, the discussion of the analysis results synthesizes the findings, providing a coherent narrative that connects the data to broader trends and implications in the field.

In conclusion, the process of conducting bibliometric research is multifaceted and requires careful consideration at each stage. From the initial selection of a high-quality article database like Dimension, to the rigorous indexing standards of Scopus, every decision contributes to the credibility and reliability of the research findings. The use of advanced tools such as Vosviewer and R Bibliometrix further enhances the analysis, allowing for a nuanced understanding of the research landscape. Ultimately, this research not only sheds light on the current state of the field but also serves as a valuable resource for future studies, paving the way for ongoing scholarly discourse and exploration.

4 RESULT AND DISCUSSIONS

4.1 Database Overview

The main information in this research can be seen in Figure 3, which depicts information on the number of articles used as 2069, data taken in the period 2017 to 2023, with 26.77 citations per article, with a total of 5287 authors, the number of single authors is 125.

The dataset analyzed in this research is extensive, comprising a total of 2069 articles published over a six-year span from 2017 to 2023. This timeframe is particularly significant as it captures a period of rapid advancement and innovation across various fields of study, making the findings of this research relevant to contemporary academic discourse. The sheer volume of articles indicates a robust engagement with the subject matter, reflecting a growing interest among scholars and researchers. Such a high number of publications suggest not only an active research community but also the potential for diverse perspectives and methodologies being applied to the topic at hand.

One of the standout metrics from this analysis is the average citation rate, which stands at 26.77 citations per article. This figure is indicative of the impact and relevance of the research being conducted. In academic circles, citations are a crucial measure of how frequently a piece of work is referenced by other scholars, thereby highlighting its influence within the field. For instance, a citation rate above 25 is generally considered to be a strong indicator of a paper's significance, suggesting that the articles in this dataset are not only being read but are also contributing meaningfully to ongoing discussions and advancements in their respective areas. This level of engagement can be illustrated by considering a hypothetical article that proposes a groundbreaking theory; if it garners numerous citations, it is likely that subsequent research builds upon its findings, further validating its importance.

Furthermore, the research identifies a total of 5287 authors contributing to these articles, which underscores the collaborative nature of contemporary research. The involvement of a large number of authors suggests that many studies are being conducted by teams rather than individuals, fostering a multidisciplinary approach that can lead to more comprehensive and innovative outcomes. For example, an article co-authored by experts in both biology and computer science could yield novel insights at the intersection of these fields, such as advancements in bioinformatics. The collaborative aspect of research not only enhances the quality of the work but also promotes the sharing of ideas and resources, ultimately driving progress in the academic community.

Among the 5287 authors, it is noteworthy that only 125 are single authors. This statistic raises interesting questions about the dynamics of authorship in academic publishing. The predominance of multi-authored articles may reflect a trend towards collaborative research efforts, where the complexity of modern problems necessitates diverse expertise. However, it also raises concerns about the challenges faced by single authors, who may struggle to gain visibility and recognition in a landscape increasingly dominated by collaborative works. The limited number of single authors could suggest a potential barrier for emerging researchers who may wish to establish their individual voices in a crowded field. This observation invites further exploration into the factors that contribute to the success of single versus multi-authored publications.

In conclusion, the analysis presented in this research, particularly as depicted in Figure 3, reveals significant insights into the academic landscape from 2017 to 2023. The considerable number of articles, high citation rate, and the extensive pool of authors highlight the active engagement and collaborative nature of contemporary research. However, the limited number of single authors raises important considerations regarding the challenges faced by individual researchers. Collectively, these findings not only illuminate the current state of academic publishing but also set the stage for future inquiries into the evolving dynamics of authorship and collaboration in research. As the academic community continues to grow and adapt, understanding these trends will be essential for fostering an inclusive and productive research environment.

Figure 2. Process diagram of the analytical

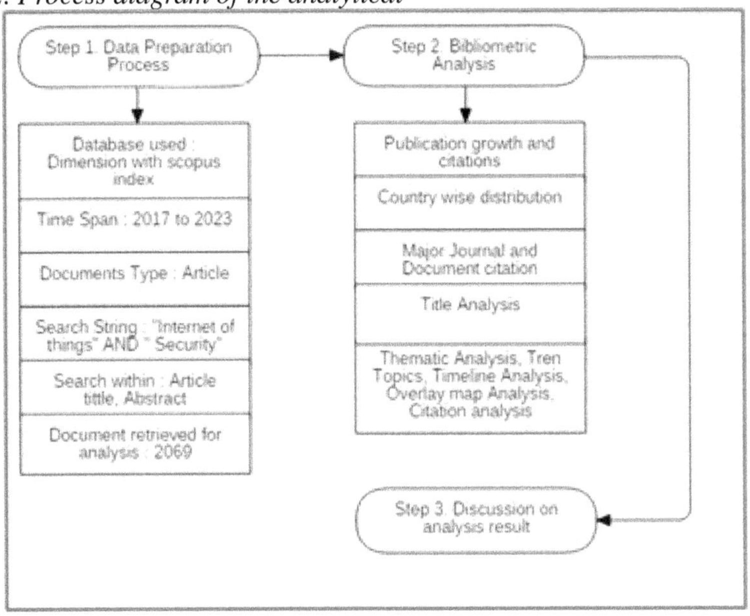

Figure 3. Main information

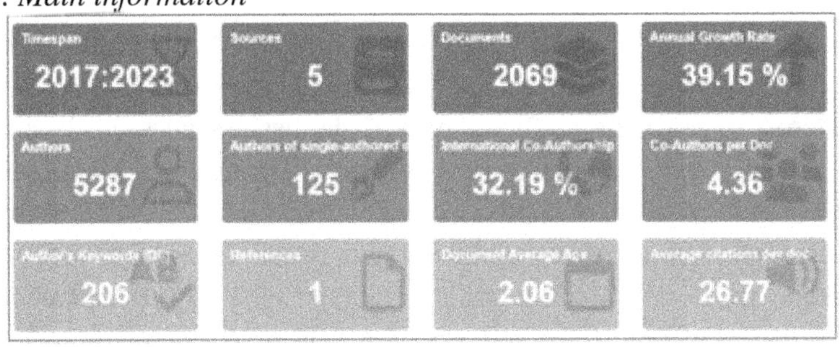

4.2 Annual increase in publications

The growth of research in the field of the Internet of Things (IoT) and security is a compelling narrative that reflects the increasing importance of these domains in our technologically driven society. As illustrated in the accompanying diagram, the upward trajectory of research publications from 2017 to 2023 indicates a burgeoning interest among scholars and practitioners alike. This trend not only underscores the critical nature of IoT and security but also highlights the evolving challenges and innovations that accompany this growth.

In 2017, the landscape of IoT research was relatively nascent, with only 62 articles published. At this stage, the focus was primarily on foundational concepts and early applications of IoT technologies. Researchers were exploring the potential of connecting everyday devices to the internet, laying the groundwork for future advancements. For instance, early studies examined how smart home devices could enhance energy efficiency and improve user convenience. The modest number of publications during this period reflects a time when the implications of IoT were just beginning to be understood, and the security concerns associated with these technologies were not yet at the forefront of academic inquiry.

As we move into 2018, the publication count surged to 138 articles, signaling a growing recognition of the significance of IoT. This year marked a pivotal moment when researchers began to delve deeper into the integration of IoT with various sectors, including healthcare, agriculture, and urban planning. For example, studies exploring the use of IoT in smart healthcare systems illustrated how connected devices could monitor patient health in real-time, thereby improving outcomes and reducing costs. The increase in publications also suggests that researchers were beginning to grapple with the security implications of these interconnected systems, recognizing that as the number of devices increased, so too did the potential vulnerabilities.

The trend continued into 2019, with 212 articles published, reflecting an even greater interest in both IoT applications and the associated security challenges. Researchers began to investigate more sophisticated security protocols and frameworks designed to protect IoT networks from cyber threats. A notable example is the development of blockchain technology as a means to enhance security in IoT environments. By decentralizing data storage and ensuring transparency, blockchain offered a promising solution to mitigate risks associated with data breaches and unauthorized access. This year marked a crucial turning point, as the dialogue around IoT began to shift from mere exploration to addressing the pressing need for robust security measures.

The year 2020 witnessed a dramatic increase in publications, with 372 articles emerging. This spike can be attributed to the global pandemic, which accelerated the adoption of IoT technologies across various sectors. As remote work and digital interactions became the norm, the demand for secure IoT solutions surged. Researchers focused on developing innovative security mechanisms to protect sensitive data transmitted through IoT devices. For instance, studies explored the implementation of artificial intelligence (AI) in threat detection, allowing for real-time monitoring and response to potential security breaches. The pandemic served as a catalyst for both the expansion of IoT applications and the urgent need for enhanced security frameworks.

In 2021, the number of publications reached 393, indicating a continued commitment to exploring the intersection of IoT and security. Researchers began to emphasize the importance of regulatory compliance and ethical considerations in the deployment of IoT technologies. The discourse around privacy concerns gained traction, with studies examining the implications of data collection and surveillance in smart cities. This year also saw an increased collaboration between academia and industry, as stakeholders sought to develop practical solutions to address security challenges. The emphasis on interdisciplinary approaches underscored the complexity of the issues at hand and the necessity for diverse expertise in tackling them.

The upward trend persisted into 2022, with 442 articles published. This year marked a significant shift towards the exploration of emerging technologies, such as edge computing and 5G networks, and their implications for IoT security. Researchers investigated how these advancements could enhance the efficiency and security of IoT systems. For instance, edge computing allows data processing to occur closer to the source, reducing latency and minimizing the risk of data exposure during transmission. This year also highlighted the importance of user education and awareness in mitigating security risks, as studies emphasized the role of end-users in maintaining secure IoT environments.

Finally, in 2023, the publication count reached 450 articles, reflecting a sustained interest in the field. As researchers continue to explore innovative solutions to security challenges in IoT, the focus has broadened to include topics such as ethical AI, machine learning for anomaly detection, and the role of government policies in shaping IoT security landscapes. The dialogue has evolved to encompass not only technical solutions but also societal implications, emphasizing the need for a holistic approach to security in an increasingly interconnected world.

In conclusion, the growth of research in IoT and security from 2017 to 2023 illustrates a dynamic and rapidly evolving field. The increasing number of publications over the years signifies a heightened awareness of the potential applications of IoT technologies, as well as the critical need to address the security vulnerabilities that accompany them. As researchers continue to explore innovative solutions and frameworks, the interplay between IoT and security will undoubtedly shape the future of technology and its impact on society. This ongoing dialogue is essential for ensuring that the benefits of IoT can be realized while safeguarding against the risks inherent in an interconnected world.

Figure 4. Security in Iot analysis used in research by year

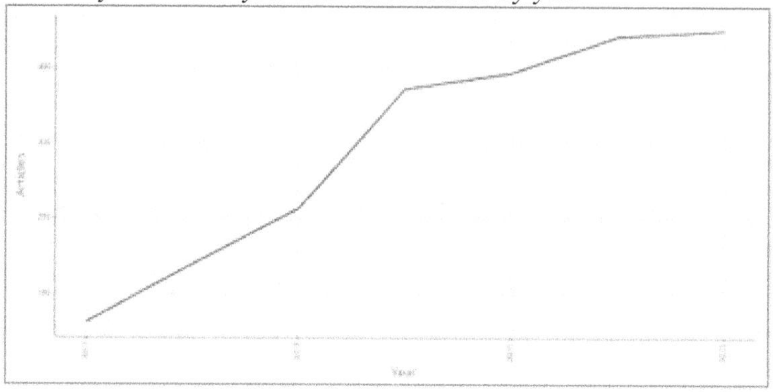

4.3 Most cited journals

The analysis of journal article authors who have received the highest citations reveals significant insights into the impact of their contributions in the field of Internet of Things (IoT). As illustrated in Figure 5, the top three authors—Lin J, Yang Y, and Chettri L—have all made their mark through their publications in the IEEE Internet of Things Journal. This journal is renowned for its rigorous peer-review process and its role in disseminating cutting-edge research, making it a pivotal platform for scholars in the IoT domain.

Lin J stands out not only for the quantity of citations but also for the quality of research presented. His articles often explore innovative applications of IoT technologies in real-world scenarios. For instance, one of his seminal papers delves into the integration of IoT with smart city infrastructures, proposing a framework that enhances urban management through data analytics. The methodologies he employs are both comprehensive and practical, demonstrating how IoT can transform urban living. This approach not only garners citations but also inspires further research and development in smart city initiatives globally.

Following closely is Yang Y, whose work focuses on security and privacy concerns in IoT systems. In an era where data breaches are increasingly common, Yang's research addresses critical vulnerabilities that could compromise user trust and system integrity. One notable study investigates the implementation of blockchain technology as a solution to enhance security protocols in IoT networks. By providing empirical evidence and case studies, Yang effectively illustrates how these technologies can work in tandem to create a more secure IoT ecosystem. His findings have not only attracted significant academic attention but have also influenced industry practices, highlighting the practical implications of his research.

Chettri L rounds out the top three with his contributions that bridge the gap between theoretical frameworks and practical applications in IoT. His research often emphasizes the importance of interoperability among various IoT devices, which is crucial for the seamless functioning of smart environments. A particularly impactful paper discusses the development of a standardized communication protocol that facilitates interaction among heterogeneous devices. By showcasing real-world implementations and the challenges faced during integration, Chettri not only enriches the academic discourse but also provides valuable insights for practitioners in the field.

The interconnectedness of these authors' works highlights a broader trend in IoT research: the necessity for collaboration between theoretical exploration and practical application. As the IoT landscape continues to evolve, the insights provided by Lin, Yang, and Chettri serve as a foundation for future innovations. Their collective contributions underscore the importance of addressing both the technical and societal implications of IoT technologies.

In conclusion, the prominence of Lin J, Yang Y, and Chettri L in the IEEE Internet of Things Journal is a testament to their significant contributions to the field. Their research not only garners citations but also influences both academic and industrial practices, shaping the future of IoT. As the field continues to grow, the foundational work laid by these authors will undoubtedly pave the way for new discoveries and advancements, reinforcing the critical role of rigorous research in driving innovation in technology.

Figure 5. Most cited journals

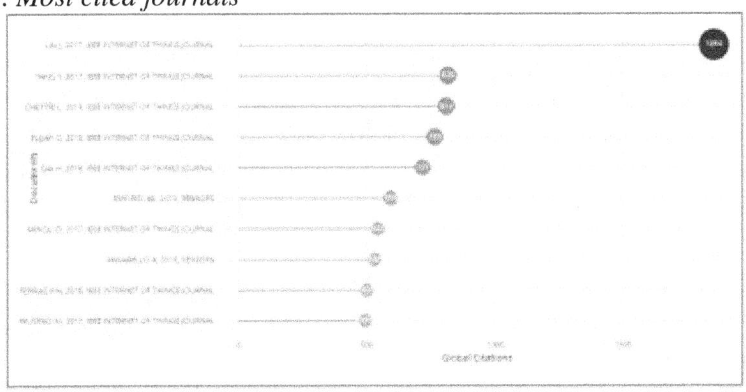

4.4 Science mapping analysis

Science mapping analysis is a technique used to display diagrams that illustrate the relationships between words in journal titles. In this context, science mapping serves as a tool that helps researchers and academics understand patterns and trends in scientific literature. Using software such as Vosviewer, researchers can easily identify relationships between key terms in relevant articles. For example, in this study, the term "Internet of Things" shows a strong association with the word "Internet", with a link strength of 329. This number shows that there are 329 article titles that use both terms together, creating a bridge between two very important concepts in today's technological world.

To better understand the significance of these numbers, we need to see how the terms "Internet of Things" (IoT) and "Internet" interact with each other in the context of research. IoT refers to the network of physical devices connected to the internet, allowing them to collect and exchange data. In this sense, "Internet" serves as the foundation that supports the IoT ecosystem. For example, in many studies discussing IoT, authors often associate this technology with the existing internet infrastructure, explaining how the speed and stability of the internet connection can affect the performance of IoT devices.

Another relevant illustration is in the smart home field, where devices such as smart thermostats, remote-controlled lights, and integrated security systems all rely on the internet to function properly. In many articles discussing this topic, "Internet" is often a keyword that appears together with "Internet of Things," underlining the importance of connectivity in the development of this technology. Thus, science mapping not only illustrates the relationship between words, but also provides deeper insight into how these concepts interact and shape each other in the context of research.

Furthermore, it is important to analyze the impact of this relationship on the development of research in the field of technology. With the increasing number of articles mentioning both terms, we can conclude that there is an increasing interest and research focus on the integration of IoT with existing internet infrastructure. This shows that researchers are not only interested in the development of IoT devices themselves, but also in how these devices can be optimized through existing internet networks. Further analysis can be conducted to explore how these trends have evolved over time, as well as how changes in internet technology have influenced innovation in the IoT field.

In conclusion, science mapping analysis provides a powerful tool for understanding the relationships between terms in the scientific literature. By highlighting the interrelationships between "Internet of Things" and "Internet", we can see a bigger picture of how these two concepts interact and influence each other in the context of research. The high link strength numbers indicate that there is a lot of interrelated research, creating a rich and complex network of knowledge. With a deeper understanding of these relationships, researchers can better direct their research focus and contribute to the development of more innovative and relevant technologies in the future.

Figure 6. Network visualization article

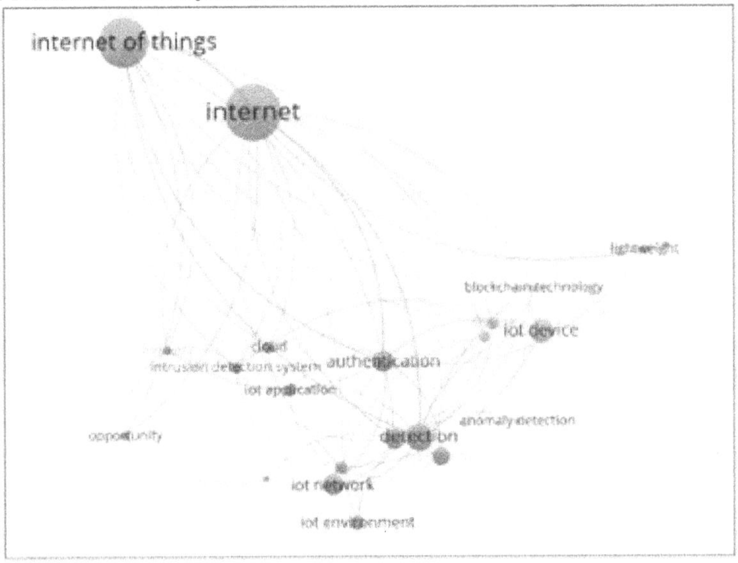

4.5 Thematic analysis

In the thematic analysis conducted in this study, four broad categories can be identified, each with a unique and relevant focus. These categories include the niche theme related to "software and physiology monitoring", the motor theme including "Computer Security, Internet of Things, and Humans", the basic theme including "Machine Learning Algorithms, Neural Networks, and Computers", and the emerging theme related to "artificial intelligence and telemedicine". A deeper analysis of each of these categories will provide greater insight into the complex interactions between technology and humans in diverse contexts.

First, in the niche category of "software and physiology monitoring", we can see how software technology plays a significant role in monitoring an individual's physiological condition. For example, health monitoring software integrated with wearable devices such as smartwatches can collect vital data such as heart rate, blood oxygen levels, and sleep patterns. Using data analysis algorithms, these devices can provide valuable insights for users and healthcare professionals. As an illustration, a study shows that continuous monitoring through these devices can help in the early detection of heart disease, allowing for faster and more effective intervention. This analysis shows the importance of software integration in improving the quality of life of individuals.

Furthermore, the motor theme category "Computer Security, Internet of Things, and Humans" highlights the challenges faced in today's digital age. As more and more devices are connected to the internet, security risks are becoming more significant. For example, IoT devices used in smart homes can become targets for cyberattacks if not properly protected. In one case, a hack of a smart home security camera exposed the owner's personal data, showing how vulnerable humans are to digital threats. Therefore, it is important to develop stronger security protocols and cybersecurity education for users. This analysis shows that computer security is not just a technical responsibility but also requires awareness and active participation from users.

In the basic category "Machine Learning Algorithms, Neural Networks, and Computers", we see how advances in artificial intelligence have pushed the boundaries of technology. Machine learning algorithms, which allow computers to learn from data and improve their performance over time, have been used in a variety of applications, from data analysis to speech recognition. For example, artificial neural networks used in image recognition have enabled applications such as increasingly accurate facial recognition. However, behind these advances come ethical challenges that need to be addressed, such as bias in the data that can influence the results of the analysis. As such, it is important to conduct careful analysis of the data used in training algorithms, to ensure fair and accurate results.

Finally, in the emerging theme of "artificial intelligence and telemedicine," we can see how this technology has revolutionized the way we access healthcare. Telemedicine allows patients to consult a doctor without having to go to the hospital, which is especially useful in emergency situations or for those living in remote areas. For example, telemedicine applications that use artificial intelligence to diagnose symptoms based on patient data can speed up the diagnosis process. However, challenges related to the privacy and security of health data must also be considered. This analysis shows that while technology offers many benefits, we also need to consider the ethical and practical implications of its application in our daily lives.

Overall, this thematic analysis shows that each of the categories identified has a significant contribution to our understanding of the interaction between technology and humans. From health monitoring to cybersecurity, to advances in artificial intelligence and telemedicine, all of these aspects are interconnected and influence each other. It is therefore important to continue research and development in this area, taking into account the ethical and practical challenges that may arise. In this way, we can ensure that technology not only develops but also provides maximum benefits to society as a whole.

Figure 7. Thematic analysis

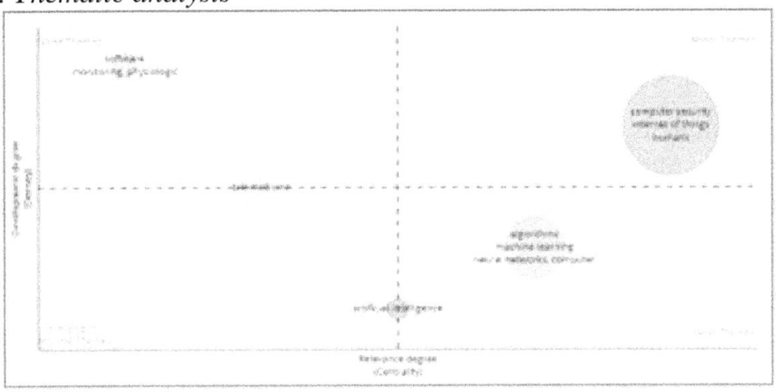

5 CONCLUSION

In this research, an analysis was carried out on reputable articles using Scopus, it was concluded that from the thematic analysis figure 7, a research mapping was produced which was developing rapidly and also became the center of discussion, namely “Computer Security, Internet of Things, Human”, while there was not much research and less researched topics are “artificial intelligence, and telemedicine.

REFERENCES

Adamova, A., Zhukabayeva, T., & Adamov, N. (2024). Machine Learning Algorithms for Intrusion Detection in IoT-enabled Smart Homes. *Procedia Computer Science, 241*, 427–432. https://doi.org/https://doi.org/10.1016/j.procs.2024.08.059. DOI: 10.1016/j.procs.2024.08.059

Brotsis, S., Limniotis, K., Bendiab, G., Kolokotronis, N., & Shiaeles, S. (2021). On the suitability of blockchain platforms for IoT applications: Architectures, security, privacy, and performance. *Computer Networks, 191*(September 2020), 108005. DOI: 10.1016/j.comnet.2021.108005

Eddine, M. S., Ferrag, M. A., Friha, O., & Maglaras, L. (2021). EASBF: An efficient authentication scheme over blockchain for fog computing-enabled internet of vehicles. *Journal of Information Security and Applications, 59*(March), 102802. DOI: 10.1016/j.jisa.2021.102802

Ge, M., Syed, N. F., Fu, X., Baig, Z., & Robles-Kelly, A. (2021). Towards a deep learning-driven intrusion detection approach for Internet of Things. *Computer Networks, 186*(December 2020), 107784. DOI: 10.1016/j.comnet.2020.107784

Gupta, S., Garg, R., Gupta, N., Alnumay, W. S., Ghosh, U., & Sharma, P. K. (2021). Energy-efficient dynamic homomorphic security scheme for fog computing in IoT networks. *Journal of Information Security and Applications, 58*(February), 102768. DOI: 10.1016/j.jisa.2021.102768

Hassan, A., Afrouzi, H. N., Siang, C. H., Ahmed, J., Mehranzamir, K., & Wooi, C.-L. (2022). A survey and bibliometric analysis of different communication technologies available for smart meters. *Cleaner Engineering and Technology, 7*, 100424. https://doi.org/https://doi.org/10.1016/j.clet.2022.100424. DOI: 10.1016/j.clet.2022.100424

Kathole, A. B., Katti, J., Lonare, S., & Dharmale, G. (2023). Identify and classify pests in the agricultural sector using metaheuristics deep learning approach. *Franklin Open, 3*, 100024. https://doi.org/https://doi.org/10.1016/j.fraope.2023.100024. DOI: 10.1016/j.fraope.2023.100024

Khashan, O. A., & Khafajah, N. M. (2023). Efficient hybrid centralized and blockchain-based authentication architecture for heterogeneous IoT systems. *Journal of King Saud University. Computer and Information Sciences, 35*(2), 726–739. DOI: 10.1016/j.jksuci.2023.01.011

Kombate, Y., & Houngue, P. (2022). A bibliometric method for analyzing IoT hacking solutions in network services. *Procedia Computer Science, 203*, 141–148.

Kumar, A., Sharma, S., Goyal, N., Singh, A., Cheng, X., & Singh, P. (2021). Secure and energy-efficient smart building architecture with emerging technology IoT. *Computer Communications, 176*(June), 207–217. DOI: 10.1016/j.comcom.2021.06.003

Kumar, R., & Sharma, R. (2022). Leveraging blockchain for ensuring trust in IoT: A survey. *Journal of King Saud University. Computer and Information Sciences, 34*(10), 8599–8622. DOI: 10.1016/j.jksuci.2021.09.004

Lee, T. F., & Chen, W. Y. (2021). Lightweight fog computing-based authentication protocols using physically unclonable functions for internet of medical things. *Journal of Information Security and Applications, 59*(701), 102817. DOI: 10.1016/j.jisa.2021.102817

Liu, X., Yu, W., Liang, F., Griffith, D., & Golmie, N. (2021). On deep reinforcement learning security for Industrial Internet of Things. *Computer Communications, 168*(December 2020), 20–32. DOI: 10.1016/j.comcom.2020.12.013

Lombardi, M., Pascale, F., & Santaniello, D. (2021). Internet of things: A general overview between architectures, protocols and applications. *Information (Basel), 12*(2), 1–21. DOI: 10.3390/info12020087

Mousavi, S. K., & Ghaffari, A. (2021). Data cryptography in the Internet of Things using the artificial bee colony algorithm in a smart irrigation system. *Journal of Information Security and Applications, 61*, 102945. DOI: 10.1016/j.jisa.2021.102945

Jayashri, N., Rampur, V., Gangodkar, D., Abirami, M., Balarengadurai, C., & Kumar, A. (2023). Improved block chain system for high secured IoT integrated supply chain. *Measurement. Sensors, 25*, 100633.

Patel, H., & Shrimali, B. (2023). AgriOnBlock: Secured data harvesting for agriculture sector using blockchain technology. *ICT Express, 9*(2), 150–159. https://doi.org/https://doi.org/10.1016/j.icte.2021.07.003. DOI: 10.1016/j.icte.2021.07.003

Paul, R., Ghosh, N., Sau, S., Chakrabarti, A., & Mohapatra, P. (2021). Blockchain based secure smart city architecture using low resource IoTs. *Computer Networks, 196*(February), 108234. DOI: 10.1016/j.comnet.2021.108234

Rahman, M. A., Zaman, N., Asyhari, A. T., Sadat, S. M. N., Pillai, P., & Arshah, R. A. (2021). SPY-BOT: Machine learning-enabled post filtering for Social Network-Integrated Industrial Internet of Things. *Ad Hoc Networks, 121*(June), 102588. DOI: 10.1016/j.adhoc.2021.102588

Rejeb, A., Rejeb, K., Abdollahi, A., Al-Turjman, F., & Treiblmaier, H. (2022). The Interplay between the Internet of Things and agriculture: A bibliometric analysis and research agenda. *Internet of Things : Engineering Cyber Physical Human Systems*, *19*(July), 100580. DOI: 10.1016/j.iot.2022.100580

Sahu, A. K., Sharma, S., Tanveer, M., & Raja, R. (2021). Internet of Things attack detection using hybrid Deep Learning Model. *Computer Communications, 176*(December 2020), 146–154. DOI: 10.1016/j.comcom.2021.05.024

Unal, D., Hammoudeh, M., Khan, M. A., Abuarqoub, A., Epiphaniou, G., & Hamila, R. (2021). Integration of federated machine learning and blockchain for the provision of secure big data analytics for Internet of Things. *Computers & Security*, *109*, 102393. DOI: 10.1016/j.cose.2021.102393

Violino, S., Figorilli, S., Ferrigno, M., Manganiello, V., Pallottino, F., Costa, C., & Menesatti, P. (2023). A data-driven bibliometric review on precision irrigation. *Smart Agricultural Technology*, *5*, 100320. https://doi.org/https://doi.org/10.1016/j.atech.2023.100320. DOI: 10.1016/j.atech.2023.100320

Virgil, P., & Diodato, P. G. (1994). *Dictionary of Bibliometrics* (1st Editio). Routledge. https://doi.org/https://doi.org/10.4324/9780203714133

Wu, M., Li, Q., Bilal, M., Xu, X., Zhang, J., & Hou, J. (2021). Multi-label active learning from crowds for secure IIoT. *Ad Hoc Networks*, *121*(February), 102594. DOI: 10.1016/j.adhoc.2021.102594

Wu, Y., Wang, Z., Ma, Y., & Leung, V. C. M. (2021). Deep reinforcement learning for blockchain in industrial IoT: A survey. *Computer Networks*, *191*(February), 108004. DOI: 10.1016/j.comnet.2021.108004

Chapter 10
Sentiment Analysis in Transportation Apps: Machine Learning for User Feedback Classification

Sunneng Sandino Berutu
https://orcid.org/0000-0002-3607-2722
Universitas Kristen Immanuel, Indonesia

Stephen Anugerah Wau
Universitas Kristen Immanuel, Indonesia

Haeni Budiati
Universitas Kristen Immanuel, Indonesia

Jatmika Jatmika
Universitas Kristen Immanuel, Indonesia

ABSTRACT

This research investigates the application of Machine Learning in the context of sentiment classification of user reviews on online transportation applications in Indonesia. This research aims to develop a model to accurately classify positive, negative, or neutral sentiments from customer reviews. Review data collected from the Gojek, Grab, and Maxim applications is utilized to train and test Machine Learning models, including natural language processing (NLP) techniques to extract significant features. Sentiment analysis results show that approximately 63.1% have negative sentiment reviews, 27.7% have positive sentiments, and 9.2% have neutral sentiments. Furthermore, the Support Vector Machine (SVM) algorithm achieved the highest level of accuracy at 92%, followed by Random Forest (RF) with an

DOI: 10.4018/979-8-3693-9846-3.ch010

accuracy level of 91%, Naive Bayes (NB) at 79%, Decision Tree (DT) at 80%, and Logistic Regression (LR) with an accuracy level of 83%. In general, the SVM model performs better than others.

1 INTRODUCTION

In recent years, Indonesia's online transportation sector has experienced tremendous expansion and change. By offering simple, quick, and dependable ride-hailing options via mobile applications, platforms like Gojek, Grab, and Maxim have completely changed how individuals access transportation services. The competition amongst providers has increased as these services continue to grow in popularity and user base nationwide, with each business vying to give the most frictionless and fulfilling user experience. Due in large part to rising smartphone adoption, better internet connectivity, and the rising need for adaptable, on-demand transportation options, this industry has emerged as one of Indonesia's fastest-growing digital industries ((Bahar et al., 2024).

User experience is a key component that makes any digital transportation platform successful. User feedback is crucial in forming the platform's public image, affecting the acquisition of new users, and directing future service improvements in a service-based sector that mostly depends on real-time interactions and consumer satisfaction. Reviews and star ratings from the Google Play Store are among the main ways people share their experiences, acting as an open line of communication between customers and service providers. Users can rank apps on the Play Store using a scale of 1 to 5 stars, with 5 denoting the highest level of satisfaction and 1 denoting the lowest. Users frequently write reviews that detail their experiences using the app and the services it offers in addition to these ratings.

For developers and service managers looking to determine the advantages and disadvantages of their products, as well as for other customers who might be thinking about downloading the app, these reviews are quite helpful. The authors in (Mustopa et al., 2020) claim that user-generated reviews serve as a social proof mechanism, influencing prospective users to choose a certain platform based on the experiences of others. For example, a review that emphasizes reliable fare estimates, courteous service, and drivers who are always on time can persuade new users to use the app. On the other hand, frequent concerns regarding app bugs or excessive costs may discourage uptake.

Due to the enormous amount of evaluations that are produced every day for popular apps like Grab and Gojek, it is not feasible to do sentiment analysis by hand. Regularly reading, analyzing, and classifying hundreds, or even millions, of evaluations is not possible for human reviewers. The procedure takes a long time and is subject to human biases and errors. The same language may be interpreted differently by various people, producing individualized and sometimes inaccurate outcomes. In research and decision-making settings where objectivity and scalability are crucial, this problem becomes much more pressing.

Researchers and practitioners are increasingly using automated sentiment analysis methods, especially those driven by machine learning, to get beyond these restrictions. According to the authors in (Brandão et al., 2025), machine learning (ML) offers strong tools that can process big datasets, learn from examples, and spot patterns in unstructured data like user evaluations. These algorithms can be used to teach computers to categorize reviews' linguistic content according to sentiments, such as favorable, negative, or neutral. This method produces more reliable and scalable results while drastically reducing the workload for human analysts.

Text-based reviews have historically been the focus of sentiment analysis. Large review text corpora are usually gathered by researchers, who then manually or semi-automatically classify the texts according to sentiment classes. They then train machine learning models like Naïve Bayes, Support Vector Machines (SVM), Random Forests (RF), or deep learning models like Long Short-Term Memory (LSTM) networks. After that, these models are trained to link particular sentiment categories to particular words, phrases, and language constructions. The models may be used to accurately classify new, unseen reviews once they have been trained.

Though text data has received a lot of attention, star ratings—an equally crucial component of user feedback—are sometimes overlooked in sentiment analysis, especially when it comes to online transportation services. Typically given with written evaluations, star ratings are structured data that provide a numerical assessment of user sentiment. Even though they are straightforward, they can offer valuable hints regarding the general tone of the related review text.

By incorporating star rating data into the sentiment analysis framework, the current study offers a fresh solution to this research gap. In this study, star ratings are specifically transformed into sentiment categories. Mapping one- and two-star ratings to negative sentiment, three-star ratings to neutral sentiment, and four- and five-star ratings to positive sentiment is a popular tactic in this procedure. Through this conversion procedure, quantitative ratings are successfully converted into categorical sentiment labels, which can subsequently serve as the basis for sentiment classification model training.

The review text is then included in the machine learning training dataset after each review is given a sentiment label determined by its star rating. Sentiment classification models that can predict the sentiment of new, unlabeled reviews based just on textual content can be developed by training different machine learning algorithms using the labeled data. This method allows for a more thorough and reliable study of user sentiment by utilizing the advantages of both structured (ratings) and unstructured (text) data.

Furthermore, the problem of manual annotation—which is frequently subjective and resource-intensive—is lessened by employing star ratings as a stand-in for sentiment labels. While drastically cutting down on the time and expense involved in creating training data, the automated labeling approach based on rating conversion guarantees increased uniformity and objectivity. Moreover, it enables the system to expand to extremely big datasets, such the thousands of reviews that are posted every day for well-known transportation apps.

In conclusion, Indonesia's online transportation sector is still thriving in a very dynamic digital environment where customer feedback is a key factor in driving innovation and improving service quality. This project attempts to create a more accurate, scalable, and efficient approach to sentiment analysis that is suited to the particular requirements of transportation platforms by utilizing machine learning techniques with both review texts and star rating data. This method has practical implications for companies looking to improve customer happiness and optimize service delivery based on real-time user feedback, in addition to supporting scholarly research in the domains of machine learning and natural language processing.

2 RELATED WORK

As a comprehensive and easily accessible source for assessing application performance, measuring user sentiment, and learning more about user behavior, Google Play Store review data has drawn the attention of many researchers in recent years. These user-generated reviews—which frequently include star ratings—provide a useful prism through which to examine actual usage trends, public opinion, and satisfaction levels in a variety of mobile applications and areas.

The study by (Mustopa et al., 2020), which investigated the public's experience with Indonesia's official COVID-19 contact tracing app, PeduliLindung, is one noteworthy example. Their study concentrated on assessing the application's alleged advantages for the community during the pandemic. They were able to identify important areas of satisfaction and worry by examining user input, providing useful information for public health officials and app developers looking to enhance digital interventions during medical emergencies.

The authors in (Kalaichelavan et al., 2020) also used evaluations from the Play Store to explore drone applications. In order to discover recurrent problems like connectivity issues, battery performance concerns, and usability barriers, their research used textual review data to extract common hurdles faced by drone app users. This study showed how, from the viewpoint of the user, app reviews may be used as a diagnostic tool to identify operational and technological limitations.

In another investigation, the authors in (Gupta & Kamthania, 2021) used sentiment analysis methods to examine user review data that was taken from the well-known open data platform Kaggle. Their research highlighted both positive and negative user experiences by examining public sentiment across a range of applications. Their method demonstrated how open datasets and natural language processing (NLP) techniques may be combined to produce insights on user engagement and satisfaction.

In order to determine behavioral changes and user attitudes toward health management, Al-Abbadey in (Al-Abbadey et al., 2021) studied user reviews from an obesity measurement application in the health area. The authors discovered behavioral patterns and motivating elements that affect the uptake and sustained usage of these health-related applications by analyzing review data. Their study demonstrated how user reviews can offer more than just technical assistance; they can also reveal social and psychological aspects of app use.

Improving on the evaluation of performance, the researcher in (Islam et al., 2023) assessed the performance of the popular Zoom video conferencing application by looking at user review data. Zoom's user base grew significantly as a result of the pandemic's transition to online employment and education, making it an excellent choice for this kind of investigation. Islam's research uncovered important user comments on user interface design, security issues, connectivity stability, and video quality. These results were crucial in determining the platform's scalability and constraints in the face of rising demand.

Sentiment-based review analysis has also been applied to sports applications. In order to determine the perceived advantages and difficulties of different sports apps, the authors in (Chembakottu et al., 2023) examined user reviews. Their research revealed which features—including live streaming, real-time score updates, and participatory elements like community forums—were most appreciated by users. These revelations shed light on how user expectations change in specialized application areas.

The authors in (Kambayo et al., 2025) conducted a more recent study that concentrated on evaluating sentiment within the Thread application, which is a social media and discussion-based platform. Through sentiment extraction from user reviews, their study examined a number of app features, such as usability, privacy, content relevancy, and community involvement. A more sophisticated comprehension of the app's perceived advantages and disadvantages resulted from this multifaceted examination.

The researcher in (Kumar et al., 2022) investigated fast commerce (q-commerce) apps, which deliver goods in minutes as opposed to days, in the context of emerging markets. Their sentiment analysis of user evaluations was useful in determining client happiness, especially with regard to order accuracy, pricing, and delivery speed. The results emphasized the logistical difficulties that q-commerce platforms encounter as well as the rising customer demand for immediate delivery.

The authors in (Khan et al., 2024), on the other hand, took a classification method to user review data, concentrating on rating sentiment classification. Their study streamlined the process of review analysis at scale by automatically classifying reviews based on sentiment polarity using machine learning techniques. Their research shows how useful it is to automate sentiment classification, particularly for applications that receive a lot of user feedback.

The authors in (Pratmanto et al., 2020) examined user sentiment in reviews of the Shopee application, a well-known e-commerce platform in Southeast Asia, in a more business-oriented setting. The purpose of their study was to find out how people reacted to several features of the platform, including customer service, pricing accuracy, delivery speed, and product availability. Their conclusions offered insightful advice for raising app functionality and user pleasure.

Analytical studies have also focused on star ratings, which are a common part of app reviews on the Play Store, in addition to text reviews. For example, Sharma and Dutta in (Sharma & Dutta, 2021) created domain-specific lexicons for a number of movie review datasets using star ratings. This method showed how unstructured text can be enhanced by structured rating data to provide sentiment analysis tools that are more accurate.

The auhors in (Schneider et al., 2021) carried out studies utilizing rating data from restaurants and movies to further examine the ramifications of rating design and how rating system design affects total consumer assessments. Their results highlighted the psychological components of rating behavior by indicating that the scale and format of ratings could have a substantial impact on user perception and feedback inclinations.

Finally, to investigate the consistency between textual reviews and star ratings, the researcher in (Sadiq et al., 2021) carried out a case study on user reviews on the Play Store. According to their research, although there is frequently alignment, there are occasionally inconsistencies—for example, a user may post a favorable comment but a negative rating, or vice versa. These discrepancies emphasize how crucial it is to examine both textual and numerical feedback in order to have a complete picture of user mood.

To sum up, an increasing amount of study shows the usefulness of Play Store review data in a variety of fields, such as entertainment, e-commerce, public health, and education. User reviews provide a comprehensive, real-time source of input that can influence future breakthroughs in mobile application development, improve user experience, and inform design decisions, whether through sentiment analysis, behavioral insights, or performance evaluation.

3 RESEARCH METHODS

This research consists of three main stages, namely data preprocessing, model development, and model evaluation. Every step is essential to the overall approach. Cleaning and preparing the data is the first step in the data preprocessing stage. Next, models are constructed and trained, and their performance is evaluated. Figure 1 provides an overview.

Figure 1. Stages of the research process

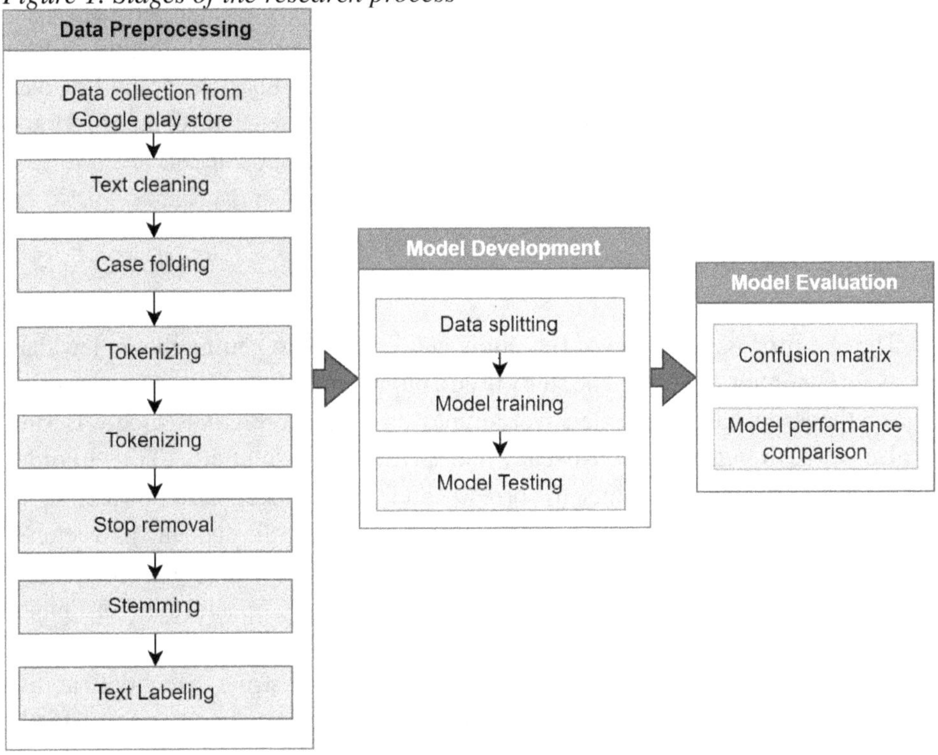

3.1 Data Collecting

With the help of several supporting libraries, the Python programming language was used to collect the data for this study. Google Colab, a cloud-based platform that offers a collaborative environment for writing and running Python code, was used for the entire process. This makes it the perfect option for managing and processing large-scale datasets. The Google Play Store, one of the most popular places to download and evaluate mobile apps, was the specific source of the review data. The dataset concentrated on Gojek, Grab, and Maxim, the three most well-known and extensively utilized online transportation apps in Indonesia. Because of their substantial user bases, steady market presence, and great relevance in the daily lives of Indonesians who depend on ride-hailing and delivery services, these three applications were chosen. The total dataset consisted of 3,000 records, with 1,000 review entries from each application. The extracted reviews offered a wealth of information for sentiment analysis since they contained both textual comments and star ratings. The extraction procedure used criteria to rank the "most relevant" reviews, which are usually based on a combination of user interaction, helpfulness votes, and recency, to make sure that the most representative and pertinent evaluations were gathered. This study attempts to provide a thorough picture of user sentiment throughout Indonesia's online transportation industry by compiling data from several well-known applications into a single dataset. Fair comparison and insightful analysis of user experiences, behaviors, and satisfaction levels across various service providers are made possible by the balanced dataset.

3.2 Data Preprocessing

The text processing phase of this study was essential to getting the review data ready for sentiment analysis. The study used a number of tried-and-true text preparation methods that have been extensively employed in earlier research, such as Berutu et al.'s work, to guarantee consistency and efficacy(Berutu et al., 2023). In order for machine learning algorithms to correctly analyze and categorize raw text input, several preprocessing processes are necessary to transform the data into a structured and analyzable format. Data cleaning, the initial stage of the preparation pipeline, entailed clearing the review texts of extraneous characters, special symbols, punctuation, and other noise. This stage aids in the removal of components that do not add to the text's semantic meaning and would otherwise obstruct the model's ability to learn. To ensure that words like "Service" and "service" are handled consistently by the algorithm, case folding was then done to change all letters to lowercase.

Following case folding and cleaning, each sentence or review was divided into discrete words or tokens as part of the tokenization process. Because it allows the algorithm to examine text at the word level, tokenization is an essential step in natural language processing. After that, stopword elimination was used. Common words that typically don't have much meaning, such "and," "is," "the," and "of," are known as stopwords. They can be eliminated to lower the dataset's dimensionality and noise.

Reducing words to their root forms, or stemming, was the last step in the text processing process. Words like "driving," "drives," and "driver," for instance, might all be stemmed to "drive." This improves the model's capacity to identify patterns and aids in grouping related words together. These thorough preprocessing procedures were used to standardize and optimize the review data for use in sentiment classification tasks based on machine learning.

Text labeling, the next step in the study technique, is essential to getting the dataset ready for sentiment analysis based on machine learning. In order for machine learning algorithms to identify and categorize sentiments based on the content of the review text, labeling entails giving each review in the dataset a sentiment class. A sentiment classification system with three categories—positive, negative, and neutral—was used in this investigation. Compared to a binary approach, this classification system covers the entire range of user experiences, enabling a more sophisticated study of user feedback. The user's star rating in the Google Play Store was used to establish the sentiment label for each review. The Play Store uses a star rating system with 1 denoting the least amount of satisfaction and 5 denoting the most. The following labeling criteria were applied in this study:

a) Positive Sentiment: Reviews with ratings higher than three stars (four or five stars) were classified as having a positive sentiment. Praise for the application, user satisfaction, or appreciation of service features are common in these reviews.
b) Negative Sentiment: Reviews that received a rating of one or two stars, or less than three stars, were classified as having negative sentiment. These evaluations typically express user discontent, grievances regarding the quality of the service, errors, or bad experiences.
c) Neutral Sentiment: Reviews with a rating of precisely three stars were classified as neutral sentiment. These reviews frequently contain conflicting comments or recommendations for improvement and are generally balanced, neither extremely favorable nor extremely negative.

With this labeling strategy, manual sentiment tagging—which can be laborious and subjective—is no longer necessary and the data can be consistently annotated automatically. By employing this method, the study guarantees that every review is assigned an objective label based on measurable user ratings, which makes it easier to create sentiment classification models that are precise and expandable.

3.3 Model Development

To guarantee that the sentiment classification models are trained and assessed efficiently, the dataset goes through a crucial division procedure during the model development stage. Eighty percent of the dataset is set aside for training, and the remaining twenty percent is set aside for testing. By enabling the algorithms to learn from a significant amount of the data while still being tested on a set they have never seen before, this often employed ratio aids in the development of strong models.

To evaluate each model's capacity for generalization and prevent overfitting, this separation is crucial.

The training step, which follows the dataset split, involves fitting multiple machine learning algorithms using the training data. Five well-known classification methods were used in this study: Logistic Regression (LR), Random Forest (RF), Support Vector Machine (SVM), Naïve Bayes (NB), and Decision Tree (DT). These algorithms were chosen because of their proven performance in tasks involving sentiment analysis and text classification. After processing the input features that were taken from the preprocessed and labeled review texts, each algorithm learns to recognize patterns that differentiate between neutral, negative, and favorable attitudes.

The 20% testing dataset is used to test the models once they have been trained. This stage assesses the models' performance on data that hasn't been seen before. A confusion matrix is used to evaluate model performance in a comprehensive manner (Shanmugavadivel et al., 2022).

3.4 Model Evaluation

A number of evaluation metrics, such as accuracy, precision, recall, and the F1-score, are obtained from the confusion matrix. These metrics shed light on how well each model handles erroneous positives and false negatives, balances accurate predictions, and improves classification efficacy overall.

Lastly, a comparison of all machine learning algorithms' performance outcomes is made. The researchers can determine which algorithm works best in terms of sentiment categorization for online reviews of transportation apps thanks to this comparison. It is anticipated that the model that performs the best will be the most appropriate for deployment in the future or more refinement in practical sentiment analysis applications.

4 RESULT AND DISCUSS

4.1 Data

The data collection results of reviews and the distribution of ratings for each online transportation apps are described in Table 1.

Table 1. Review data based on rating stars

Online transportation apps	Rating stars					Total
	1(lowest)	2	3	4	5(highest)	
Gojek	615 (61.5%)	217 (21.7%)	86 (8.6%)	53 (5.3%)	119 (11.9%)	1000
Grab	490 (49.0%)	103 (10.3%)	95 (9.5%)	66 (6.6%)	246 (24.6%)	1000
Maxim	381 (38.1%)	143 (14.3%)	120 (12%)	94 (9.4%)	262 (26.2%)	1000

Three well-known online transportation apps—Gojek, Grab, and Maxim—have rating distribution data that offers insightful information about consumer perception and happiness. 1,000 users have rated each app, with 1 denoting the lowest rating and 5 denoting the highest. The distribution of these evaluations shows divergent patterns in app performance and user experience, which could be the result of variations in customer service, pricing, service quality, and app usability.

The vast majority of users, or 61.5% of all respondents, rated Gojek with the lowest possible rating of one star. Among the three platforms, this one has the largest percentage of 1-star evaluations, indicating that Gojek users are generally unhappy. A variety of concerns, including technical difficulties, subpar customer support, service delivery delays, or recently implemented price or policy changes that were not well accepted, could be the cause of the large number of unfavorable reviews. Interestingly, 11.9% of users gave Gojek 5-star evaluations, suggesting that a tiny percentage of users had extremely pleasant experiences despite the overwhelming unfavorable opinion. Nevertheless, it seems that Gojek's user experience is highly divided, with only 53 users (5.3%) rating it with four stars and less than 10% rating it with three or four stars. There wasn't much of a middle ground; users either had extremely bad experiences or, less frequently, very good ones.

Grab, on the other hand, offers a more evenly distributed distribution, albeit with a sizable percentage of negative ratings. Despite being high, Grab's 1 star rating from 49% of users is 12.5 percentage points worse than Gojek's. With 246 users (24.6%), Grab notably garnered a comparatively high percentage of 5-star ratings—more than twice Gojek's share. This implies that Grab is more adept at providing a consistently satisfying user experience, or at the very least, satisfies the needs of a greater percentage of its consumers. The mid-range evaluations for Grab likewise exhibit a more uniform distribution: 6.6% awarded it four stars, while 9.5% gave it three. These numbers show a wider range of user viewpoints, which may be a reflection of differing experiences in other locations, service categories, or eras. The comparatively high amount of top ratings suggests that Grab has a sizable base of happy consumers, even though there is still a sizable percentage of unhappiness.

At 38.1%, Maxim, on the other hand, is the app with the lowest percentage of 1-star reviews. Even though this still shows a noticeable degree of discontent, it is far less than that of Grab and Gojek. Furthermore, 262 consumers (26.2%) gave Maxim the largest percentage of 5-star evaluations, indicating high satisfaction. Furthermore, Maxim does better in the middle, with 9.4% of customers rating it with four stars and 12% with three. According to this comparatively more balanced rating profile, users may find Maxim to offer a more reliable or agreeable experience. Additionally, particularly in emerging or regional markets, it can suggest less technical or customer service problems or possibly more competitive pricing and service availability.

In contrast to Gojek, which has a significant degree of user dissatisfaction, Maxim seems to have the most positive user perception when compared to the other two platforms. Strong overall satisfaction is indicated by the fact that Maxim has the largest percentage of 5-star ratings in addition to the lowest percentage of 1-star ratings. Grab is positioned in the middle; while it receives less 1-star reviews than Gojek, it still receives a lot of criticism, which it offsets with a high number of top ratings. On the other hand, Gojek has received a lot of unfavorable reviews, which can indicate that important user experience areas require major enhancements.

To sum up, the rating data reveals glaring disparities in the ways that consumers view and interact with these online transit options. Maxim is the platform with the highest user satisfaction, Grab shows promise with mixed but improved results, and Gojek may need to immediately address the reasons behind its high number of low ratings in order to win back user trust and remain competitive in the market, even though each of the three apps has advantages and disadvantages.

4.2 Sentiment Measurement

The rating stars in Table 1 are grouped into positive, negative, and neutral sentiment categories. The 1-3 rules established in the methodology section are applied. The results of grouping ratings into sentiment categories are described in Figure 2.

Figure 2. The sentiment categories percentage

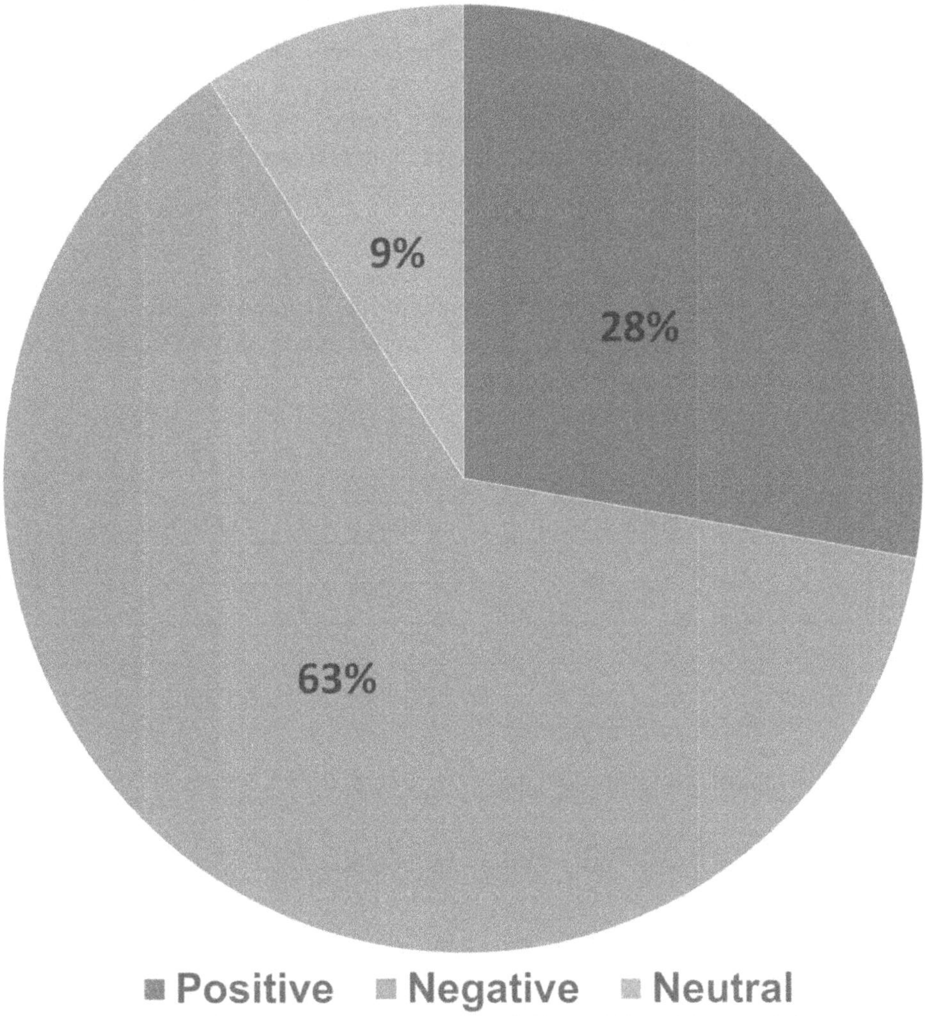

The figure above illustrates that negative category have the highest percentage (63%) with 1893 reviews, followed by positive ratings (28%) with 831 reviews and neutral category (9%) with 276 reviews. According to the findings, consumer satisfaction with online transportation applications in Indonesia is still poor.

The word occurrence calculation in the positive sentiment dataset shows that the words with the highest occurrences are “application,” “driver,” “maxim,” “good,” “friendly,” “fast,” and “service,” “assist,” “point,” “help,” “street,” and “easy.”

For negative sentiment dataset, the calculation of word occurrences shows the words with the highest occurrences are "application," "driver," "order," "help," "already," "order," "times," "pay," "use," "really," and "disappointed".

On the other side, the words with highest occurrences in the neutral sentiment dataset are "driver," "application," "please," "order," "good" "point," "promo," "maxim," "already," "pay," "address," "driver," and "times".

4.3 Classification Model of RF

The confusion matrix is employed to obtain the RF model's parameter values of accuracy, recall, precision, and f1-score. Table 2 and Figure 3 describes the RF model's performance at the testing stage.

Table 2. Performance of the RF model

Sentiment	Precision (%)	Recall (%)	F1-Score (%)	
Positive		**84**	**92**	**87**
Negative		99	91	94
Neutral		91	89	89

Figure 3. The bar chart of the RF model performance

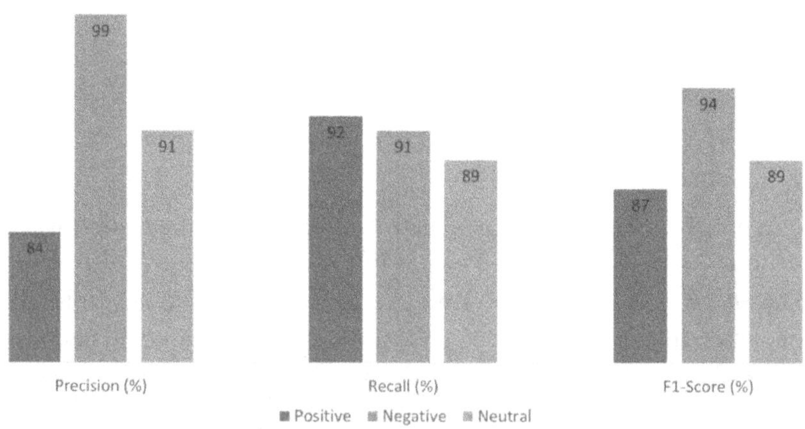

From Table 2, the model performs admirably in sentiment classification, with noteworthy outcomes in each of the three sentiment categories such as positive, negative, and neutral. The model performs best at detecting negative sentiment out of all of these; its maximum precision of 99% shows that it can accurately detect negative instances with few false positives. High F1-score and Recall values of 94% and 91%, respectively, which show balanced accuracy and coverage, are also present.

The model's maximum Recall for positive opinion is 92%, indicating that it effectively captures most positive cases. The F1-score of 87% shows strong overall performance in this category as well, notwithstanding Precision's somewhat lower score of 84%. The model consistently performs well in F1-score and Recall, both at 89%, and reaches its peak Precision score of 91% in the event of neutral mood. This suggests that neutral expressions can be reliably identified without suffering from serious misclassification.

In conclusion, the model is most effective at identifying negative sentiment, which is followed by positive sentiment and neutral sentiment. The model appears to be well-balanced and efficient in sentiment classification tasks, as indicated by the consistently high metrics in every category.

4.4 Classification Model of SVM

The parameter values, such as accuracy, recall, precision, and f1-score, of the SVM model are obtained from the confusion matrix. The performance of the SVM model is shown in Table 3 and Figure 4.

Table 3. Performance of the SVM model

Sentiment	Precision (%)	Recall (%)	F1-Score (%)
Positive	84	96	89
Negative	99	94	96
Neutral	96	87	91

Figure 4. The bar chart of the SVM model performance

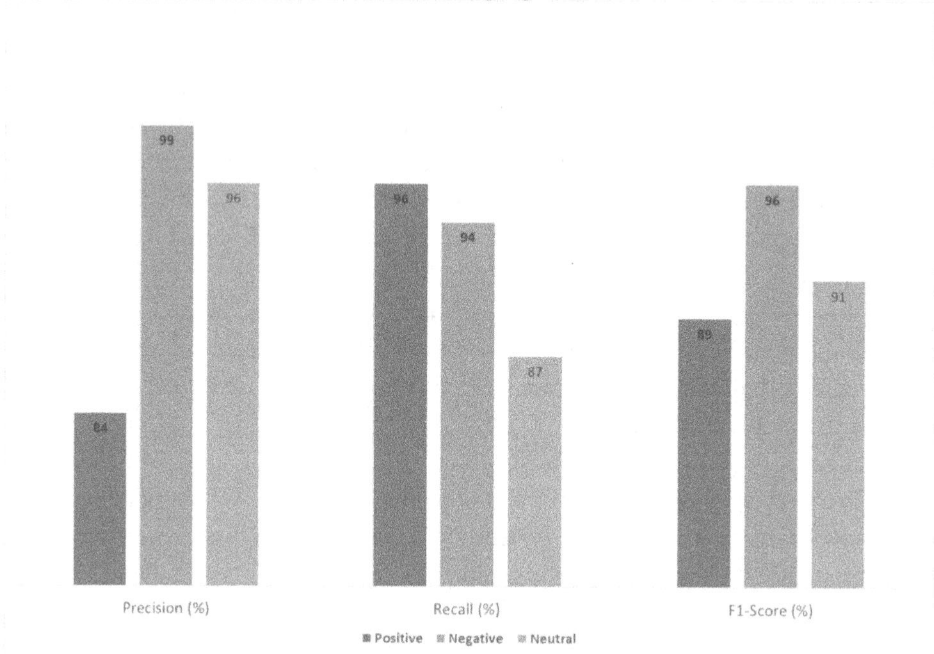

Based on Table 3 above, all three sentiment categories—positive, negative, and neutral—show excellent and balanced performance from the model, with each category emphasizing a particular classifier strength.

The model's high Recall of 96% for positive sentiment shows that it can accurately identify almost all positive cases. The F1-score of 89% shows a good balance between accuracy and completeness in identifying positive attitudes, even though precision is a little lower at 84%. This implies that, although occasionally misclassifying other situations as positive, the model is marginally more likely to forecast positives, capturing the majority of pertinent occurrences.

The model reaches 99% precision in the case of negative sentiment, indicating very accurate predictions with very few false positives. This category exhibits the most stable and dependable performance, as seen by its remarkable F1-score of 96% and Recall of 94%. The model is very good at accurately and reliably detecting negative opinion.

With a precision of 96% for neutral sentiment, the model demonstrates its great accuracy in predicting a neutral sentiment. However, with a somewhat lower recall of 87%, some neutral cases might go unnoticed. With a slightly more cautious approach in identifying neutral utterances, the F1-score of 91% nevertheless shows good overall performance in this category.

In conclusion, the SVM model is well implemented in identifying sentiments. The result signifies that negative sentiment has the top score, followed particularly by neutral and positive sentiment.

4.5 Classification Model of NB

The confusion matrix is employed to obtain the NB model's parameter values of accuracy, recall, precision, and f1-score. Table 4 presents the performance of the NB model.

Table 4. Performance of the NB model

Sentiment	Precision (%)	Recall (%)	F1-Score (%)
Positive	85	62	71
Negative	69	96	80
Neutral	89	78	83

According to Table 4, the model's performance across sentiment categories shows that each sentiment type behaves differently, with various precision, recall, and F1-score trade-offs and strengths.

The model's 85% precision for positive sentiment indicates that it is typically accurate when predicting a positive opinion. However, the model misses a large number of real positive events, as evidenced by the significantly lower recall of 62%. This imbalance is reflected in the F1-score of 71%, which indicates that the model is conservative in classifying feelings as positive, putting accuracy ahead of coverage.

On the other hand, the model has a high Recall of 96% for negative sentiment, capturing almost all negative cases. Precision is lower at 69%, though, meaning that even though it detects the majority of negatives, it also contains a higher number of false positives. These two factors are balanced by the F1-score of 80%, which indicates that the model is more inclusive of negative sentiment at the expense of precision.

With a great Recall of 78% and a strong Precision of 89%, the neutral sentiment findings are the most balanced, yielding an F1-score of 83%. This implies that neutral statements are regularly recognized by the model with comparatively few errors in both directions.

In conclusion, the model has the most balanced indicates and does the best overall at identifying neutral opinion. It is quite accurate in identifying positive sentiment but lower sensitive, while it is very sensitive to negative opinion but is less precise. Depending on whether the task emphasizes recall, precision, or a combination of both for each sentiment category, these patterns can guide further model adjustment.

4.6 Classification Model of DT

The performance of the DT model is described in Table 5.

Table 5. Performance of the DT model

Sentiment	Precision (%)	Recall (%)	F1-Score (%)
Positive	78	73	75
Negative	85	86	85
Neutral	78	81	79

Based on Table 5, positive, negative, and neutral sentiment are the three sentiment categories in which the sentiment analysis model performs consistently and fairly, with the ability to identify negative sentiment showing especially good results.

When it comes to negative opinion, the model performs best, with Precision and Recall closely matching at 85% and 86%, respectively. With a robust F1-score of 85% as a result of this balance, the model is shown to be accurate and thorough in detecting negative expressions.

The model performs moderately when positive opinion is present, with a 78% precision and a 73% recall rate. An F1-score of 75% results from this, indicating that although predictions are mostly accurate, the model does miss some real positive examples. The ability to increase recall while preserving precision might be improved.

With an F1-score of 79%, Precision at 78%, and Recall at 81%, the model remains consistent for neutral opinion. This suggests a strong and consistent performance, with a modest preference for recall, which aids in the model's ability to catch more neutral events with little accuracy loss.

In conclusion, the model exhibits excellent balance and accuracy, performing best when identifying negative sentiment. While its performance for neutral opinion is consistent and dependable, positive sentiment detection might use some improvement to increase memory and overall efficacy.

4.7 Classification Model of LR

The performance of the linier regression model is shown in Table 6.

Table 6. Performance of the LR model

Sentiment	Precision (%)	Recall (%)	F1-Score (%)
Positive	84	72	77
Negative	81	93	86
Neutral	86	85	85

Positive, negative, and neutral opinion are the three sentiment categories in which the model performs well and consistently, with negative and neutral sentiments exhibiting the most efficacy.

The model successfully detects the great majority of negative cases, achieving its best Recall of 93% for negative opinion. The high F1-score of 86% indicates a good balance between precision and recall, albeit with a little lower Precision of 81%. This implies that, although occasionally including non-negative examples, the model is very good at capturing negative sentiment.

With an F1-score of 85%, recall of 85%, and precision of 86%, the neutral sentiment values are remarkably stable. This suggests that the model has low trade-offs between false positives and false negatives and is both accurate and dependable in detecting neutral material.

With an 84% precision rate for positive opinion, the model shows that it is typically right when it predicts something to be positive. Recall falls to 72%, though, indicating that some favorable occurrences are overlooked. The model is more careful when awarding positive opinion, giving correctness precedence over completeness, according to the resulting F1-score of 77%.

In conclusion, the model exhibits great recall and precision and excels at identifying neutral and negative feelings. Although the detection of positive sentiment is also rather robust, recollection should be strengthened to guarantee that more positive examples are accurately detected. All things considered, the model works well for tasks that call for accurate and balanced sentiment classification.

4.8 Performance Comparison Between Models

The authors employ the average parameters value of the Precision, Recall, and F1-score for comparing the performance between models. The result is shown in Table 7.

Table 7. Performance of the machine learning model

Model	Accuracy (%)	Precision (%)	Recall (%)	F1-Score (%)
RF	91	91	90	90
SVM	92	93	92	92
NB	79	81	78	78
DT	80	80	80	79
LR	83	83	83	82

The diagram in the following Figure 5 shows the performance comparison based on Table 7.

Figure 5. The bar chart of performance comparison between machine learning models

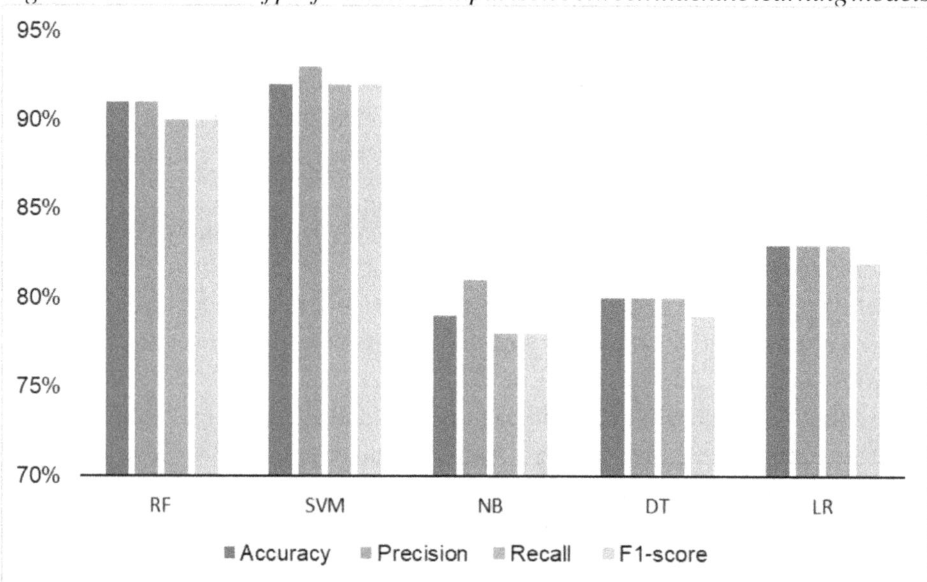

Five machine learning models were evaluated: Random Forest (RF), Support Vector Machine (SVM), Naïve Bayes (NB), Decision Tree (DT), and Logistic Regression (LR). The results show significant performance disparities, with SVM being the most successful model across all metrics.

Accuracy (92%), Precision (93%), Recall (92%), and F1-Score (92%) were the highest scores attained by Support Vector Machine (SVM), suggesting that it not only produces extremely accurate predictions but also strikes a good balance between accurately identifying pertinent instances and reducing false classifications. These findings imply that SVM, with its excellent discriminative power and generalization capabilities, is the most dependable model for the specified classification problem.

The Random Forest (RF) model is second in the graph, and it likewise showed strong performance with constant Precision (91%), Recall (90%), and F1-Score (90%), in addition to Accuracy (91%). RF still produces good results even though it lags behind SVM by a small margin, especially when it comes to stability and handling intricate patterns in the data. Its ability to capture nonlinear interactions and its resistance to overfitting are probably influenced by its ensemble structure.

On the other hand, Naïve Bayes (NB) and Decision Trees (DT) perform significantly less well. NB exhibits insufficient recall (78%), presumably as a result of its strong independence assumptions between features, which may not hold true for the dataset, with accuracy at 79% and F1-score at 78%. Similarly, with an F1-Score of 79% and 80% in Accuracy, Precision, and Recall, DT indicates a comparatively simple fit to the data. Compared to their more sophisticated equivalents, these models could have trouble handling more complicated or noisy patterns.

The performance of Logistic Regression (LR) is slightly higher compared to the results of NB and DT, with an F1-Score of 82% and accuracy, precision, and recall all at 83%. This shows a more balanced performance and implies that LR can be a trustworthy baseline, particularly where interpretability and simplicity are valued more highly than maximum accuracy.

In summary, SVM and RF perform significantly better than the other models, with SVM having a slight advantage on every parameter. SVM seems to be the most appropriate approach for applications requiring high predicted accuracy along with small error. While NB and DT could be preferred for basic or beginning investigation, LR offers an adequate balance between interpretability and performance. These findings emphasize how crucial it is to choose a model according to the particular needs and difficulty of the work at hand.

5 CONCLUSION

For the purposes of automatic sentiment analysis of reviews of online transportation applications in Indonesia, user-provided star ratings from the Google Play Store were transformed into sentiment categories. A systematic set of guidelines guided the conversion process, classifying mid-range ratings as neutral sentiment, low ratings as negative sentiment, and high ratings as positive sentiment. The mood distribution that results shows that Indonesian users are still not very satisfied with online transportation services. This conclusion is derived from the fact that, of the total reviews examined, 63% are of a negative nature. On the other hand, only 28% of evaluations express favorable feeling, while 9% are neutral. According to these findings, a sizable percentage of users are not happy with the platforms' dependability, quality of service, or other features. This study utilized and assessed a number of popular machine learning techniques, including Random Forest (RF), Support Vector Machine (SVM), Naïve Bayes (NB), Decision Tree (DT), and Logistic Regression (LR), to analyze and categorize the sentiment of user evaluations. Preprocessed text data and tagged sentiment categories based on user evaluations were used to train these models. Based on measures such as accuracy, precision, recall, and F1-score, SVM and RF were found to be the best-performing models. Both showed strong performance across a variety of evaluation measures and good classification accuracy, which qualified them for use in upcoming online review sentiment classification tasks.

In order to improve the research methods going forward, the authors suggest classifying textual reviews according to sentiment polarity values, which are determined through the use of natural language processing algorithms. To evaluate consistency and enhance comprehension, these polarity-based classifications will thereafter be contrasted with sentiment categories based on ratings. To further enhance the sentiment classification model, the study also intends to investigate deep learning methods, specifically BERT (Bidirectional Encoder Representations from Transformers). In comparison to conventional machine learning techniques, BERT's capacity to discern contextual subtleties in text promises more precise sentiment prediction and deeper understanding of user viewpoints.

ACKNOWLEDGMENT

The authors would like to thank the ministry of research technology high education (RISTEK DIKTI) and Informatics department, Faculty of Science and Computer (FISKOM), Universitas Kristen Immanuel Yogyakarta, Indonesia for supporting this research activity.

REFERENCES

Al-Abbadey, M., Fong, M. M. W., Wilde, L. J., Ingham, R., & Ghio, D. (2021). Mobile health apps: An exploration of user-generated reviews in Google Play Store on a physical activity application. *Digital Health, 7*, 1–12. DOI: 10.1177/20552076211014988 PMID: 34017609

Bahar, A., Prananta, A. W., & Afifudin, M. (2024). *Mobile App Usability, Data Privacy, Brand Loyalty, and Customer Retention in Indonesia ' s Online Transportation Services Industry. 5*(2), 2562–2572.

Berutu, S. S., Budiati, H., & Gulo, F. (2023). Data preprocessing approach for machine learning-based sentiment classification. *INFOTEL, 15*(4), 317–325. DOI: 10.20895/infotel.v15i4.1030

Brandão, J. G., Castro Junior, A. P., Pacheco, V. M. G., Rodrigues, C. G., Belo, O. M. O., Coimbra, A. P., & Calixto, W. P. (2025). Optimization of machine learning models for sentiment analysis in social media. *Information Sciences, 694*(October 2024), 121704. DOI: 10.1016/j.ins.2024.121704

Chembakottu, B., Li, H., & Khomh, F. (2023). A large-scale exploratory study of android sports apps in the google play store. *Information and Software Technology, 164*(June), 107321. DOI: 10.1016/j.infsof.2023.107321

Gupta, A., & Kamthania, D. (2021). Study of Sentiment on Google Play Store Applications. SSRN *Electronic Journal*. DOI: 10.2139/ssrn.3833926

Islam, M. J., Datta, R., & Iqbal, A. (2023). Actual rating calculation of the zoom cloud meetings app using user reviews on google play store with sentiment annotation of BERT and hybridization of RNN and LSTM. *Expert Systems with Applications, 223*(March), 119919. DOI: 10.1016/j.eswa.2023.119919 PMID: 36969371

Kalaichelavan, K., Malik, H., Husnu, N., & Sreenath, S. (2020). What Do People Complain about Drone Apps? A Large-Scale Empirical Study of Google Play Store Reviews. *Procedia Computer Science, 170*, 547–554. DOI: 10.1016/j.procs.2020.03.124

Kambayo, A. N., Berutu, S. S., & Nshimiyimana, A. (2025). *Aspect-Based Sentiment Analysis Using Latent Dirichlet Allocation (LDA) and DistilBERT on Threads App Reviews. 9*(01), 25–34.

Khan, M. Q., Malik, F., Rahman, N., Pakhtunkhwa, K., & Arabia, S. (2024). *Optimized Sentiment Classification of Google Play Store App Ratings Using Advanced Machine Learning Models*. 252–266. DOI: 10.21015/vtse.v12i4.1997

Kumar, P. M., Jeswanth, A. R., & Manani, J. (2022). Sentimental Analysis on Quick Commerce Platform Based on Customer Reviews in Google Play Store. [IJRISE]. *International Journal of Research in Science & Engineering, 2*(06), 29–42.

Mustopa, A., Pratama, E. B., Hendini, A., & Risdiansyah, D. (2020, November). Analysis of user reviews for the pedulilindungi application on google play using the support vector machine and naive bayes algorithm based on particle swarm optimization. In *2020 Fifth International Conference on Informatics and Computing (ICIC)* (pp. 1-7). IEEE.

Pratmanto, D., Rousyati, R., Wati, F. F., Widodo, A. E., Suleman, S., & Wijianto, R. (2020). App Review Sentiment Analysis Shopee Application in Google Play Store Using Naive Bayes Algorithm. *Journal of Physics: Conference Series, 1641*(1), 012043. Advance online publication. DOI: 10.1088/1742-6596/1641/1/012043

Sadiq, S., Umer, M., Ullah, S., Mirjalili, S., Rupapara, V., & Nappi, M. (2021). Discrepancy detection between actual user reviews and numeric ratings of Google App store using deep learning. *Expert Systems with Applications, 181*(June), 115111. Advance online publication. DOI: 10.1016/j.eswa.2021.115111

Schneider, C., Weinmann, M., Mohr, P. N. C., & vom Brocke, J. (2021). When the stars shine too bright: The influence of multidimensional ratings on online consumer ratings. *Management Science, 67*(6), 3871–3898. DOI: 10.1287/mnsc.2020.3654

Shanmugavadivel, K., Sampath, S. H., Nandhakumar, P., Mahalingam, P., Subramanian, M., Kumaresan, P. K., & Priyadharshini, R. (2022). An analysis of machine learning models for sentiment analysis of Tamil code-mixed data. *Computer Speech & Language, 76*(January), 101407. DOI: 10.1016/j.csl.2022.101407

Sharma, S. S., & Dutta, G. (2021). SentiDraw: Using star ratings of reviews to develop domain specific sentiment lexicon for polarity determination. *Information Processing & Management, 58*(1), 102412. Advance online publication. DOI: 10.1016/j.ipm.2020.102412

Chapter 11
Vesting Smart Irrigation–Farming Infusing Blockchain and Internet of Agricultural Things (IoAT):
Focusing Ground Water Scarcity Ensuring Space-Age Sustainable Farming

Bhupinder Singh
https://orcid.org/0009-0006-4779-2553
Sharda University, India

Christian Kaunert
https://orcid.org/0000-0002-4493-2235
Dublin City University, Ireland & University of South Wales, UK

ABSTRACT

Smart irrigation systems enabled by blockchain and the Internet of Agricultural Things (IoAT), agriculture is no longer under existential threat due to groundwater scarcity; sustainable farming has become a possibility. Over-pumping and depleted aquifers are depleting the vital irrigation water source, ground-water as warmer planet sows floods by storms rampage. Smart Irrigation uses real-time data from the Internet of Things (IoAT) sensors to improve water use efficiency and savings. The secure and transparent nature of the blockchain technology guarantees that water

DOI: 10.4018/979-8-3693-9846-3.ch011

distribution is impartial, sustainable farming practices are enforced through smart contracts. Farmers can use these state-of-the-art technologies to make informed decisions, which would help in minimal water wastage and getting a better crop output. Not only does it combat the issue of over-extraction from aquifers, but also serves as a tool to drive sustainable agriculture practices into the future, ensuring water secure food production in an environmentally challenged world.

1. INTRODUCTION

The biggest challenges experienced by agriculture in the world are water scarcity. As the world population escalates, so does competition for increasingly strained water resources especially in areas where groundwater needs to be pumped rapidly. Groundwater is a principal supplier of irrigation, but it has been decreasing because of over-extraction and poor management along with the influence caused by climate change. Increasing demands for freshwater resources have improved the need to tailor traditional irrigation systems that were proving inefficient in conserving this paramount resource and could be supplemented with modern technological interventions that are urgently needed to ensure sustainable agricultural practices (Pullo et al., 2023). That is where smart irrigation systems combined with Blockchain and the Internet of Agricultural Things (IoAT) can take a significant part in solving groundwater depletion while enabling space age, sustainable agriculture.

The Internet of Agricultural Things (IoAT) is the transformation of traditional agriculture methods into Digital farming by using advanced technologies from a subset of broader IoT trend. IoAT helps to connect all the devices, sensors and platforms together for controlling as well monitoring of agricultural process from far. IoAT helps in smart irrigation by enabling accurate and real-time monitoring of water levels, state of soil and well-being of crops which help farmers to take decisions based on data obtained through these devices regarding when they should irrigate their fields. By using this precision agriculture method each drop of water is used, thus addressing the immediate problem of rapidly depleting groundwater. The wireless irrigation system is also the age-old difference between traditional and smart water management in agriculture. These conventional irrigation systems, however, can thus waste water and create inefficiencies in the watering process by being relatively inflexible they are not capable of adapting to changing minute-to-minute or crop-to-crop moisture needs. While smart irrigation leverages sensors, the cloud, artificial intelligence (AI), and Internet of Agricultural Things to gather insights on moisture in soils or air, pigmentation intensities inside leaves. This allows saving water to the maximum, implementing real-time requirements so minimum wastage is lined with higher efficiency of using water (Latif et al., 2021).

Although blockchain technology is usually thought of in terms of cryptocurrency and finance, it has made its way into agriculture by providing a secure and transparent platform for handling the functions necessary to grow crops. Blockchain can increase transparency and traceability in water usage as well e.g., smart irrigation. Blockchain can use smart contracts to automate irrigation schedules as defined by conditions, so that the water flows simply and justly. It also brings transparency by creating an unchangeable proof about used water, in order to control the consumption of groundwater and prevent over-extraction. Such a synergy between blockchain and IoAT makes it a reliable, authenticable and accessible scheme; promoting the sustainable water management (Udutalapally et al., 2020). A blend of these high technology is incorporated in to irrigation systems that face various problems. For one, it guarantees the water is only used as much as necessary something that any region dealing with water scarcity and unpredictable rainfall patterns can attest to its significance. Secondly, it offers farmers feedback from the field in real-time and data-driven insights that inform them to make well-thought-out decisions on irrigation, crop health as a whole or resource management. Third Sustainability: It helps in decreasing the wastage of water and also to recharge the groundwater. In the end, blockchain infused IoAT smart irrigation acts as a space-age solution which suits sustainable farming in 21st century (Sendros et al., 2022).

Figure 1. Dimensions of introduction section (Original)

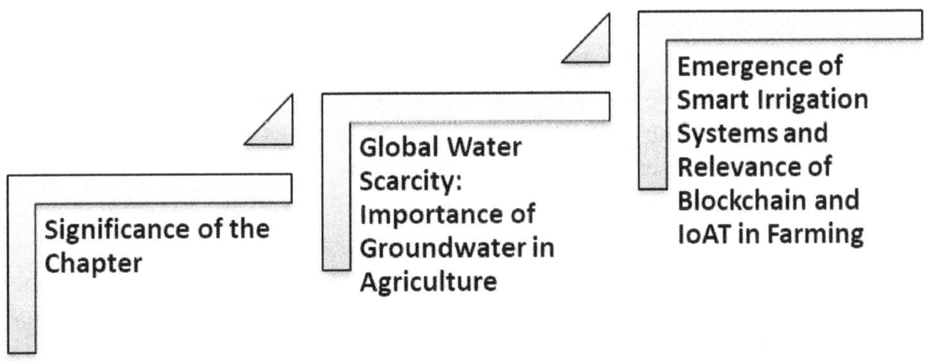

1.1 Significance of the Chapter

This study can be highly important as it deals with one of the major challenges in global agriculture, groundwater scarcity by providing a novel solution. Groundwater depletion is a worldwide problem that endangers the global food supply, particularly in arid and dry regions heavily reliant on irrigation for agriculture. It directly affects the yield of crops, as well as causes land degradation, loss of biodiversity in both flora and fauna species etc. This might be caused by over abstraction/falling ground water level, which again results into environmental damage also disrupting natural ecosystems. This research aims at reducing those effects caused by blockchain, with the utilization of smart irrigation systems empowered by block-chaining and IoAT (Internet of Agriculture Things) and this can lead to more sustainable practices in water application performance for agriculture (Chen et al., 2020).

The integration of blockchain and IoAT in smart irrigation allows for a transparent & auditable water management system which is one of the most significant value propositions. Water, particularly ground water is one of the resources that are most mismanaged as there is no proper monitoring and enforcement mechanism. This data can be independently recorded and stored on the blockchain using one of many tamper-proof systems available. It also will ensure that local governments, farmers and other stakeholders have access to information on water usage which can be used for monitoring reported status of groundwater resources. That this happens is essential for fair shares of water by all and to avoid over-extraction.

Smart irrigation systems enabled with IoAT provide an advanced capability of water management which was unimaginable. Through the continuous monitoring of soil moisture content, weather projections and plant conditions via sensors, IoATT lets farmers fine-tune how they irrigate according to the precise needs of individual fields (Al-Shareeda et al., 2023). This granularity of control thereby helps to prevent over-irrigation, a common issue seen in traditional irrigation systems that adhere to set schedules rather than incorporating real-time data. That means, for an example as the scenario described earlier on a much larger scale rather than only being assigned to monitor irrigation over one field and alerting if it miscalculates that through this smart technique of IoAT coupled with AI there is no chance of any inaccuracy since live data would be provided from various sensors placed within topsoil et cetera giving real time information prompting less usage or none at all just when needed. Though it saves water is by protecting crops from being either parched or over-soaked, resulting greater crop health and yield.

This chapter also reflects on the long-range sustainability improvements that arise from deploying these technologies in agriculture. Smart irrigation systems can also save water and encourage better groundwater utilization to recharge depleted aquifers, as well playing a cooperative role in sustaining the quality of this vital resource

(Sharma & Singh, 2022). As a result, the collaborative integration of blockchain and IoAT in smart irrigation systems can contribute to sustainable agriculture practices over time to make sure that generations come will have enough water resources for food production as well as environmental conservation.

This chapter makes the case within a broader discussion on sustainable agriculture for technology in relation to environmental threats. They will also increase the burden on natural resources especially water in this age of food production and require renewed efforts to promote sustainable agriculture. The study highlights that highly advanced technologies such as blockchain and IoAT are practical solutions to feed the growing population without exhausting our scarce water resources (Kathar et al., 2021). It will provide a template for the role of technology in developing an agricultural system that is both more adaptable to impending environmental stressors and increasingly sustainable. This aimed to have a holistic approach toward the study of blockchain and IoAT for these WSN based smart irrigation systems to combat groundwater scarcity so that it can be benchmarked as such in literature. With an emphasis on the bundles of these technologies, there are practical recommendations that could help sustainably farm, judiciously manage water and secure food for future generation as well. These results can be used to inform policy makers, growers and technology developers alike smart irrigation is not's just a neat gimmick irrigators need it in order to get their groundwater depletion under control before sustaining fresh water use till the 21st century (Upadhyaya et al., 2022).

1.2 Overview of Global Water Scarcity: Importance of Groundwater in Agriculture

Groundwater, which makes up some 30% of thc world´s fresh water resources has a vital role in food production;- especially in dry and semi-dry areas. It is an essential source of irrigation, which sustains almost 40 percent of the world's food supply. Due to increased extraction primarily for agricultural purposes, replenishment of groundwater needs far more time than what has been taken out. This overabundance endangers food quality because of lost oceans that give crops the water they have to develop, smaller yields and residue soil health (Bhushan et al., 2021). The main concerns of the global scarcity of water are agriculture, economy and ecosystem. Multiple factors such as population growth, climate change and unsustainable water management are stressing freshwater resources. As of today, a little over one-third of all people on Earth live in countries with significant water stress and this number is projected to more than double within the next ten years.

Groundwater is paramount to agriculture, and it specifies the food production, livelihoods and rural economies. This degradation means that groundwater dependence will only increase as surface water might become more unreliable due to climate variability. It is thus critical to manage it sustainably so as not only for food security, but also to have adequate water in the aquifer system (Fernández-Carames & Fraga-Lamas, 2018).

1.3 Emergence of Smart Irrigation Systems and Relevance of Blockchain and IoAT in Farming

The advent of smart irrigation systems represents a groundbreaking change in agricultural water management designed to deal with increased global water scarcity by the efficient and precise use of water. Smart irrigation, on the other hand employs sensors and data analytics to feed real-time information, like soil moisture levels or weather conditions directly from a field back to an irrigator; using that information helps decide when and where water should be applied. Also see table below. Water use is fine-tuned based on data, avoiding waste and boosting crop productivity.

Smart irrigation also gains the upper hand with Blockchain and IoAT (Internet of Agricultural Things) (Radeva & Popchev, 2022). Through tamper-proof records and smart contracts, blockchain guarantees the fairness of water distribution while allowing people to share this valuable resource as transparently secure. Together, this network of devices and sensors called IoAT gathers real-time information on the environmental conditions so that farmers know when to turn on irrigation systems or how much water is being used across fields. Using both, the conservation of groundwater has made while encouraging sustainable farming practices and increased resilience in agriculture to water scarcity & climate variability.

2. GROUNDWATER SCARCITY AND AGRICULTURAL CHALLENGES

Worldwide, the shortage of groundwater is considered to be one of the most important issues that seriously limit agricultural activities in regions affected by water deficit. Utilized primarily in the agricultural sector, approximately 70% of global terrestrial water withdrawals come from groundwater management practices (Singh et al., 2023). But the problem is there and climate change, alongside unsustainable extraction of water, is causing an unprecedented depletion in aquifers right around India. Which is starting to lead already crop yields are falling, water supplies shrinking and the entire socio-economic pattern of this area begins a slide into entropy. In this category, there are articles on reasons and ramifications of groundwater depletion; climate change effects on hydrology with a focus on water stress in agriculture; the importance of groundwater for resilience farming besides contrast upon traditional irrigation practices to modern technologies. The socio-economic impacts of groundwater scarcity on farming communities are discussed at last.

2.1 Causes and Consequences of Groundwater Depletion

The major reason is over-extraction of water for agriculture thus making a decline in groundwater. Farmers around the world depend on groundwater to irrigate fields, in large part because surface water sources like rivers and lakes are either not available or severely overdrawn. This eventually causes the water table to gradually sink (Shi et al., 2019), decreasing groundwater recharge. It does not help that water is badly managed here wasteful irrigation systems, no laws about withdrawing groundwater. The consequences of groundwater depletion are many. Aquifers are being eaten up, other wells are running dry and farmers everywhere find it harder to access the water they need for crops. This can decrease agricultural productivity and even make farmland defunct in extreme cases. Groundwater levels are also plummeting, leading to land subsidence which in turn damages construction infrastructure and changes ecosystems. Over-extraction of groundwater can also lead to the intrusion of salt water into coastal aquifers, rendering irrigation and domestic drinking water supplies unusable (Sengupta et al., 2021). These challenges have a compounding affect in agricultural areas, such that groundwater depletion becomes an issue of food security.

2.2 Climate Change and Water Stress in Agriculture

Climate change will also lead to changing patterns of weather and therefore water demand for irrigation may increase adding pressure on already over-stressed groundwater systems. Hotter temperatures also evaporate more moisture from the ground and water, leaving less to nourish crops. The reduced availability of surface water impacts negatively on the usage and subsequent extraction of ground water, which further accelerates its depletion (Prakash et al., 2023). Apart from climate change, the water stress in agriculture is a collateral poor groundwater management making ground more scarcity.

More exotic, the water resources are becoming unpredictable due to rising global temperatures, shifting precipitation patterns and an increase in extreme events related with weather such as droughts or floods. Longer drots in many areas have reduced surface water supplies, forcing farmers to rely more heavily on groundwater for irrigation. At the same time, rainfall patterns are increasingly erratic and unreliable which disturbs normal planting and harvesting cycles that add up to limited water sources (Akram et al., 2020). Global replenishment of underground reservoirs has a strong effect on sustainability and future water use for agriculture in areas where climate change is predicted to intensify drought.

2.3 Groundwater use in Sustainable Agriculture

In parts of Australia where it is essential groundwater sustains viable farming e.g. regions with unreliable rainfall or limited surface water resources. It acts as a water reservoir during rainless times, serving the irrigation needs and makes sure that plants do not go completely dry. In regions with water scarcity, irrigation by groundwater is the main driver of food production and livelihoods in rural communities. Responsible management of groundwater is important for the sustenance farmers. This includes (a) reducing appropriations, for example through better irrigation practices and water-use efficiency in industrial processes; the adoption of drought-resistant crops or other renewable energy sources that are less water-intensive than current supplies from aquifers/wells/springs etc., as well increasing conservation efforts on protecting recharge areas so high-quality clean groundwater deposits don't dry up due excessive pumping rates (Bathalapalli et al., 2021). The sustainable management of groundwater will help keep this essential natural resource for generations to come, and it helps agriculture tackle the challenges posed by climate change and other environmental pressures.

2.4 Traditional Irrigation vs. Modern Irrigation

All conventional methods of irrigation like flood irrigation are inefficient and hence depletive to the groundwater levels. Unlike a flood irrigation where water is applied evenly across the entire area resulting in unnecessary evaporation, runoff and leaching of soil to far below root depth. However, the above method is water intensive and in heavily groundwater reliant areas leads to over-extraction of ground-water.

Modern irrigation alternatives, e.g. Drip and sprinkler Irrigation systems actually lessen water use while also simplifies in managing of water agriculture. Drip irrigation is the method of providing water to plant roots through a network of tubes and emitters which serves two main purposes, reducing evaporation loss from air along with minimal runoff (Vangipuram et al., 2023). Instead, sprinkler systems spread water evenly over fields so no excess water needs to be applied. Sensors are available, that can be installed in these modern systems and which help automated controls to provide water according to real-time conditions consequently even more improving the already-existing efficiency.

Farming has seen a complete reinvention in water management, brought about by increasingly smart irrigation technologies under the umbrella of internet-enabled agriculture (IoAT) and blockchain. Such systems utilize input from soil moisture sensors, weather predictions and crop health monitoring to fine-tune irrigation timing. For example, blockchain technology could provide water usage records that are secure and transparent in relation to the amount of groundwater being used sustainably without exploitation. They can help farmers reduce their water footprint and slowly begin to replenish groundwater levels while saving on critical resource usage, assisting more in making agriculture a sustainable ecosystem (Haque et al., 2023).

2.5 Effects of Groundwater Scarcity on Farming Socioeconomic

The shortage of groundwater has severe social and economic implications for farmers, especially smallholder farmers in developing countries. Irrigation rises in cost as groundwater becomes more difficult to access. The other is to dig deeper wells in hopes of reaching the water table, which requires both energy and money energy because you have drill down to reach that far or a well pump at least twice as powerful on average. Many small-scale farmers, who have a limited capital base and rely on their crop sales to survive from one growing season to the next often simply give up farming altogether or switch over to less water-intensive but lower value crops with fewer price risks (Tanwar et al., 2022).

3. SMART IRRIGATION SYSTEMS: OVERVIEW AND BENEFITS

Modern irrigation is revolutionizing agriculture, while smart systems are making this water use more efficient in many ways. With increasing water scarcity and the impacts of climate change, smart irrigation presents a sustainable solution to manage our valuable resource more effectively. They use sophisticated sensors, data analytics and automation to monitor precisely when they need water for their crops (Amoah & Oh, 2020). The section touches upon the concept of smart irrigation, technologies and sensors used in smart irrigation, water efficiency strategies for successful irrigational processes as well as data-driven decision-making in relation to irrigation along with examples from different successful projects on how they were able to introduce Smart Iot based solutions.

Figure 2. Smart irrigation systems (Original)

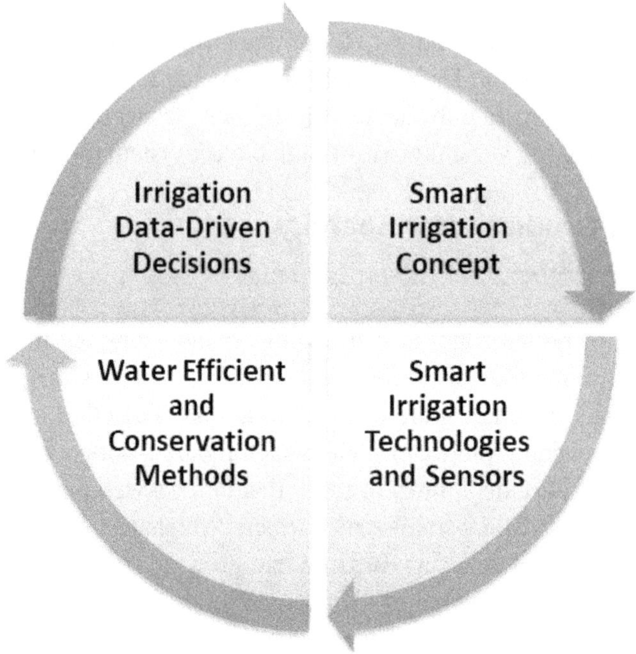

3.1 Smart Irrigation Concept

Smart irrigation: It is an application of sensors, control systems and hydraulic machines as well as real-time data to manage water use in agriculture. Unlike normal irrigation practices that apply water evenly whether or not the requirement arises; smart irrigation system has been built keeping in mind to granulate and decide the amount of required water based on different parameters like Soil moisture, Weather conditions, Crop type etc. This reduces waste of water thereby ensuring that crops are adequately hydrated, enhancing both the efficiency and healthiness (Kose et al., 2022).

With the help of sensors smart irrigation systems save water, they are automatic and work with remote configured smartphones or computers using which farmers can manage their watering schedules by time. Such systems use sensors and data analysis tools to help growers make decisions with better knowledge of what really going on in the field, rather than relying exclusively upon manual monitoring or guesswork when deciding what should be irrigated how often (Singh, 2023). Consequently, smart irrigation conserves the water resources of nature as well as it enhances sustainable farming, minimizes over-extraction from ground reserves and further strengthens agricultural resilience for coping up with dramatic changes in climatic trends leading toward water scarcity.

3.2 Smart Irrigation Technologies and Sensors

Smart irrigation systems heavily depend on various types of environmental conditions for which advanced technology and sensors are applied that sense data in real-time. Materials used during design and construction consist of soil moisture sensors, weather stations, flow meters etc. The soil moisture sensors are used to measure the water content in the soil, which helps farmers know when irrigation is needed and how much they should pour (Ordonez et al., 2023). Taking weather conditions (such as temperature, humidity, rainfall and wind speed) into account along with forecasts at every level enables the irrigation systems to change water delivery accordingly. Including flow meters in the design of irrigation systems, they are able to monitor water usage and quantify how much are being applied allowing farmers if there face any inefficiencies or leaks along with educated management options.

Many smart irrigation systems deploy controllers or automated valves to manage the amount of water circulating through based on data from their sensors. Therefore, these systems are connected to the Internet of Agricultural Things (IoAT) which easily connect other sensors and devices in various places on a farm. Smart irrigation technologies make use of real-time data to decrease water wastage, making for a more sustainable way to farm and providing better resource efficiency (Wahid et al., 2022).

3.3 Water Efficient and Conservation Methods

Smart irrigation systems can also be set to water in the early morning or late evening, when its cooler out. In addition, they include a feature to detect rainfall and stop irrigation when there is enough moisture in the soil. Smart irrigation helps prevent groundwater depletion and uses a lot less water, which is important in the face of severe global drought issues. Increased water use efficiency and working towards Water conservation in agriculture is one of the highlighted advantages of smart irrigation. These systems eliminate wasteful overwatering and minimize evaporation losses to ensure water is only applied exactly when plants need it most. Systems such as drip irrigation and micro-sprinkler deliver water directly to the root zone of plants reducing wastage on both occasions and allowing only necessary supply (Hassija et al., 2021).

3.4 Irrigation Data-Driven Decisions

These systems use data collected from sensors to help farmers know when and how much they need to irrigate, thereby making better-informed decisions about their irrigation scheduling and water management. This system allows farmers to adapt their irrigation practices on the fly according to real-time factors including soil moisture, crop growth stages and weather conditions. The irrigation system can lessen or even turn off those sprinklers (Pikatza et al., 2023). Weather forecasts again can provide information to hold off on irrigation if rain is expected, thus saving more water. The optimization of water use renders benefits such as increased crop yields and reduced operating costs by reducing both water and energy incorporation in the process.

4. BLOCKCHAIN IN AGRICULTURE: ENHANCING TRANSPARENCY AND EFFICIENCY

Blockchain technology is becoming increasingly popular in a number of industries, one being agriculture as it has been used to improve transparency, traceability and efficiency. As an example, blockchain presents innovative solutions to the optimization of water use and management guaranteeing a fair distribution common in smart irrigation & other implementations aimed at promoting sustainability (Kumari & Muthulakshmi, 2023). It also explains the basics of blockchain, how it is used in agriculture production and a percentage about smart contracts for water management purposes as well as study which show that real time monitoring using IoT sensors supplies transparency to manage water sources while examining Blockchain applicability into this industry.

A blockchain is a decentralized, distributed ledger technology that allows secure keeping of records in transparent and un-refutable way. Chaining] It is a block with multiple data blocks chained together using the timestamp for each data and transaction. Any data added to the blockchain stays there forever and cannot be modified which makes it a trusted method for recording transactions (Panda et al., 2020). Blockchain is a technology that leverages a network of peers, which means multiple participants (nodes) validate the transactions among themselves based on consensus without central administration. Solutions for blockchain applications in agriculture include providing a record and tracking methods to the global supply chain of, such as the movement of goods, water use efficiency & resource distribution. The technology offers a transparent means of monitoring and achieving agricultural processes without compromising any aspect, promoting the authentic accountability between farmers as well as suppliers even regulators (Monteiro et al., 2021).

The blockchain has a role to play in water resource management as well that we will look into more detail as when combined with smart irrigation system. This way, it is possible to have the real-time recording of water usage, which makes tracking irrigation practices also transparent and efficient. This would enable, for example, farmers to use blockchain and adequately audit their water consumption and control authorities or policy-makers could monitor the equitable distribution of a sensitive natural resource such as water. Blockchain also can assist in certifying environmentally friendly and organic agriculture by capturing an undeniable permanent record of adherence to ecological standards (George & Jayashree, 2022). There are many areas in agriculture where blockchain technology can be leveraged such as improving supply chain transparency, maintaining sustainable resource management. Traceability is a key area to apply blockchain. Using a blockchain to track every stage of the agricultural supply chain helps stakeholders trace products from farm-to-fork. Not only does this guarantee the safety and quality of food, but it also gives consumers peace-of-mind that their products were produced sustainably and ethically.

4.1 Smart Contracts for Water Use and Distribution

Smart contracts for water usage and distribution are one of the best blockchain applications that can be applied in agriculture. A smart contract is a self-executing contract where the terms/conditions are directly written into code. It triggers transaction implementation without the requirement of notaries upon fulfillment of pre-set conditions. For instance, in the water industry these smart contracts can be utilized to control and modulate the distribution of this asset across different regions using real-time information taken from smart agriculture systems. A smart contract can also distribute water to various farms, for example considering soil moisture levels, crop requirements and weather conditions (Karlsson et al., 2018). It will make distribution of water both efficient and just, thus minimizing wastage and exhaustion of underground water. Smart contracts also provide the opportunity for farmers to trade water rights in a transparent and secure way which improves water utilization throughout regions.

4.2 Transparency in Water Resource Management

Transparency in Water Resource Management One of the key advantages of blockchain that can be used extensively in agriculture practices is a better water resource management. Water is a contentious and scarce good, the fair distribution of which presents an enormous challenge. This track and audit trail security of transaction details is exactly where blockchain technology can alleviate the challenges with a tamper-proof account concerning how much water stakeholders are allowed to take, along with verification about how it was used (Alzoubi et al., 2022). Recording every water use transaction on a blockchain would allow farmers, regulators and policymakers to view the up-to-the-minute in transparent information about consumption. This level of transparency helps avoid disputes over water allocation, aids compliance with policies for managing water and encourages more sustainable practices. Blockchain can promote transparency and collaboration among stakeholders in areas where water is a critical resource.

Despite the potential, there are various challenges to implement blockchain in agriculture. The nature of blockchain requires that farms have internet access and reliable, probably not in developing worlds. Furthermore, most farmers especially smallholders might be unable to use blockchain systems because they are not technically efficient. The other problem is that blockchain solutions are expensive, and in a lot of cases they might be too much to handle by individual farmers (Mitra et al., 2022). Building and owning up blockchain platforms is a costly affair so there will always be resistance to new technology primarily due to cost or understanding of what they are getting in return.

Data privacy concerns may serve as a barrier. However, even though blockchain is secure and tamper-proof the data must be shared amongst multiple parties due to its decentralized nature. Farmers may not want to share data on things as sensitive water use with third parties poses important questions about ownership and control of such data (Singh & Kaunert, 2024). Blockchain technology provides several opportunities for agriculture by promoting water management transparency and practices efficiency. Yet, key determinants for large-scale deployment will be combating infrastructure issues and providing affordable solutions while integrating data privacy measures. In a modern day world with evolving digital technologies, there is also hope for blockchain to be used as an instrument that could foster more sustainable farming and create the fair distribution of limited resources such as water (Farooq et al., 2022).

5. INTERNET OF AGRICULTURAL THINGS (IOAT): REVOLUTIONIZING FARMING

For the agriculture industry, innovation in farming practices and sustainability has been aided by IoT- enabled technology of The Internet of Agricultural Things (IoAT). With the ability to collect and evaluate data in real time, IoAT provides insights that allow farmers to make more informed decisions which will enhance productivity while optimizing outputs. Read more for the role of IoAT in groundwater monitoring, precision farming, water efficiency etc. IoAT includes a variety of technologies used to link farm machinery, sensors and data analytics platforms (Voulgaris et al., 2019). Those multiple device integrated products gather critical information about soil conditions, the health of crops, weather patterns and water availability. Through this data, farmers are able to take a real-time look at their fields and make more informed decisions leading to increase in crop yield while minimizing environmental degradation. This is a substantial change to make farming more intelligent, as the IoAT can now be integrated in agricultural processes (Kaur et al., 2023).

5.1 IoAT Ground-water Monitoring Sensors

Groundwater Shortage is a challenge for agriculture, in particular in arid and semi-arid zones. The sensors from IoAT are essential for the monitoring of groundwater accumulations and providing accurate information on water levels, quality, and flow rates. Adopting these sensors helps farmers to monitor changes in soil moisture and allows them to change the practices of irrigation accordingly. As the ground water levels are continuously monitored and controlled, IoAT contributes to avoid over extraction of water from earthly reserves making conservation a reality (Gupta et al., 2020).

5.2 IoAT Integrated Precision Farming

Precision farming is all about maximizing yields with minimum resources wastage; IoAT significantly extends the potential of precision farming. Using IoAT Farming technologies they can easily know the amount of water available in soil and number of nutrients level are there, Crop status is healthy or not. This data opens the way to tailored water and nutrient delivery as well pest prevention strategies such that each crop has everything it needs without excess of anything. The outcome is an agriculture that requires fewer resources and has a better yield with less environmental impact. Wireless communication is the foundation of IoT, facilitating smooth data flow between numerous sensors and devices. This enables automatic changes in irrigation schedules due to factors such as weather events or soil moisture levels. The providing of farmers with real-time data, they can improve their irrigation practices to reduce water wastage and make sure that crops are receiving the necessary moisture (Siwach et al., 2022).

5.3 Water Require Effective Use through IoAT

It demonstrated the critical role of IoAT in increasing water use efficiency for agricultural purposes. Farmers can use data collected from sensors and other devices to guide irrigation strategies that are more directly matched with the needs of a given crop. The result drastic cuts to water waste and reductions in energy costs, leading to a more sustainable agriculture (Rudrakar & Rughani, 2022). With embedding IoAT in agricultural ecosystems, food production is yielded that can withstand the trials of climate change and water scarcity a crucial step for our future planet (Rakhee & Elayidom, 2023).

6. INTEGRATION OF BLOCKCHAIN AND IOAT IN SMART IRRIGATION

When combined with the Internet of Agricultural Things (IoAT) and Blockchain, new benefits arise that could change current smart irrigation systems. This will provide improved resource management to farmers, especially in relation of the use of groundwater because we are putting together a decentralized and transparent blockchain that combines with real-time data operating on IoAT. This synergy allows for both sustainable agricultural practices and higher operational efficiency. A blockchain is a decentralized and transparent ledger that improves the traceability of transactions or data. It can enhance the trustworthiness of data that is being collected from all types sensors, some even deployed in field when integrated with IoAT (Kumari & Sharma, 2023). Alternatively, IoAT sensors track soil moisture, weather conditions and crop health the data of which can be stored securely on a blockchain. This integration also provides an immutable ledger of environmental conditions and resource utilization, giving farmers and stakeholders access to reliable objective data. This allows both parties to have an understanding of the irrigation activity and water use so that more informed decisions can be made for better resource availability and less waste.

6.1 Blockchain-based on IoAT Solutions in Groundwater Management

Blockchain Usage: It gives a transparent interface for locating and governing water usage so it can be used in the management of groundwater (Akhtar et al., 2021). Through the use of smart contracts a self-executing contract with the terms directly written into code farmers are able to automate and optimize irrigation schedules by connecting these IoAT sensors to their land. For instance, a smart contract can be designed to trigger irrigation based on the detection of soil moisture levels maximizing efficiency and efficacy with every drop (Flores et al., 2021) . The use of sophisticated automation helps to reduce the risk of over-irrigation and leads us on a path towards sustainable groundwater resource management. The blockchain can also enable other branches in the agricultural supply chain to work together. Farmers, water authorities and agricultural experts can then monitor the groundwater resources better by using a platform where they all have access. By working together in this way, we can address the challenges of resource rights and allocation around water to ensure a fair distribution that is more resilient (Mistry et al., 2020).

6.3 Integrated System in Data Privacy and Security

The data's security and privacy are essential, as we move into the more connected world. By integrating blockchain with IoAT, the security of agriculture data is further improved and secured from unauthorized access or alteration (Gaba et al., 2023). Climate-resilient solutions use blockchain technology to protect farmers' water, soil and crop health data through encryption and decentralized storage. For farmers who are worried about competitive advantage or regulatory scrutiny it matters all the more. Blockchain transparency enables stakeholders to trust the data they receive while ensuring that individual privacy is maintained if everybody can see and verify near real-time what sort of information receives or has received. Farmers might select specific data sets they're comfortable sharing to relevant parties, and retain control of their information. Consequently, it is an important balance of security and transparency for enabling smart irrigation practices to foster collaboration in innovation (Singh, 2023).

6.4 Applying Blockchain-IoAT to Reduce Water Waste

Using real-time data monitoring, farmers have a better understanding to their irrigation practices and areas of inefficiencies. Using the IoAT technology, farmers obtain data from sensors placed throughout fields that let them know water is not needed in certain areas and that they can preserve larger swaths of land by updating irrigation approaches (Ahamed et al., 2023). The automation of smart contracts allows water irrigation schedules to be followed precisely as per need and not routine. Especially in times of drought or water scarcity, this adaptability is very helpful and every drop literally counts. With the power of both blockchain and IoAT combined, in particular to water systems created by farmers using rainwater sources individually cured for each farm can create irrigation systems that are not only sustainable or conserve but also offer security as part of their long-term agriculture solution (Guven & Parlak, 2022).

The combination of blockchain and IoAT in irrigation systems is a great step forward for water resources control. Through the unique attributes of both technologies, farmers can utilize irrigation practices more efficiently and secure their data facilitating collaboration among key players thus laying a robust foundation for an upcoming sustainable agricultural future.

7. GROUNDWATER SCARCITY THROUGH SMART SOLUTIONS: GROUNDWATER MONITORING VIA IOAT SENSORS

The challenge of rapidly depleting groundwater, which leads to a decline in agricultural productivity and sustainability particularly problematic for regions where irrigated agriculture is one of few choices. The Internet of Agricultural Things (IoAT) is one such smart solution to tackle this growing concern and has brought in advanced sensor technologies that provide an efficient way for groundwater monitoring. Using real-time data from IoAT sensors, this technology could notify farmers and water managers about fluctuations in groundwater levels/quality to promote conscientious irrigation practices (Ravindran, 2022).

With the strategic deployment of sensors across agricultural fields, IoAT allows stakeholders to keep a close eye on soil moisture levels and any changes in ground water availability as well as enabling them to measure the impact of irrigation practices over time (Strobel, 2020). Monitoring soil moisture, using sensors and taking important weather data into account our method is designed to help farmers perfectly adjust their irrigation schedules so that just the right amount of water gets delivered where it needs with least waste (Vangipuram et al., 2022). If the level of soil moisture stays healthy, then water stripped from underground doesn't need to be used, preserving an invaluable source. Through the use of its sensors IoAT can detect abnormalities which may be due to pollution or salinity in water that if not identified at an early stage would lead to a major problem. This way of early warning not only increases the reliability but also supports interventions that can lead to a sustainable groundwater management.

IoAT sensors are extremely beneficial in tackling groundwater prolongation by allowing citable monitoring and management of water balance. These are standout smart solutions that enable sustainable irrigation practices and save water, while ensuring the sustenance of underground freshwater sources to drive forward a more reliable agricultural system in light of growing water scarcity trends. IoAT based sensors integrated with data analytics platforms supplement decision support by creating situational context using historical trends and patterns related to groundwater resources. These data allows farmers to design practices that change with the weather and maintain water storage for generations (Chattopadhyay et al., 2022).

8. CASE STUDIES- SPACE-AGE TECHNOLOGIES IN SUSTAINABLE FARMING: DRONES, SATELLITE IMAGING AND REMOTE SENSING FOR WATER MONITORING

The high-flying drones and satellite-imaging technology: farmers are gearing up for space-age tech in sustainable farming practices transitioning to utilizing high-tech tools such as drone monitoring, satellite imaging with remote sensing systems for water monitoring (Lamtzidis & Gialelis, 2018). Space-age technologies such as drones, satellite imaging and remote sensing have transformed the way it approach farming sustainably offering better pain-points in terms of monitoring and managing water. The data produced by these technologies allow farmers to decide what when, where, how much to irrigate; which areas of their farm need more or less water application and nutrient supply all assuring sustainable agricultural practices (Bera et al., 2023).

Drones with high-resolution cameras and sensors provide real time surveillance of these fields helping the providers to do an effective job in case pests are detected. For example, they produce detailed pictures from which crop health, soil moisture and irrigation patterns can be measured. In essence, by analyzing this imagery farmers are able to perform health checks of their crop finding areas that need assistance (drought-stressed crops or over-watered regions) and adjusting them accordingly (Dinc et al., 2019). Targeted delivery means literal pinpoint irrigation, making water use more economical and less wasteful which is an important consideration where freshwater resources are stretched to their limit.

Satellites enhance drone technology by providing larger area, more detailed knowledge over broad regional scales. Multispectral and hyperspecteral sensors on satellites can detect soil moisture, vegetation health and other phenomena from space. This information is critical to evaluate the groundwater availability and effects of various agricultural practices on water resources. This information can be used by farmers in real time and help them adapt their farming practices so that water is utilized effectively while sustaining ably (Bodkhe et al., 2020). Remote sensing tools provide additional information about climatic aspects such as rainfall and temperature profiles which supplement water monitoring applications. Incorporating this data into predictive models, farmers can get an idea of when water is needed and amend their irrigation schedules (Bayih et al., 2022). Understanding how the environment affects water resources is critical to help farmers be proactive in saving this precious commodity.

Drones, satellite imaging and remote sensing these advances mark a significant change in farming towards sustainability (Verma, et al., 2022). Using space-age technologies, farmers are able to efficiently track water resources they can tighten up their irrigation practices ideally leading them away from unsustainable groundwater abstractions, and make informed data-driven decisions that build sustainability and resilience against increasing pressures on global freshwater availability.

9. CHALLENGES AND LIMITATIONS IN IMPLEMENTING SMART IRRIGATION

The need to solve water scarcity and escalate agricultural sustainability requires deploying smart irrigation systems leveraging cutting-edge technologies like the Internet of Agricultural Things (IoAT) acting together with blockchain. On the other hand, issues also exist with respect to some challenges and limits that impair good representation of these novelties (Ferrag & Shu, 2021).

9.1 Technological Barriers

The major technology challenges to overcome for smart irrigation is system complexity. Integrating multiple IoAT devices, sensors and blockchain platforms might be challenging for unskilled farmers. Furthermore, the interoperability between technologies may be difficult to achieve, as traditional irrigation infrastructure might not work with newer generation smart solutions (Phang et al., 2023).

9.2 Expenses Loss in IoAT and Blockchain Adoption

Implementing IoAT and blockchain technologies on numerous farms is costly, which can be a barrier to many farmers making the change particularly smallholders (Bala & Kaur, 2022). For someone who is working with a tight budget, buying all the equipment like sensors, drones and other software solutions can be expensive. Maintenance and software service charges, on the other hand will continue to cost money.

9.3 Data Privacy and Security Issues

Smart irrigation systems are fundamentally dependent on data sharing and collection, which raises concerns about data privacy and security. Farmers are understandably reluctant to take up the technology for fear of potential abuse and unauthorized access of sensitive data, such as water usage patterns or crop health

(Dwivedi et al., 2021). The risk of data breach and misuse triggers concerns regarding security systems in place, as well affecting the farmers' confidence.

9.4 Resistance from Farmers to Adopting Technology

The farmers may simply not be aware of the benefits that theoretically smart irrigation brings. Most traditional farmers may be reluctant to adopt new techniques as they feel these are either complex or of not much use. This resistance can be overcome through education and training programs that emphasize the long-term sustainability of smart irrigation practices (Podder et al., 2021).

9.5 Development of Legal and Regulatory Frameworks

The lack of legal and institutional frameworks regulating applications possibly results in IoAT and blockchain that might obstruct development (Chauhan & Dwivedi, 2022). Farmers must be assured that such technologies meet the necessary data, water-rights and environmental regulations. Having robust policies in place may drive smarter irrigation systems into safe and effective operational frameworks (Namdech et al., 2025). It can be concluded that although smart irrigation has the potential for sustainable agriculture there is a need to address technological challenges, cost-effectiveness and other issues like data privacy related costs, change resistance by farmers and regulations which act as a bottleneck in its implementation (Vilas Boas et al., 2023).

Figure 3. Challenges and limitations in implementing smart irrigation (Original)

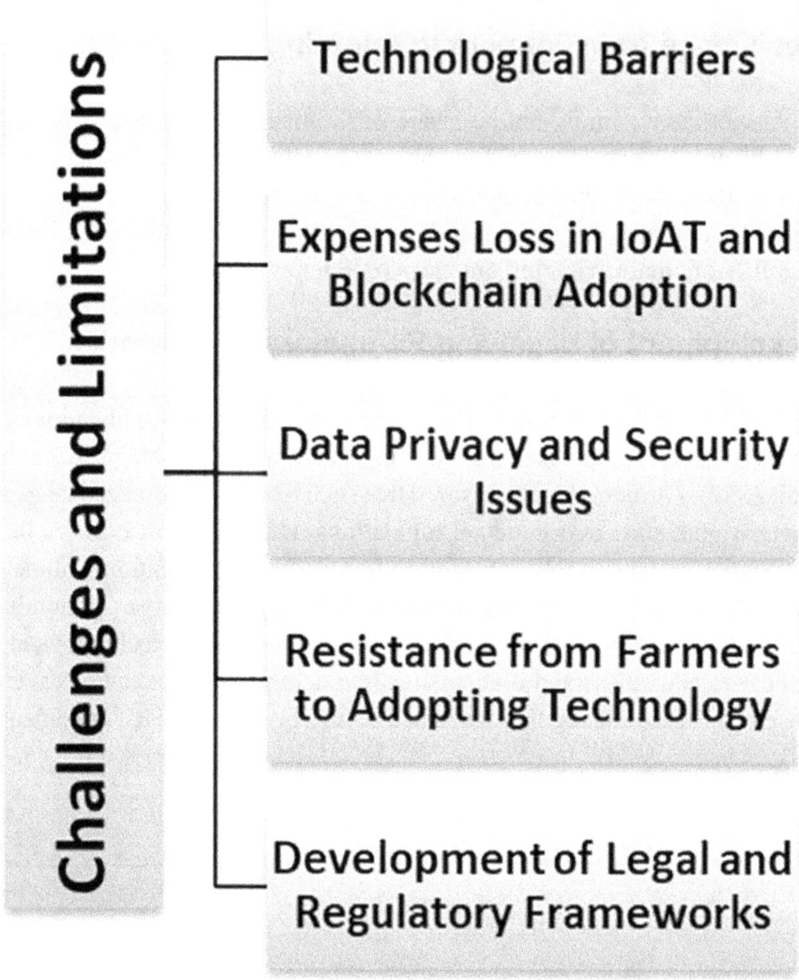

10. CONCLUSION AND FUTURE SCOPE

The reduction of agricultural yields under conditions where groundwater resources need to be mobilized is also affecting food security and the economy. Lower income loss from lower crop yields is a follow-on effect that can ultimately affect local economies with agriculture as the main source of employment and economic activity. Water shortages also deepen social inequalities as only richer farmers have resources and are able to invest in technology which reduces water consumption or entitles them for new sources of support, while poor part has no solutions how else they can irrigate their harvest.

In farming dependent regions however, water scarcity can escalate migration and land is unable to produce forcing the farmers into alternative employment in urban areas or abroad. This disrupts rural communities, displaces the accumulated knowledge of generations in traditional agriculture and impoverishes those who most require our support. Socio-economic aspects of groundwater depletion point to the imperative for sustainable water management practices to guarantee farming and rural livelihoods in times that lie ahead. Groundwater scarcity is a profound and enduring issue impacting agriculture worldwide with wide-reaching ecological, societal, and economic outcomes. This can be effectively addressed by devising a harmonized approach through updated irrigation technology, prevention of over-extraction from groundwater via sustainable water table management and developing policies for equitable access to the resource. Through innovative technologies such as smart irrigation systems and the utilization of blockchain, farmers can help reduce Groundwater depletion and transition towards more resilient, sustainable agriculture methods.

REFERENCES

Ahamed, N. N., Karthikeyan, P., Velliangiri, S., & Gunjan, V. K. (2023). Blockchain Security for the Internet of Things. In *Modern Approaches in IoT and Machine Learning for Cyber Security: Latest Trends in AI* (pp. 75–92). Springer International Publishing.

Akhtar, M. M., Rizvi, D. R., Ahad, M. A., Kanhere, S. S., Amjad, M., & Coviello, G. (2021). Efficient data communication using distributed ledger technology and iota-enabled internet of things for a future machine-to-machine economy. *Sensors (Basel), 21*(13), 4354. DOI: 10.3390/s21134354 PMID: 34202173

Akram, S. V., Malik, P. K., Singh, R., Anita, G., & Tanwar, S. (2020). Adoption of blockchain technology in various realms: Opportunities and challenges. *Security and Privacy, 3*(5), e109. DOI: 10.1002/spy2.109

Al-Shareeda, M., Ali, M., & Manickam, S. (2023). The blockchain internet of things: Review, opportunities, challenges, and recommendations. *Indonesian Journal of Electrical Engineering and Computer Science, 31*(3), 1673–1683. DOI: 10.11591/ijeecs.v31.i3.pp1673-1683

Alzoubi, Y. I., Al-Ahmad, A., Kahtan, H., & Jaradat, A. (2022). Internet of things and blockchain integration: Security, privacy, technical, and design challenges. *Future Internet, 14*(7), 216. DOI: 10.3390/fi14070216

Amoah, E., & Oh, J. Y. (2020). BLOCKCHAIN IN IoT AND PROJECT MANAGEMENT. *Issues in Information Systems, 21*(3).

Bala, K., & Kaur, P. D. (2022, January). Changing trends of blockchain in IoT: Benefits and challenges. In *2022 12th International Conference on Cloud Computing, Data Science & Engineering (Confluence)* (pp. 324-329). IEEE.

Bathalapalli, V. K., Mohanty, S. P., Kougianos, E., Yanambaka, V. P., Baniya, B. K., & Rout, B. (2021, December). A puf-based approach for sustainable cybersecurity in smart agriculture. In *2021 19th OITS International Conference on Information Technology (OCIT)* (pp. 375-380). IEEE. DOI: 10.1109/OCIT53463.2021.00080

Bayih, A. Z., Morales, J., Assabie, Y., & de By, R. A. (2022). Utilization of internet of things and wireless sensor networks for sustainable smallholder agriculture. *Sensors (Basel), 22*(9), 3273. DOI: 10.3390/s22093273 PMID: 35590963

Bera, S., Dey, T., Mukherjee, A., & Buyya, R. (2023). E-CropReco: A dew-edge-based multi-parametric crop recommendation framework for internet of agricultural things. *The Journal of Supercomputing, 79*(11), 1–35. DOI: 10.1007/s11227-023-05131-4

Bhushan, B., Sahoo, C., Sinha, P., & Khamparia, A. (2021). Unification of Blockchain and Internet of Things (BIoT): Requirements, working model, challenges and future directions. *Wireless Networks, 27*(1), 55–90. DOI: 10.1007/s11276-020-02445-6

Bodkhe, U., Tanwar, S., Parekh, K., Khanpara, P., Tyagi, S., Kumar, N., & Alazab, M. (2020). Blockchain for industry 4.0: A comprehensive review. *IEEE Access : Practical Innovations, Open Solutions, 8*, 79764–79800. DOI: 10.1109/ACCESS.2020.2988579

Chattopadhyay, P., Patel, H. P., & Parmar, V. (2022, August). Internet of Things (IoT) in Smart Agriculture. In *2022 3rd International Conference on Electronics and Sustainable Communication Systems (ICESC)* (pp. 536-540). IEEE.

Chauhan, N., & Dwivedi, R. K. (2022, April). Designing a secure smart healthcare system with blockchain: a review. In *2022 1st International Conference on Informatics (ICI)* (pp. 214-219). IEEE. DOI: 10.1109/ICI53355.2022.9786895

Chen, F., Xiao, Z., Cui, L., Lin, Q., Li, J., & Yu, S. (2020). Blockchain for Internet of Things applications: A review and open issues. *Journal of Network and Computer Applications, 172*, 102839. DOI: 10.1016/j.jnca.2020.102839

Dinc, E., Kuscu, M., Bilgin, B. A., & Akan, O. B. (2019). Internet of everything: A unifying framework beyond internet of things. In *Harnessing the Internet of Everything (IoE) for Accelerated Innovation Opportunities* (pp. 1–30). IGI Global. DOI: 10.4018/978-1-5225-7332-6.ch001

Dinc, E., Kuscu, M., Bilgin, B. A., & Akan, O. B. (2019). Internet of everything: A unifying framework beyond internet of things. In *Harnessing the Internet of Everything (IoE) for Accelerated Innovation Opportunities* (pp. 1–30). IGI Global. DOI: 10.4018/978-1-5225-7332-6.ch001

Dwivedi, S. K., Roy, P., Karda, C., Agrawal, S., & Amin, R. (2021). Blockchain-based internet of things and industrial IoT: A comprehensive survey. *Security and Communication Networks, 2021*, 1–21. DOI: 10.1155/2021/7142048

Farooq, M. S., Javid, R., Riaz, S., & Atal, Z. (2022). Internet of Things (IoT) Technologies in Greenhouse Farming: A Systematic Literature Review. *IEEE Access : Practical Innovations, Open Solutions.*

Fernández-Caramés, T. M., & Fraga-Lamas, P. (2018). A Review on the Use of Blockchain for the Internet of Things. *IEEE Access : Practical Innovations, Open Solutions, 6*, 32979–33001. DOI: 10.1109/ACCESS.2018.2842685

Ferrag, M. A., & Shu, L. (2021). The performance evaluation of blockchain-based security and privacy systems for the Internet of Things: A tutorial. *IEEE Internet of Things Journal, 8*(24), 17236–17260. DOI: 10.1109/JIOT.2021.3078072

Flores, A., Morales, A., Campos, G., & Gelso, J. (2021, December). Energy Efficiency Using IOTA Tangle for Greenhouse Agriculture. In *Annual International Conference on Information Management and Big Data* (pp. 122-138). Cham: Springer International Publishing.

Gaba, S., Khan, H., Almalki, K. J., Jabbari, A., Budhiraja, I., Kumar, V., Singh, A., Singh, K. K., Askar, S. S., & Abouhawwash, M. (2023). Holochain: An agent-centric distributed hash table security in smart iot applications. *IEEE Access : Practical Innovations, Open Solutions, 11*, 81205–81223. DOI: 10.1109/ACCESS.2023.3300220

George, G. M., & Jayashree, L. S. (2022). Fusion of Blockchain-IoT network to improve supply chain traceability using Ethermint Smart chain: A Review. *KSII Transactions on Internet and Information Systems, 16*, 3694–3722.

Gupta, D., Bhatt, P., & Bhatt, S. (2020, December). A game theoretic analysis for cooperative smart farming. In *2020 IEEE International Conference on Big Data (Big Data)* (pp. 2350-2359). IEEE. DOI: 10.1109/BigData50022.2020.9377935

Guven, I., & Parlak, M. (2022, November). Blockchain, AI and IoT Empowered Swarm Drones for Precision Agriculture Applications. In *2022 IEEE 1st Global Emerging Technology Blockchain Forum: Blockchain & Beyond (iGETblockchain)* (pp. 1-6). IEEE.

Haque, M. A., Haque, S., Zeba, S., Kumar, K., Ahmad, S., Rahman, M., & Ahmed, L. (2023). Sustainable and efficient E-learning internet of things system through blockchain technology. *E-Learning and Digital Media*, •••, 20427530231156711.

Hassija, V., Zeadally, S., Jain, I., Tahiliani, A., Chamola, V., & Gupta, S. (2021). Framework for determining the suitability of blockchain: Criteria and issues to consider. *Transactions on Emerging Telecommunications Technologies, 32*(10), e4334. DOI: 10.1002/ett.4334

Karlsson, K., Jiang, W., Wicker, S., Adams, D., Ma, E., van Renesse, R., & Weatherspoon, H. (2018, July). Vegvisir: A partition-tolerant blockchain for the internet-of-things. In *2018 IEEE 38th International Conference on Distributed Computing Systems (ICDCS)* (pp. 1150-1158). IEEE. DOI: 10.1109/ICDCS.2018.00114

Kathar, S., Hardel, H., Alam, Z., Phansalkar, S., & Sajjan, R. (2021, December). Auction System for Agricultural Trade Using Blockchain Technology: A Survey with Proposed Framework. In *2021 IEEE Pune Section International Conference (PuneCon)* (pp. 1-10). IEEE. DOI: 10.1109/PuneCon52575.2021.9686543

Kaur, U., Kaur, M., & Khang, A. (2023). Futuristic Technologies in Agriculture Challenges and Future Prospects. *Handbook of Research on AI-Equipped IoT Applications in High-Tech Agriculture*, 347-364.

Kose, U., Prasath, V. S., Mondal, M. R. H., Podder, P., & Bharati, S. (Eds.). (2022). *Artificial Intelligence and Smart Agriculture Technology*. CRC Press. DOI: 10.1201/9781003299059

Kumari, A., & Sharma, I. (2023, October). Augmentation of Internet of Things Security and Privacy with IOTA Tangle Network. In *2023 4th IEEE Global Conference for Advancement in Technology (GCAT)* (pp. 1-5). IEEE. DOI: 10.1109/GCAT59970.2023.10353394

Kumari, S., & Muthulakshmi, P. (2023). Artificial Intelligence—Blockchain Enabled Technology for Internet of Things: Research Statements, Open Issues, and Possible Applications in the Near Future. *Privacy Preservation of Genomic and Medical Data*, 433-480.

Lamtzidis, O., & Gialelis, J. (2018, December). An IOTA based distributed sensor node system. In *2018 IEEE Globecom Workshops (GC Wkshps)* (pp. 1-6). IEEE. DOI: 10.1109/GLOCOMW.2018.8644153

Latif, S., Idrees, Z., Huma, Z., & Ahmad, J. (2021). Blockchain technology for the industrial Internet of Things: A comprehensive survey on security challenges, architectures, applications, and future research directions. *Transactions on Emerging Telecommunications Technologies*, *32*(11), e4337. DOI: 10.1002/ett.4337

. Leduc, G., Kubler, S., & Georges, J. P. U,"Innovative Blockchain-based farming marketplace and smart contract performance evaluation,". *J of C. Production, Elsevier*.

Mistry, I., Tanwar, S., Tyagi, S., & Kumar, N. (2020). Blockchain for 5G-enabled IoT for industrial automation: A systematic review, solutions, and challenges. *Mechanical Systems and Signal Processing*, *135*, 106382. DOI: 10.1016/j.ymssp.2019.106382

Mitra, A., Singhal, A., Mohanty, S. P., Kougianos, E., & Ray, C. (2022). eCrop: A novel framework for automatic crop damage estimation in smart agriculture. *SN Computer Science*, *3*(4), 319. DOI: 10.1007/s42979-022-01216-8

Monteiro, E. S., da Rosa Righi, R., Barbosa, J. L. V., & Alberti, A. M. (2021). APTM: A model for pervasive traceability of agrochemicals. *Applied Sciences (Basel, Switzerland)*, *11*(17), 8149. DOI: 10.3390/app11178149

Namdech, P., Rotjanawasuthorn, S., Singh, B., & Kaunert, C. (2025). Harnessing Artificial Intelligence For Lifelong Education: Opportunities And Challenges. *Asian Education and Learning Review*, *3*(2), 4–4.

Ordóñez, J., Alexopoulos, A., Koutras, K., Kalogeras, A., Stefanidis, K., & Martos, V. (2023). Blockchain in Agriculture: A PESTELS Analysis. *IEEE Access : Practical Innovations, Open Solutions*.

Panda, S. S., Mohanta, B. K., Dey, M. R., Satapathy, U., & Jena, D. (2020, July). Distributed ledger technology for securing IoT. In *2020 11th international conference on computing, communication and networking technologies (ICCCNT)* (pp. 1-6). IEEE. DOI: 10.1109/ICCCNT49239.2020.9225333

Phang, S. K., Chiang, T. H. A., Happonen, A., & Chang, M. M. L. (2023). From Satellite to UAV-Based Remote Sensing: A Review on Precision Agriculture. *IEEE Access : Practical Innovations, Open Solutions*, *11*, 127057–127076. DOI: 10.1109/ACCESS.2023.3330886

Pikatza, A., Sainz, N., González, I., & Emaldi, M. (2023, July). Development of an Open Source IoT-Blockchain Platform for Traceability of Fresh Products from Farm to Fork. In *International Congress on Blockchain and Applications* (pp. 487-497). Cham: Springer Nature Switzerland. DOI: 10.1007/978-3-031-45155-3_46

Podder, A. K., Al Bukhari, A., Islam, S., Mia, S., Mohammed, M. A., Kumar, N. M., & Abdulkareem, K. H. (2021). IoT based smart agrotech system for verification of Urban farming parameters. *Microprocessors and Microsystems*, *82*, 104025. DOI: 10.1016/j.micpro.2021.104025

Prakash, C., Singh, L. P., Gupta, A., & Lohan, S. K. (2023). Advancements in smart farming: A comprehensive review of IoT, wireless communication, sensors, and hardware for agricultural automation. *Sensors and Actuators. A, Physical*, *362*, 114605. DOI: 10.1016/j.sna.2023.114605

Pullo, S., Pareschi, R., Piantadosi, V., Salzano, F., & Carlini, R. (2023, December). Integrating IOTA's Tangle with the Internet of Things for Sustainable Agriculture: A Proof-of-Concept Study on Rice Cultivation. [). MDPI.]. *Informatics (MDPI)*, *11*(1), 3. DOI: 10.3390/informatics11010003

Radeva, I., & Popchev, I. (2022). Blockchain-Enabled Supply-Chain in Crop Production Framework. *Cybernetics and Information Technologies*, *22*(1), 151–170. DOI: 10.2478/cait-2022-0010

Rakhee, M., & Elayidom, S. (2023, January). Blockchain Technology as a Solution for Vulnerabilities in Internet of Things-A Survey. In *2023 Advanced Computing and Communication Technologies for High Performance Applications (ACCTHPA)* (pp. 1-6). IEEE. DOI: 10.1109/ACCTHPA57160.2023.10083337

Ravindran, M. (2022). Energy Efficient Clustering for Precision Agriculture using LEACH Protocol. *NeuroQuantology : An Interdisciplinary Journal of Neuroscience and Quantum Physics*, *20*(7), 2702.

Rudrakar, S., & Rughani, P. (2022). *IoT based Agriculture (IoTA)*. Architecture, Cyber Attack, Cyber Crime and Digital Forensics Challenges.

Sendros, A., Drosatos, G., Efraimidis, P. S., & Tsirliganis, N. C. (2022). Blockchain applications in agriculture: A scoping review. *Applied Sciences (Basel, Switzerland)*, *12*(16), 8061. DOI: 10.3390/app12168061

Sengupta, A., Gill, S. S., Das, A., & De, D. (2021). Mobile edge computing based internet of agricultural things: a systematic review and future directions. *Mobile Edge Computing*, 415-441.

Sharma, A., & Singh, B. (2022). Measuring Impact of E-commerce on Small Scale Business: A Systematic Review. *Journal of Corporate Governance and International Business Law*, *5*(1).

Shi, X., An, X., Zhao, Q., Liu, H., Xia, L., Sun, X., & Guo, Y. (2019). State-of-the-art internet of things in protected agriculture. *Sensors (Basel)*, *19*(8), 1833. DOI: 10.3390/s19081833 PMID: 30999637

Singh, B. (2023). Blockchain Technology in Renovating Healthcare: Legal and Future Perspectives. In *Revolutionizing Healthcare Through Artificial Intelligence and Internet of Things Applications* (pp. 177-186). IGI Global.

Singh, B. (2023). Federated Learning for Envision Future Trajectory Smart Transport System for Climate Preservation and Smart Green Planet: Insights into Global Governance and SDG-9 (Industry, Innovation and Infrastructure). *National Journal of Environmental Law*, *6*(2), 6–17.

Singh, B., & Kaunert, C. (2024). Integration of Cutting-Edge Technologies such as Internet of Things (IoT) and 5G in Health Monitoring Systems: A Comprehensive Legal Analysis and Futuristic Outcomes. *GLS Law Journal*, *6*(1), 13–20. DOI: 10.69974/glslawjournal.v6i1.123

. Singh, P., Kumar, A., & Chopra, M. (2023). Real-World Applications of BC Technology in Internet of Things. *Machine Learning Applications: From Computer Vision to Robotics*, 97-122.

Siwach, V., Sehrawat, H., & Dahiya, O. (2022, October). Hybrid Blockchain Security Mechanism for protection of data related to Precision Agriculture. In *2022 10th International Conference on Reliability, Infocom Technologies and Optimization (Trends and Future Directions)(ICRITO)* (pp. 1-7). IEEE.

. Strobel, G. (2020). Agriculture in the Era of the Internet of Things: An Information System Architecture for Smart Agriculture.

Tanwar, R., Chhabra, Y., Rattan, P., & Rani, S. (2022, September). Blockchain in IoT Networks for Precision Agriculture. In *International Conference on Innovative Computing and Communications* [Singapore: Springer Nature Singapore.]. *Proceedings of ICICC*, *2*, 137–147.

Udutalapally, V., Mohanty, S. P., Pallagani, V., & Khandelwal, V. (2020). sCrop: A novel device for sustainable automatic disease prediction, crop selection, and irrigation in Internet-of-Agro-Things for smart agriculture. *IEEE Sensors Journal*, *21*(16), 17525–17538. DOI: 10.1109/JSEN.2020.3032438

Upadhyaya, A., Singh, Y., & Anand, P. (2022, May). Blockchain for IoT: A Comprehensive Review for Precision Agricultural Networks. In *The International Conference on Recent Innovations in Computing* (pp. 787-802). Singapore: Springer Nature Singapore.

Vangipuram, S. L., Mohanty, S. P., & Kougianos, E. (2023, October). CroPAiD: Protection of Information in Agriculture Cyber-Physical Systems Using Distributed Storage and Ledger. In *IFIP International Internet of Things Conference* (pp. 375-394). Cham: Springer Nature Switzerland.

Vangipuram, S. L., Mohanty, S. P., Kougianos, E., & Ray, C. (2022). G-DaM: A Distributed Data Storage with Blockchain Framework for Management of Groundwater Quality Data. *Sensors (Basel)*, *22*(22), 8725. DOI: 10.3390/s22228725 PMID: 36433322

Verma, A., Bhattacharya, P., Madhani, N., Trivedi, C., Bhushan, B., Tanwar, S., Sharma, G., Bokoro, P. N., & Sharma, R. (2022). Blockchain for industry 5.0: Vision, opportunities, key enablers, and future directions. *IEEE Access : Practical Innovations, Open Solutions*, *10*, 69160–69199. DOI: 10.1109/ACCESS.2022.3186892

Vilas-Boas, J. L., Rodrigues, J. J., & Alberti, A. M. (2023). Convergence of Distributed Ledger Technologies with Digital Twins, IoT, and AI for fresh food logistics: Challenges and opportunities. *Journal of Industrial Information Integration, 31*, 100393. DOI: 10.1016/j.jii.2022.100393

Voulgaris, S., Fotiou, N., Siris, V. A., Polyzos, G. C., Jaatinen, M., & Oikonomidis, Y. (2019). Blockchain technology for intelligent environments. *Future Internet*, *11*(10), 213. DOI: 10.3390/fi11100213

Wahid, A., Juliady, I., Zain, S. G., & Parenreng, J. M. (2022). Secure Wireless Sensor Network using Cryptography for Smart Farming Systems. *Internet of Things and Artificial Intelligence Journal*, *2*(4), 248–262. DOI: 10.31763/iota.v2i4.554

Chapter 12
YOLO-based CCTV Surveillance Advanced Techniques for Cheating Detection

Mizanul Ridho Aohana
https://orcid.org/0009-0006-2228-6718
University of Mataram, Indonesia

Fitri Bimantoro
https://orcid.org/0000-0003-1813-0279
University of Mataram, Indonesia

I. Gede Pasek Suta Wijaya
https://orcid.org/0000-0002-3813-013X
University of Mataram, Indonesia

ABSTRACT

Digitalization has become essential for most educational institutions to measure students' knowledge and abilities through examinations. However, cheating remains a persistent issue, with students using various ways such as exchanging answer sheets, using hidden notes, or giving codes. Human monitoring is often inconsistent and limited by focus. This research proposes a solution for solving the cheating detection problem using computer vision, specifically by monitoring the suspicious behavior of students (physical) during the exam through CCTV. The method used to solve the problem is the YOLO (You Only Look Once) algorithm, comparing three versions—YOLOv5, YOLOv6, and YOLOv7. In this study, the accuracy results for each algorithm variation are 43%, 37%, and 51%, respectively. The existence of

DOI: 10.4018/979-8-3693-9846-3.ch012

imbalanced classes in the dataset is the main factor that affects the model performance. Consequently, an extended experiment is undertaken by adding 10% data to the imbalanced classes. The highest accuracy recorded is 60%, given by YOLOv7, reflecting a noteworthy 9% increase in accuracy.

1 INTRODUCTION

Technology has been proven to optimize the learning process (Cai & King, 2020) and has been shown to improve the quality of student understanding (Al-Labadi & Sant, 2021). In addition, teaching staff can also improve the learning experience provided with the help of technology (Alahmari, 2023). An examination is one of the essential aspects in educational institutions to measure how far the ability and knowledge of students related to the material that has been given (Sletten, 2021). Written exams (Cambron-Goulet et al., 2019) are among the most widely used methods (Ellis et al., 2023). In this modern era, there are many ways to automate traditional exams (Hoque et al., 2020), including technological assistance (Ferdig et al., 2020).

The main thing that needs to be considered during the exam is the integrity of students in taking the exam (Comas-Forgas et al., 2021). In more specific terms, academic integrity is described as "a dedication to six core values: honesty, trust, fairness, respect, responsibility, and courage... these fundamental values serve to guide and enhance ethical decision-making abilities and behavior." (https://academicintegrity.org/, 2021). Integrity is influenced by various factors (Ossai et al., 2023) that can affect personality, especially the habituation of the learning environment at home (Hutton et al., 2020).

However, in the implementation of the exam, there are often indications of cheating that disrupt the level of academic integrity (Janke et al., 2021). Cheating is a dishonest behavior exhibited by students when completing assignments or examinations (Septiana, 2022). In most cases, this behavior is typically driven by a low sense of self-efficacy (Fadillah, 2019). Cheating can be done using various methods (Zulkifli N, 2022), which aim to increase exam scores (Chen et al., 2020) and avoid failing the exam (Ifeagwazi et al., 2019). This is undoubtedly very dangerous in educational institutions because it affects the quality of student's outcome (Siddhpura & Siddhpura, 2020).

Computer vision has been widely applied across various fields, including education, where its integration can enhance learning experiences and improve safety measures (Mahajan et al., 2020). Installing Closed Circuit Television (CCTV) in classrooms not only serves security purposes but also facilitates the application of advanced computer vision algorithms, enabling real-time monitoring and analysis of classroom activities (Nishchal et al., 2020).

However, the implementation of this technology also raises critical ethical concerns, particularly in relation to privacy and personal data protection. Video recordings collected during examination monitoring may contain sensitive personal information, which necessitates strict compliance with data protection regulations. In the context of educational institutions, the use of such data must consider legal frameworks such as Indonesia's Personal Data Protection Law (UU PDP Indonesia) (Dewan Perwakilan Rakyat Republik Indonesia, 2022) the European Union's General Data Protection Regulation (GDPR) (European Union Lex, 2016). Accordingly, any CCTV-based examination monitoring system should be developed following the principles of *privacy by design* and *data minimization*, ensuring that only essential data is collected and processed with adequate safeguards.

Among the array of computer vision techniques, one of the most popular algorithms is YOLO (You Only Look Once), known for its speed and accuracy in object detection (Diwan et al., 2023; Wu et al., 2022). The YOLO algorithm has seen diverse applications in numerous domains, including mask detection to ensure compliance with health guidelines, monitoring social distancing during the pandemic, detecting clothing attraction in retail settings, analyzing traffic accidents for safety improvements, and counting the number of vehicles in traffic management systems (Senthilkumar et al., 2021)(Ijjina et al., 2020)(Srinivasan et al., 2020)(Chhadikar et al., 2019). Based on some information from the references above, this algorithm is also very possible to detect student cheating when taking an exam (Nishchal et al., 2020).

The YOLO algorithm has evolved over time, giving rise to multiple variants, each designed to improve performance in object detection tasks (Cong et al., 2023). These variants, such as YOLOv5, YOLOv6, and YOLOv7, offer unique architectural improvements and optimizations, making them increasingly efficient in detecting objects across different types of images and videos. The availability of these variants presents an exciting opportunity for researchers to experiment with and test each version individually, as their performance can differ significantly based on several factors. These factors include the algorithm's underlying architecture, the quality and size of the dataset used for training, and the complexity of the objects or scenes being detected (Institute of Electrical and Electronics Engineers. Bangalore Section & Institute of Electrical and Electronics Engineers, n.d.).

Therefore, this research will examine detailly some YOLO algorithms and compare their performance in identifying student cheating during written exams using CCTV system monitoring. The overall research aims to optimize cheating detection in the exam environment and contribute to developing exam security solutions based on computer vision technology.

2 RELATED WORK

Examinations have an important role in measuring a student's understanding and reasoning, and they also play a crucial role in measuring the success of the learning process in educational institutions. Cheating during exams has become a classic and significant problem for educational institutions. During physical exams, students often use unfair means to pass the exam. Students are always looking for ways to cheat, mainly due to the negligence of invigilators.

Frequently used exam proctoring methods usually rely on the consistency of human proctors, which can be tedious, time-consuming, labor-intensive, and ineffective in preventing students from cheating. Therefore, educational institutions need to streamline these problems by implementing computer vision technology so that supervision can be done automatically.

In recent years, various studies have been conducted to create technologies that can overcome the solution to these problems. The use of deep learning algorithms has proven effectively for detection and classification problems. Previous research (Masud et al., 2022) has made an intelligent exam supervisor system that can detect student cheating based on the movement of examinees. The methods used are Random Forest, Logistic Reg, LSTM, 2DCNN, CNN-LSTM, Conv-LSTM, and CNN-BiGRU. From the comparison of these methods, the highest accuracy was obtained at 97.7% in the Random Forest and LSTM methods.

Previous research (Fang et al., 2020), developed an intelligent monitoring system that uses image recognition monitored live on streaming video from the exam room. The system divides the video frame into predefined candidate seat areas and alerts the invigilator if the limits are exceeded.

In previous research, a model was developed using spatial-temporal features to detect abnormal behaviour from students (Liu et al., 2020). The proposed model can identify movements such as turning around and raising hands with an accuracy rate of 93.3%.

CCTV modelling has also detected various suspicious activities during exams (Adil et al., 2019). The model recognizes activities such as hand contact and stealing classmates' exam sheets based on a certain threshold. That project converts video input from a camera (CCTV) into image frames. That study uses a Gaussian Filter to remove background, foreground, and noise for pre-processing. Suspicious activity detection is performed using the Haar-like feature algorithm developed by Viola and Jones. The proposed model successfully identified heads with 70% accuracy, hands with 72% accuracy, and faces with 84% accuracy.

Using the concept of deep learning (Kaddoura & Gumaei, 2022), a cheating detection model was successfully created during online exams using the Deep CNN method. Algorithm implementation was carried out on the front and rear cameras with accuracy achieved 99.83% and 99.81%, respectively.

Another study using CCTV (Nishchal et al., 2020) tried to automatically detect abnormal or cheating activities in the exam using the Open Pose method. The model was developed by detecting the body posture of students during the exam using CCTV. If the number of cases detected exceeds a predetermined threshold, it will be detected as cheating, and a report will be sent to the examiner. The accuracy obtained in that study was 63%.

Trying to dig more deeply (Hussein et al., 2022), there is research using BRISK, HOG, MSER, SURF and SURF&HOG feature extraction to predict five classes of student cheating actions in the classroom. That research resulted in 91% accuracy.

The research discusses the utilization of computer vision methods through CCTV to detect anomalous activities of students during exams. That research centres on monitoring students' suspicious behaviour during physical exams through CCTV cameras. In making the model, that research used YOLOv3 and got an accuracy of 88.03% (Malhotra & Chhabra, 2021).

Research on an electronic cheating detection model that allows supervisors to perform Student/Person Detection and Tracking, Detect suspicious activity, Create alerts, and Mark attendance. Students are detected from videos using YOLOv7, and DeepSort trackers are used to track people detected by the YOLOv7 algorithm. The accuracy obtained in that study is 81.67% (Zaffar et al., 2023).

Based on some of the references described previously, that research will focus on using the YOLO algorithm, as it has been proven to have a pretty good level of accuracy in predicting student activity. The selection of that algorithm is also based on its significant level of existence today in the domain of computer vision problems, especially in the context of object detection and object classification.

3 RESEARCH METHODS

3.1 Proposed Work

Deep learning has many important factors that can affect its success, so it is essential to do careful planning (Ball et al., 2017). Therefore, before starting this research, it is necessary to design a comprehensive process so that the steps can be carried out in a structured manner to achieve optimal results. In the context of this research, there is a model development flow that starts from data collection to model evaluation. The following is a series of stages that we apply.

Figure 1. Proposed work plan

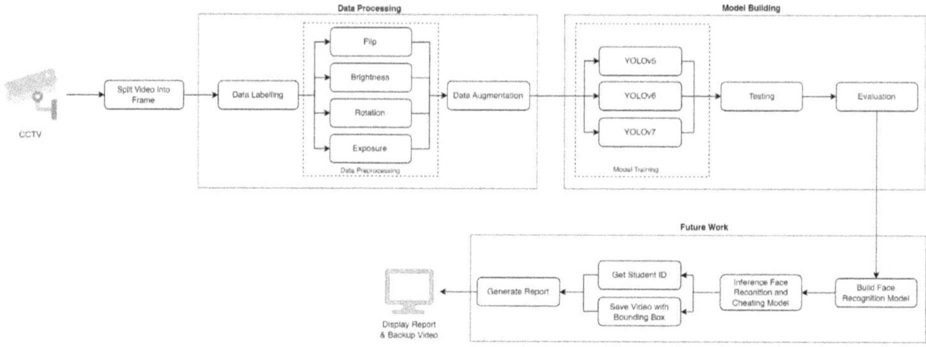

Based on Figure 1, the model development starts with data collection from CCTV and then proceeds with the separation of video into photo frames. The data obtained will be processed through the data preprocessing stage before modelling. Once the dataset has been preprocessed, the research moves into the modeling phase. The proposed model will be developed using the YOLO algorithm, a popular and efficient real-time object detection method. YOLO is chosen for its ability to simultaneously detect and classify objects in a single pass, making it suitable for identifying multiple objects and behaviors in the CCTV frames. After the model is successfully created, the following process includes a testing and evaluation stage to measure the model's performance. At this stage, the model will be tested using data not involved in the training process to ensure its reliability and effectiveness. The results of this evaluation will help refine the model, if necessary, before full implementation. Finally, after the testing and evaluation process is complete, the best model for recognizing problems will be obtained.

In future work, there is potential to develop the system further using face recognition. Through this model, the system can automatically identify students suspected of cheating during exams. Every suspicious gesture or action will be meticulously recorded and stored as part of the exam report. The recorded outcomes will be integrated into the exam report data and presented to the respective course instructors. Consequently, the utilization of facial recognition technology not only enhances security during exams but also provides valuable information to instructors for further evaluation. This development can be a positive step towards creating a fair and trustworthy exam environment, supporting effective exam management, and offering innovative solutions to potential cheating activities.

3.2 Data Collection

The dataset in this research was taken using CCTV from Hikvision (DS-2CD1121-I) with a resolution of 1920x1080, 25 fps in room A3-01, located in Building A, Informatics Engineering Dept., Faculty of Engineering, University of Mataram. The room has a capacity of 35 people and has an area of about 40 m^3.

The dataset used in this study consists of 17 students who voluntarily participated in a controlled simulated exam environment, demonstrating five distinct behaviors: normal, cheating, exchanging papers, giving coded signals, and looking around for assistance from friends. The entire dataset was collected over a span of approximately one hour and thirty minutes, capturing a variety of student actions that simulate real-world examination scenarios.

Figure 2. Dataset collecting

Figure 2 illustrates the process of collecting the dataset, which resulted in high-quality images suitable for training machine learning models. The dataset was recorded using a CCTV camera in video format, capturing the activities of the students throughout the exam simulation. However, to effectively utilize this video-based dataset for further processing and model development, it was necessary to convert the recorded videos into individual frames or images. This step is crucial because most computer vision models, particularly those designed for action recognition or behavior detection, work more effectively with image sequences rather than raw video data.

The video-to-frame extraction process was performed using Python, with the OpenCV library. OpenCV, a widely-used library for computer vision tasks, allows for precise control over the frame extraction process. In this study, frames were extracted at regular intervals of every 7 seconds. The decision to extract frames at this interval was made to balance the need for sufficient data coverage while avoiding redundancy and ensuring the model could capture variations in student behavior effectively.

The extracted frames are then used as the input for the training process of machine learning models. Each frame serves as a static representation of a moment within the video, capturing a student's posture, movements, or interactions that could indicate one of the predefined behaviors (e.g., cheating, exchanging papers). This method of converting video into frames not only simplifies the data structure but also helps enhance the computational efficiency of model training.

Additionally, this frame-based approach enables the use of various image processing and feature extraction techniques, which are crucial for improving the accuracy of the model in recognizing specific behaviors. For instance, techniques such as background subtraction, object detection, or even pose estimation can be applied to each frame to better understand the actions depicted, thus enriching the dataset for more accurate prediction of student behaviors during exams.

3.3 Data Pre-Processing

From the dataset that has been collected, it is necessary to do data preprocessing, such as removing unclear images and increasing the brightness of the image. In this case, the author uses the Roboflow platform to annotate the dataset to separate one label from another. Figure 3 is an example of the annotation result.

Figure 3. Dataset annotation

Figure 3, a dataset has undergone an annotation process. This annotation process involves grouping the data into five different classes, namely Cheat Sheet, Exchange Paper, Giving Code, Looking Friend and Normal. These data were thoroughly annotated, starting from the examinee's head to hands, to show the body gestures the examinee exhibited clearly.

Once the process of annotating the dataset was complete, the authors proceeded to apply several preprocessing steps to enhance the data quality and improve model performance. The preprocessing involved operations such as horizontal flipping, rotation (between -5° to 5°), brightness enhancement (between -25% to 25%), and exposure enhancement (between -10% to 10%) (Setiyono et al., 2021)(Shandilya et al., 2023)(Park et al., 2020).

Horizontal flipping was used to create mirror images of the original images, which helps the model learn to recognize objects from different orientations, ultimately improving its robustness to variations in object positioning.

Rotation within a small angle range of -5° to 5° was implemented to simulate slight variations in viewpoint, mimicking real-world scenarios where CCTV camera (objects) may not always be perfectly aligned. This helps the model generalize better by learning to recognize objects despite minor rotational differences.

Brightness enhancement adjusted the image brightness by up to 25% in both directions, allowing the model to become more resilient to varying lighting conditions that may occur in different environments. Similarly, exposure enhancement within a range of -10% to 10% was applied to account for fluctuations in image exposure, further ensuring that the model can accurately detect objects under various lighting scenarios.

In addition, the dataset was augmented three times to improve the performance of the model in object detection (Zoph Barretand Cubuk, 2020), bringing the total amount of annotated data to 10,252 (Sharma et al., 2020). This data is divided into various classes, with the following distribution: Normal class has 3,975 annotations, Looking Friend has 3,897 annotations, Cheat Sheet has 1,890 annotations, Giving Code has 372 annotations, and the last class, Exchange Paper, has 136 annotations.

3.4 Object Detection

Real-time object detection is essential in computer vision systems (Zaffar et al., 2023). Object detection involves classifying and locating potential objects in an image (Zhao et al., 2019). Many algorithms can be used in object detection, such as Faster R-CNN, SSD, RetinaNet, and YOLO (Zaffar et al., 2023).

Faster R-CNN (Region-Based Convolutional Neural Networks) is one of the most accurate object detection algorithms, known for achieving state-of-the-art results in various tasks. It operates through a two-stage process: first, it generates region proposals areas of interest where objects might be located using a Region Proposal Network (RPN), and then it classifies these regions to identify objects and their corresponding positions. This two-stage detection process allows Faster R-CNN to excel in detecting small objects and handling complex scenes with multiple, overlapping objects (Dewi et al., 2024). However, despite its high accuracy, Faster R-CNN is computationally intensive and relatively slow compared to other models, which makes it less suitable for real-time applications, especially when processing continuous video streams, such as CCTV footage for cheating detection (Magalhães et al., 2023).

The Single Shot Multibox Detector (SSD), on the other hand, is a faster object detection algorithm that performs detection in a single stage (Dewi et al., 2024). Instead of first generating region proposals, SSD directly predicts both the object class and the bounding boxes at multiple scales using a single convolutional network. This one-stage detection approach makes SSD significantly faster than Faster R-CNN, with the added benefit of being able to detect objects at different sizes and aspect ratios. However, SSD tends to trade off some accuracy for speed, particularly in detecting smaller objects (Kim et al., 2020).

RetinaNet introduces a novel focal loss function to address the class imbalance issue common in object detection. This algorithm can maintain a balance between speed and accuracy, offering competitive performance when compared to Faster R-CNN and SSD. It combines a feature pyramid network with a ResNet backbone, providing good detection capabilities across a wide range of object sizes. However, like SSD, RetinaNet can struggle with detecting small objects and requires more computational power than YOLO, which is often a crucial factor in real-time applications (Lin et al., 2020) (Zhou et al., 2021).

This research will focus on using the YOLO algorithm to build a cheating detection model due to its emphasis on speed and efficiency. Unlike Faster R-CNN's two-stage process or SSD's focus on multiple scales, YOLO treats object detection as a single regression problem, predicting both the object class and location in one pass through the network. This design allows YOLO to process images extremely fast, making it highly suitable for real-time detection tasks, such as monitoring live CCTV footage for cheating behavior (Lavanya & Pande, 2023)(Cao et al., 2023).

There are several variants of the YOLO algorithm that will be used in the model series, namely YOLOv5s, YOLOv5m, YOLOv5l, YOLOv6s, YOLOv6m, YOLOv6l, and YOLOv7. It is important to note that Ultralytics developed YOLOv5 and YOLOv7, and YOLOv6 was developed by Meituan (Meituan, 2022). The objective of selecting these models is to compare their performance, enabling the identification of the most suitable model for the specific case at hand.

In general, the YOLO architecture can be described as an object detection approach that processes the image and generates object predictions in a single feedforward pass through the neural network. This approach enables fast and efficient object detection. Using various variants of YOLO, we will evaluate the performance and accuracy of each model in detecting suspicious behavior during exams. The following is the architecture of YOLO (Olorunshola et al., 2023).

Figure 4. YOLO architecture (Barba-Guaman et al., 2021)

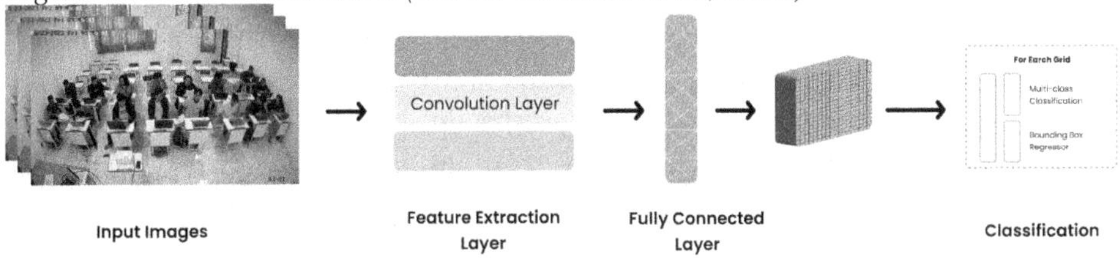

Figure 5. YOLO layers (Olorunshola et al., 2023)

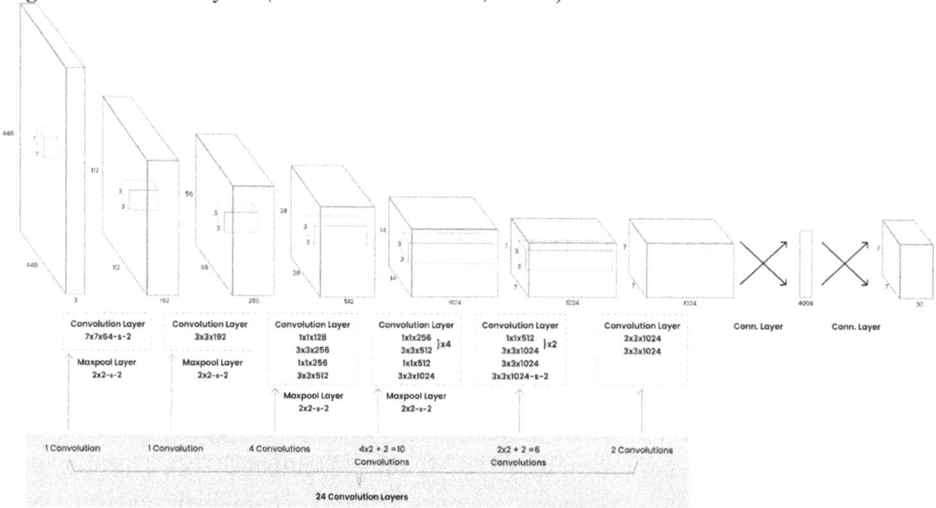

In Figure 4 and Figure 5, the YOLO algorithm starts with image input as the first step. This image then goes through a series of feature extraction stages in a convolutional neural network (CNN). This stage aims to identify unique features present in the image, such as object edges, textures, colours, and other patterns. This feature extraction stage results in a high-dimensional data representation that describes the image more abstractly.

Once the features have been successfully extracted, the next step is sending them to the fully connected layer in the neural network. This fully connected layer serves to combine and process the previously extracted features. Here, the processing involves more complex calculations and the formation of more abstract representations.

The output of the fully connected layer will be the final determinant in the classification process in the YOLO algorithm. This output includes information about the location of the object (bounding box coordinates), the detected object's class, and the detection's confidence level. In other words, the YOLO algorithm uses the features that have been processed through the fully connected layer to produce complete information about the objects in the image, including what the object is, where it is located, and the extent of its confidence.

With this approach, YOLO can efficiently and accurately detect objects in images by combining advanced feature extraction with information processing in the fully connected layer to provide robust classification results.

3.5 Object Tracking and Cropping

Object tracking is a fundamental task in computer vision, focusing on detecting and continuously tracking objects across a sequence of images or video frames. The primary goal of object tracking is to accurately locate and follow a designated object in subsequent frames after it has been identified in the initial frame (Soleimanitaleb et al., 2019). In its implementation, this is crucial in scenarios where monitoring the movement or behavior of objects over time is required, such as in surveillance or activity recognition (Reddy et al., 2015).

In this research, object annotation and cropping are performed using Roboflow, a powerful web-based tool specifically designed for image annotation and dataset management. Roboflow provides an intuitive interface for labeling objects within images and generating datasets that are optimized for training machine learning models. Its role is critical in the object detection pipeline, as it allows for precise annotation and efficient management of the data used in training, validation, and testing phases of the model development process.

The output dataset from Roboflow is structured into three main subsets: training data, validation data, and test data. This division ensures that the model is trained on one portion of the data, validated during training to fine-tune its parameters, and tested on unseen data to evaluate its generalization ability. Each subset contains two key components: a text file (txt) and an image file. The text file includes the label annotations, which specify the bounding box coordinates and class labels for each object in the image, while the image file contains the corresponding visual data. This structured dataset enables the model to learn from the labeled examples and detect objects accurately in future unseen frames.

3.6 Metrics Evaluation

In machine learning, evaluating the performance of a model is a crucial step to ensure its effectiveness in making accurate predictions. Various metrics are used to assess a model's performance, particularly in supervised learning tasks. Among these, Confusion Matrix is one of the most frequently used evaluation methods in machine learning, especially supervised learning (Hasnain et al., 2020). The confusion matrix contains four evaluation values, namely True Positive (*TP*), True Negative (*TN*), False Positive (*FP*) and False Negative (*FN*) (Ilahi et al., 2022) (Purwita Sary et al., 2023).

Table 1. Confusion matrix theory

		Prediction	
		Positive	**Negative**
Actual	Positive	TP	FN
	Negative	FP	TN

Based on Table 1, the evaluation matrix can be calculated in accuracy, precision, and recall. The following equation is used to calculate the evaluation value (Tripathi & Kumar, 2020).

$$Acc = \frac{TP + TN}{TP + FN + FP + TN} \tag{1}$$

$$P = \frac{TP}{TP + FP} \tag{2}$$

$$R = \frac{TP}{TP + FN} \tag{3}$$

Accuracy (*Acc*) is one of the testing methods based on the overall level of closeness between the predicted value and the actual value. Precision (*P*) is the accuracy of predictions with correct labels with positive results. At the same time, Recall (*R*) represents the quantity of quality in the entire prediction process (Farokhah, 2019).

In the YOLO algorithm, one of the leading indicators to assess model performance is calculating the *mAP* (Mean Average Precision) value (Lee & Hwang, 2022). The *mAP* metric is a commonly used metric in object detection tasks to measure the extent to which the model can detect and classify objects with high accuracy (Durve et al., 2021).

Conceptually, *mAP* will calculate the average precision for each object class (Gros et al., 2018). Precision measures the extent to which the detection made by the model is the intended object, and it is calculated by counting the number of true positives (*TP*) divided by the total number of positive predictions made by the model.

$$AP = \sum_{i=1}^{N} P(i)\,\Delta\,\mathrm{R(i)} \tag{4}$$

$$mAP = \frac{1}{N}\sum_{i=1}^{N} AP(img_i) \tag{5}$$

Where *AP* defines the average precision, *N* defines the number of classes, and img_i represents the *i*-th image iteration. The *AP* and *mAP* values range from 0-1, while the ideal value for both benchmarks is 1.

4 EXPERIMENTS AND RESULTS

4.1 Result and Discussion

In this research, experiments have been conducted using Python programming language using pyTorch on an NVIDIA V100 device (Google Collab). The model has been trained with a learning rate of 0.005, batch size 16, 100 epoch, and image size 640x640 (Jiang et al., 2021)(Xu et al., 2022)(Afis Julianto et al., 2022; Garg et al., 2018). This time, the dataset division in training is divided into 70% of the total dataset, with 20% of the data for validation and 10% for testing (Nguyen et al., 2021), This is also based on research conducted previously (Varma et al., 2019) and uses a similar composition in dealing with unbalanced datasets, 70% of the data is used to train, 10% for validation and 20% for testing.

The algorithms selected for testing include YOLOv5s, YOLOv5m, YOLOv5l, YOLOv6s, YOLOv6m, YOLOv6l, and YOLOv7. Each of these models represents a different variant of the YOLO architecture, which is widely used in object detection tasks due to its real-time performance and high accuracy. The variations in the model size (small, medium, and large) allow for a trade-off between speed and accuracy, where smaller models are typically faster but may compromise on precision, while larger models offer higher accuracy at the cost of increased computational demand.

It is important to note that for the YOLOv7 algorithm, the medium and large variants could not be evaluated due to resource constraints. This limitation is primarily related to the significant computational power and memory required to train and test these larger models, which often exceeds the available resources in standard testing environments. Despite this, YOLOv7 is still considered the most advanced of the YOLO series, offering improved performance through enhanced architecture and optimization techniques. Future research may explore these larger variants when sufficient computational resources become available, which could lead to further insights into the trade-offs between accuracy, speed, and resource usage across different YOLO models.

In the conducted experiments, it was observed that certain datasets exhibit class imbalances, particularly in the Exchange Paper and Giving Code categories. As illustrated in Figure 6, the dataset distribution is uneven, with a significantly higher representation of one class compared to the other. This imbalance poses a notable challenge for the model during the training process, as it may lead to biased predictions favoring the majority class. Class imbalances are a common issue in machine learning and can adversely affect the model's ability to generalize, particularly for the underrepresented class. Such conditions force the model to deliver optimal performance despite the lack of proportionate data for each class.

The imbalance depicted in Figure 6 is not uncommon in real-world datasets, and it poses a critical challenge in machine learning tasks, particularly in classification problems. When the model is trained on such an imbalanced dataset, it may fail to adequately learn the features of the minority class, thus resulting in biased predictions. Without proper handling, the model might prioritize optimizing accuracy for the majority class, often at the expense of recall and precision for the underrepresented class.

In this study, no specific balancing techniques such as oversampling, undersampling, or synthetic data generation (e.g., SMOTE) were applied to the dataset. Instead, the raw performance of the algorithms was evaluated on this naturally imbalanced data to assess how well each model can cope with class imbalances under realistic conditions. This approach helps reveal the inherent strengths and limitations of the algorithms in handling imbalanced data without intervention.

Figure 6. Dataset detail

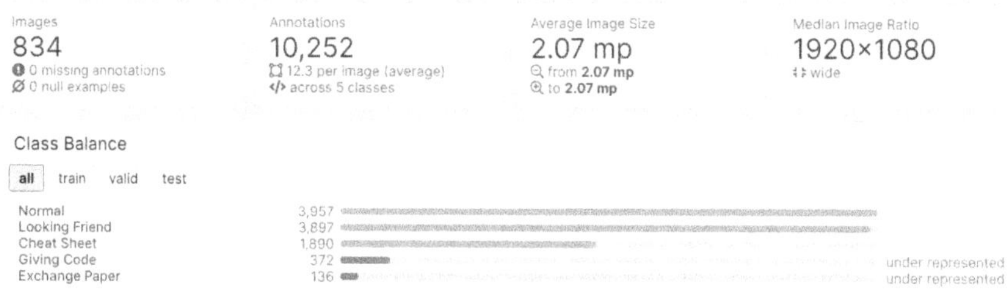

The Imbalanced dataset as shown in Figure 6, the model will be forced to provide the best performance with a disproportionate dataset. Furthermore, regarding the results of the analysis that has been done, it is found that each algorithm has varying performance in modelling with the same dataset. Moving forward, future research could explore the application of various techniques to address class imbalance, such as cost-sensitive learning or data augmentation strategies specifically designed for the minority class.

The implementation of this system can significantly reduce the workload of human proctors, allowing them to concentrate on more strategic tasks. Furthermore, it enables the automatic and objective detection of suspicious behaviors, minimizing the potential for human subjectivity or bias during exam monitoring. In addition, the digital documentation produced through recordings and detection results serves as authentic evidence that can support academic audits, violation investigations, and evaluations of exam quality. Therefore, this technology not only enhances operational efficiency but also strengthens academic integrity within educational institutions.

4.2 Comparative Analysis

In this section, we present a comparative analysis of the models trained during the experiments by examining their performance across several critical parameters. These parameters are essential for understanding both the computational efficiency and the accuracy of each model in addressing the object detection task. Specifically, the analysis focuses on the following metrics: the total number of parameters (Param.), the computational complexity measured in GFLOPs (Giga Floating Point Operations), the mean average precision at IoU threshold 0.50 (mAP:50), and the more stringent mean average precision across multiple IoU thresholds from 0.50 to 0.95 (mAP50:95).

These metrics offer a comprehensive view of each model's capabilities, allowing us to assess the trade-offs between computational cost and detection accuracy. Models with fewer parameters and lower GFLOPs may be more efficient, but their performance in terms of mAP could be impacted, particularly in challenging datasets with imbalances or varied object sizes. On the other hand, models with higher mAP scores may require more computational resources, which is crucial to consider in real-time applications or when resource constraints are a concern.

The results of this analysis are summarized in Table 2, where each model's performance is compared based on these parameters. This table provides an invaluable resource for understanding the differences between algorithms and selecting the most appropriate model for deployment, depending on the specific needs of the application, such as speed, accuracy, or resource efficiency.

Table 2. Model result

Algorithm	Indicators			
	Param.	**GFLOPs**	**mAP:50**	**mAP50:95**
YOLOv7	36,50 M	103.2	0,559	0,344
YOLOv6l	59,54 M	150.51	0,535	0,333
YOLOv6m	34,81 M	85.63	0,505	0,311
YOLOv6s	18,50 M	45.17	0,504	0,305
YOLOv5l	46,13 M	107.7	0,555	0,363
YOLOv5m	20,87 M	47.9	0,497	0,337
YOLOv5s	7,02 M	45,15	0,549	0,351

From Table 2, it can be seen that YOLOv7 has a fairly low GFLOPs rate and a relatively high *mAP* in both metrics (*mAP*:50 and *mAP* 50:95). This indicates that the model performs well in detecting objects with fairly good accuracy over a wide variety of object sizes.

YOLOv6l has a higher GFLOPs rate than YOLOv7, but its *mAP* is slightly lower. This may indicate that this model requires more computation power for similar performance to YOLOv7.

Meanwhile, the YOLOv6m has a lower GFLOPs rate than larger models (such as the YOLOv6l) but has a lower *mAP*. Opting for YOLOv6m could be advantageous when faced with computational power limitations. YOLOv6s, a more lightweight model with significantly lower GFLOPs, entails a trade-off with detection performance compared to larger models. This model may be a suitable choice when prioritizing faster execution with minimal compromise on accuracy.

YOLOv5l has a moderate level of GFLOPs and a relatively high *mAP*. It may be a good choice if looking for a balance between performance and computation power. YOLOv5m is a lighter model, which can run faster but has a lower *mAP* than the larger model. YOLOv5s is the lightest model in terms of GFLOPs but still has a competitive *mAP*.

The selection of a model is contingent on specific requirements. Given sufficient computational resources, opting for YOLOv7 or YOLOv6l could be advantageous for achieving high performance. Conversely, when faced with limitations in computation power, the consideration of lighter models like YOLOv6s or YOLOv5s may be warranted.

In addition to evaluating models based on accuracy and computational efficiency, execution time is a critical factor to consider, especially in real-time applications or environments with limited computational resources. The evaluation of execution time is divided into three main stages: pre-processing time, inference time, and Non-Maximum Suppression (NMS) time.

Pre-processing time refers to the duration required to prepare the data before it is fed into the model, such as image normalization, resizing, or augmentation. Inference time, on the other hand, measures how quickly the model processes input data and generates predictions, making it one of the key indicators for assessing model performance in real-time scenarios. The final stage, NMS, is responsible for eliminating redundant predictions and selecting the best prediction for each detected object, which also plays a role in the overall efficiency of the object detection system.

Table 3 summarizes the execution time comparisons of the different algorithms tested. The parameters recorded include the average pre-processing time, average inference time, and average NMS time. These results offer a comprehensive view of each algorithm's relative speed and help identify which models are the most time-efficient, particularly in applications requiring rapid responses, such as CCTV-based surveillance systems or autonomous vehicle navigation.

Table 3. Model execution time

Algorithm	Time		
	Average pre-process	Average inference	Average NMS
YOLOv7	0.25 ms	112.9 ms	1.0 ms
YOLOv6l	0.18 ms	45.68 ms	1.97 ms
YOLOv6m	0.18 ms	27.43 ms	1.51 ms
YOLOv6s	0.18 ms	12.72 ms	2.52 ms
YOLOv5l	0.4 ms	40.6 ms	1.6 ms
YOLOv5m	0.4 ms	22.0 ms	1.6 ms
YOLOv5s	0.4 ms	9.1 ms	1.2 ms

YOLOv7 performs well in terms of accuracy (*mAP*) and requires a reasonable inference time (112.9 ms). The low pre-processing time (0.25 ms) indicates that the model processes the input quickly before going through the inference process.

YOLOv6l performs reasonably well in terms of accuracy, although it requires more computational power than YOLOv7. The slower inference time (45.68 ms) indicates that this model is slower in detecting objects. YOLOv6m is a more efficient model in terms of computation power than YOLOv6l but with a slight sacrifice in accuracy. The faster inference time (27.43 ms) could be viable when working with restricted computational resources. YOLOv6s is the lightest model regarding computation power, but accuracy and inference time are lower. The high NMS time (2.52 ms) may be the reason for the longer inference time.

YOLOv5l has good performance in terms of accuracy and reasonable inference time. However, the slightly higher pre-processing time (0.4 ms) and slower inference time (40.6 ms) should be considered. YOLOv5m is a lighter model with lower performance compared to the larger model. The faster inference time (22.0 ms) can be a good choice for real-time applications. YOLOv5s is a very lightweight model with a fast inference time (9.1 ms), though slightly lower accuracy.

4.3 Model Evaluation

To further assess the performance of the models, we conducted a thorough evaluation using a confusion matrix, which serves as the foundation for calculating several key metrics: Accuracy, Average Precision (AP), and Average Recall (AR). These metrics are essential for gaining a deeper understanding of the models' behavior, especially in scenarios where class imbalances and other complexities might affect their ability to generalize well.

Accuracy provides an overall measure of how often the model's predictions are correct, making it a straightforward metric for evaluating model performance. However, accuracy alone is often insufficient, particularly when dealing with imbalanced datasets. Therefore, we also employ Average Precision (AP), which focuses on the model's ability to correctly identify positive instances, considering both precision and recall at varying confidence thresholds. This metric is particularly useful in object detection tasks where detecting small or occluded objects is critical. Average Recall (AR) complements AP by measuring the model's capacity to capture all relevant instances, ensuring that the model doesn't miss too many true positives.

By combining these metrics, we obtain a more comprehensive evaluation of the models, providing insights into not just their correctness but also their robustness in detecting objects across a wide range of conditions. The results of these evaluations are presented in Table 4, offering a detailed comparison of each model's performance based on the testing data. This table is crucial in understanding how each algorithm balances accuracy, precision, and recall, allowing us to determine which model is best suited for real-world deployment depending on the task's requirements.

Table 4. Model evaluation metrics

Algorithm	**Metrics Evaluation**		
	Accuracy	**AP**	**AR**
YOLOv7	0,51	0,500	0,513
YOLOv6l	0,37	0,379	0,371
YOLOv6m	0,33	0,399	0,326
YOLOv6s	0,37	0,401	0,365
YOLOv5l	0,39	0,401	0,391
YOLOv5m	0,39	0,406	0,385
YOLOv5s	0,43	0,492	0,431

Based on Table 4. evaluation metrics, YOLOv7 has high accuracy and good performance in terms of average precision (*AP*) and average recall (*AR*). However, despite its fast inference time, it requires higher computational power (GFLOPs: 103.2) than other models.

YOLOv6l has lower accuracy than YOLOv7 and lower *AP* and *AR* performance. However, its inference time is faster than that of YOLOv7. YOLOv6m has lower accuracy but slightly higher *AP* performance than YOLOv6l. The inference time is faster than YOLOv6l but still requires significant computation power. YOLOv6s has a similar accuracy level to YOLOv6l but with a faster inference time. It is a good choice when seeking a balance between accuracy and speed of inference.

YOLOv5l has a relatively high accuracy rate and good *AP* and *AR* performance. Its moderate inference time can also be a good choice if it has sufficient resources. YOLOv5m has fairly good accuracy and faster inference time compared to other models. It can be a good choice for real-time applications. While the YOLOv5s have good accuracy and very fast inference time. This model can be a strong choice for real-time applications that require high-speed object detection.

4.4 Best Model Evaluation

Based on the extensive experiments and thorough analysis conducted in this study, many factors contribute to determining the most effective model, including accuracy, computational efficiency, execution time, and the ability to handle complexities such as class imbalance and small object detection. After further observation, the two best models are YOLOv5s and YOLOv7.

The selection of YOLOv5s is based on a good level of accuracy (Accuracy: 0.43) with a competitive *mAP* (*mAP*@50: 0.549). Very fast inference time (Average inference: 9.1 ms), making it suitable for real-time applications. It also has a very low GFLOPs rate (45,15), which means it requires low computation power.

Meanwhile, YOLOv7 has a higher accuracy rate (Accuracy: 0.51) with a good *mAP* (*mAP*@50: 0.559). Reasonable inference time (Average inference: 112.9 ms), which may still be usable for real-time applications depending on the requirement. Although it requires higher computation power (GFLOPs: 103.2), it is a good choice if accuracy is a top priority.

The model evaluation is more detailed, using a matrix of confusion, which is more closely related to the best model. The following Table 5 shows the comparison results of the test.

Table 5. Confusion matrix YOLOv5s

	Predicted						
Actual		CS	EP	GC	LF	N	BG
	CS	0,69	0	0	0,11	0,14	0,06
	EP	0,06	0,25	0,03	0,56	0	0,11
	GC	0,03	0	0,11	0,57	0,17	0,13
	LF	0,04	0	0	0,78	0,11	0,05
	N	0,08	0	0,01	0,14	0,75	0,03
	BG	0,25	0,03	0,02	0,39	0,31	0

From Table 5. confusion matrix, there are five active classes, namely Cheat Sheet (CS), Exchange Paper (EP), Giving Code (GC), Looking Friend (LF) and Normal (N). At the same time, BG is a Background (BG) class that describes the part of the image that does not contain objects of any class identified by the model. It is a specialized class used to indicate areas without significant or important objects in the image.

Table 6. Confusion matrix YOLOv7

				Predicted			
Actual		CS	EP	GC	LF	N	BG
	CS	0,72	0	0	0,09	0,16	0,02
	EP	0,04	0,52	0,04	0,17	0,13	0,09
	GC	0,06	0	0,28	0,34	0,29	0,03
	LF	0,05	0,02	0,04	0,74	0,13	0,03
	N	0,07	0	0,01	0,1	0,81	0,01
	BG	0,21	0,04	0,08	0,36	0,31	0

Based on the analysis of the confusion matrix in Table 5 and Table 6, it can be concluded that the YOLOv7 model performs better than the YOLOv5s model in detecting the Exchange Paper and Giving Code classes in the imbalance dataset case. In the Exchange Paper class, YOLOv7 has a True Positive (*TP*) value of 0.52, while YOLOv5s only has a *TP* of 0.19. Similarly, in the Giving Code class, YOLOv7 has a *TP* of 0.28, while YOLOv5s only has a *TP* of 0.06.

Nonetheless, it is important to note that the performance of both models is far from ideal. This may be due to several factors. Firstly, the dataset's imbalance may affect the model's performance, as the model may tend to focus on the majority class and ignore the minority class. Secondly, the parameters and hyperparameters of the model may play an important role in performance. In further research, parameter tuning, and more careful dataset processing may be required to improve model performance. In addition, the performance of cheating detection can also be affected by environmental factors, such as the quality of CCTV images and the intelligence level of the model used.

4.5 Adding Dataset Experiment

After obtaining YOLOv7 as the best-performing model in the previous experiments, the researcher endeavored to further train the YOLOv7 model by adding 10% data for two labels that exhibited data imbalance, namely "Giving Code" (592 data) and "Exchange Paper" (525 data). Consequently, a new dataset was acquired as follows:

Figure 7. Add giving code and exchange paper dataset

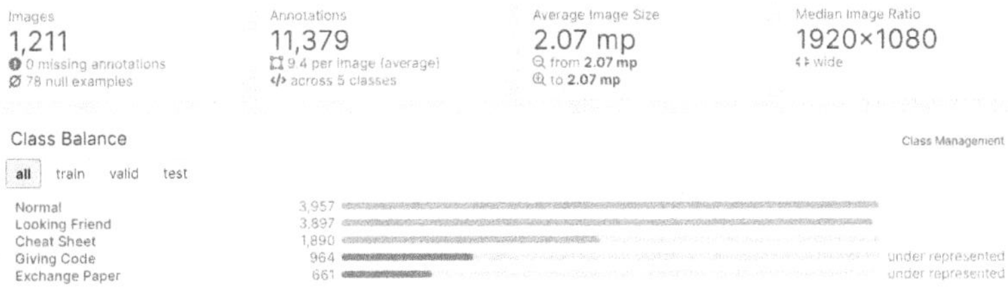

The training process followed the same methodology as the previous training, with no additional changes to the training or testing scenarios. Upon completion of the training, an accuracy value of 0.60 and a *mAP* of 0.66 were obtained. This indicates an improvement in accuracy of approximately 9% and an increase in *mAP* of around 10% compared to the previous experiment.

From this experiment, it is evident that augmenting the dataset for classes with data imbalance can enhance the accuracy performance of the model, resulting in improved outcomes. The increased dataset size contributes to a more comprehensive representation of the specific classes, allowing the model to learn better and generalize patterns associated with Giving Code and Exchange Paper.

4.6 Privacy and Surveillance Ethics

Although the use of CCTV-based behavior detection technology offers numerous benefits in educational settings, its implementation also presents several ethical challenges, particularly concerning privacy. Video recordings produced during exam monitoring may contain sensitive personal data such as facial features, expressions, and individual behaviors, which—if misused—could lead to violations of privacy rights. Therefore, it is essential to apply data protection principles from the system design stage (privacy by design), including techniques such as face blurring or the use of encrypted identifiers for anonymization.

In addition, strict limitations must be imposed on data retention, ensuring that recordings are stored only for the necessary duration and are promptly deleted afterward. The deployment of such systems should also involve full transparency with all participants through explicit informed consent, and must comply with legal frameworks such as Indonesia's Personal Data Protection Law or the General Data Protection Regulation (GDPR) at the international level. By considering these aspects, the use of digital surveillance systems can deliver maximum benefits without compromising individuals' fundamental rights to privacy.

5 CONCLUSION

The best models to use are YOLOv5s and YOLOv7, according to the tests carried out in this study. The YOLOv5s model is a very lightweight model with a very fast inference time (9.1 ms) and does not require large computations in its implementation. On the other hand, this model has a significant shortcoming when detecting classes with dataset deficiencies, namely Giving Code and Exchange Paper. However, this deficiency can be recognized better in YOLOv7 with better accuracy of 51%, an increase of about 8% higher than the YOLOv5s model, so the YOLOv7 model is the best option in the first scenario.

In the second scenario, models were constructed by adding the dataset by 10%, focusing on addressing the imbalance in two classes, Giving Code and Exchange Paper. Utilizing YOLOv7 yielded an accuracy of 60%, indicating a notable improvement of 9% compared to the previous model. This observation underscores the significant impact of data imbalance on YOLO performance to detect objects accurately, as well as the importance of addressing this issue in order to optimize model performance. The results also underscore the broader implications of class imbalance in machine learning tasks, demonstrating that even a relatively small increase in training data can lead to significant improvements in model accuracy, especially when focused on underrepresented classes.

Future development can be done by adding datasets to classes with deficiencies to obtain a balanced dataset, as well as incorporating synthetic data methods, such as SMOTE or similar techniques, to further address data imbalance and improve model performance. Additionally, exploring alternative research methods, such as experimenting with different model architectures and incorporating advanced algorithms like Fast-RCNN and SSD, presents opportunities for novel insights into improving detection accuracy and robustness. Fine-tuning model hyperparameters and systematically optimizing factors like learning rates and anchor box sizes are crucial for enhancing adaptability. Furthermore, evaluating models in more realistic scenarios, including diverse environmental conditions and complex backgrounds, ensures their applicability beyond controlled experiments. This multifaceted approach aims to propel YOLOv7 and similar models toward heightened accuracy, adaptability, and practical utility in real-world contexts.

REFERENCES

https://academicintegrity.org/. (2021). *THE FUNDAMENTAL VALUES OF ACADEMIC INTEGRITY*. International Center for Academic Integrity. https://academicintegrity.org/images/pdfs/20019_ICAI-Fundamental-Values_R12.pdf

Adil, M., Simon, R., & Khatri, S. K. (2019). Automated Invigilation System for Detection of Suspicious Activities during Examination. *2019 Amity International Conference on Artificial Intelligence (AICAI)*, 361–366. DOI: 10.1109/AICAI.2019.8701263

Al-Labadi, L., & Sant, S. (2021). Enhance Learning Experience Using Technology in Class. *Journal of Technology and Science Education, 11*(1), 44–52. DOI: 10.3926/jotse.1050

Alahmari, M. (2023). Exploring the Influential Factors Affecting Staff Willingness to Adopt Augmented Reality. *International Journal of Information and Education Technology (IJIET), 13*(7), 1078–1084. DOI: 10.18178/ijiet.2023.13.7.1907

Ball, J. E., Anderson, D. T., & Chan, C. S. (2017). Comprehensive Survey of Deep Learning in Remote Sensing: Theories, Tools, And Challenges for the Community. *Journal of Applied Remote Sensing, 11*(04), 1. DOI: 10.1117/1.JRS.11.042609

Barba-Guaman, L., Naranjo, J. E., Ortiz, A., & Gonzalez, J. G. P. (2021). Object Detection in Rural Roads Through SSD and YOLO Framework. In H. and D. G. and M. F. and R. C. A. M. Rocha Álvaro and Adeli (Ed.), *Trends and Applications in Information Systems and Technologies* (pp. 176–185). Springer International Publishing. DOI: 10.1007/978-3-030-72657-7_17

Cai, H., & King, I. (2020). Education Technology for Online Learning in Times of Crisis. *Proceedings of 2020 IEEE International Conference on Teaching, Assessment, and Learning for Engineering, TALE 2020*, 758–763. DOI: 10.1109/TALE48869.2020.9368387

Cambron-Goulet, É., Dumas, J. P., Bergeron, É., Bergeron, L., & St-Onge, C. (2019). Guidelines for Creating Written Clinical Reasoning Exams: Insight from a Delphi Study. *Health Profession Education, 5*(3), 237–247. DOI: 10.1016/j.hpe.2018.09.001

Cao, F., Xing, B., Luo, J., Li, D., Qian, Y., Zhang, C., Bai, H., & Zhang, H. (2023). An Efficient Object Detection Algorithm Based on Improved YOLOv5 for High-Spatial-Resolution Remote Sensing Images. *Remote Sensing (Basel), 15*(15), 3755. DOI: 10.3390/rs15153755

Chen, B., Azad, S., Fowler, M., West, M., & Zilles, C. (2020). Learning to Cheat: Quantifying Changes in Score Advantage of Unproctored Assessments Over Time. *L@S 2020 - Proceedings of the 7th ACM Conference on Learning @ Scale*, 197–206. DOI: 10.1145/3386527.3405925

Chhadikar, N., Bhamare, P., Patil, K., & Kumari, S. (2019). Image Processing Based Tracking and Counting Vehicles. *2019 3rd International Conference on Electronics, Communication and Aerospace Technology (ICECA)*, 335–339. DOI: 10.1109/ICECA.2019.8822070

Comas-Forgas, R., Lancaster, T., Calvo-Sastre, A., & Sureda-Negre, J. (2021). Exam Cheating and Academic Integrity Breaches During The COVID-19 Pandemic: An analysis of internet search activity in Spain. *Heliyon*, *7*(10), e08233. Advance online publication. DOI: 10.1016/j.heliyon.2021.e08233 PMID: 34722942

Cong, X., Li, S., Chen, F., Liu, C., & Meng, Y. (2023). A Review of YOLO Object Detection Algorithms based on Deep Learning. *Frontiers in Computing and Intelligent Systems*, *4*(2), 17–20. DOI: 10.54097/fcis.v4i2.9730

Dewan Perwakilan Rakyat Republik Indonesia, P. R. I. (2022, October 17). *UU No. 27 Tahun 2022 Perlindungan Data Pribadi*. BPK RI. https://peraturan.bpk.go.id/Details/229798/uu-no-27-tahun-2022

Dewi, A. L., Pardede, H. F., Suryawati, E., Pratiwi, H., Heryana, A., Yuliani, A. R., & Ramdan, A. (2024). Two-Stage Object Detection for Autonomous Vehicles With VGG-16 Based Faster R-CNN. *Jurnal Elektronika dan Telekomunikasi (JET)*, *24*(1), 25. DOI: 10.55981/jet.551

Diwan, T., Anirudh, G., & Tembhurne, J. V. (2023). Object Detection Using YOLO: Challenges, Architectural Successors, Datasets and Applications. *Multimedia Tools and Applications*, *82*(6), 9243–9275. DOI: 10.1007/s11042-022-13644-y PMID: 35968414

Durve, M., Bonaccorso, F., Montessori, A., Lauricella, M., Tiribocchi, A., & Succi, S. (2021). Tracking Droplets in Soft Granular Flows with Deep Learning Techniques. *The European Physical Journal Plus*, *136*(8), 864. DOI: 10.1140/epjp/s13360-021-01849-3 PMID: 34458055

Ellis, R., Cleland, J., Scrimgeour, D. S. G., Lee, A. J., Hines, J., & Brennan, P. A. (2023). Establishing The Predictive Validity of the Intercollegiate Membership of the Royal Colleges of Surgeons Written Examination: MRCS Part A. *The Surgeon*, *21*(6), 323–330. Advance online publication. DOI: 10.1016/j.surge.2023.07.004 PMID: 37544852

European Union Lex. (2016, May 4). *Regulation - 2016/679 - EN - GDPR- EUR-Lex*. European Union. https://eur-lex.europa.eu/eli/reg/2016/679/oj

Fadillah, A. (2019). Hubungan Antara Efikasi Diri Dengan Perilaku Menyontek Pada Mahasiswa. *Psikoborneo: Jurnal Ilmiah Psikologi*, *7*(4), 657–664. DOI: 10.30872/psikoborneo.v7i4.4846

Fang, Y., Ye, J., & Wang, H. (2020). Realization of Intelligent Invigilation System Based on Adaptive Threshold. *2020 5th International Conference on Computer and Communication Systems (ICCCS)*, 201–205. DOI: 10.1109/ICCCS49078.2020.9118581

Farokhah, L. (2019). Implementasi K-Nearest Neighbor Untuk Klasifikasi Bunga Dengan Ekstraksi Fitur Warna RGB. [JTIIK]. *Jurnal Teknologi Informasi Dan Ilmu Komputer*, *7*(6), 1129–1135. DOI: 10.25126/jtiik.2020722608

Ferdig, R. E., Baumgartner, E., Hartshorne, R., Kaplan-Rakowski, R., & Mouza, C. (2020). Teaching, Technology, and Teacher Education During the COVID-19 Pandemic: Stories from the Field. *AACE-Association for the Advancement of Computing in Education*, 125–128.

Garg, D., Goel, P., Pandya, S., Ganatra, A., & Kotecha, K. (2018). A Deep Learning Approach for Face Detection using YOLO. *2018 IEEE Punecon*, 1–4. DOI: 10.1109/PUNECON.2018.8745376

Gros, D., Habermann, T., Kirstein, G., Meschede, C., Ruhrberg, S. D., Schmidt, A., & Siebenlist, T. (2018). Anaphora Resolution. *International Journal of Information Retrieval Research*, *8*(3), 33–45. DOI: 10.4018/IJIRR.2018070103

Hasnain, M., Pasha, M. F., Ghani, I., Imran, M., Alzahrani, M. Y., & Budiarto, R. (2020). Evaluating Trust Prediction and Confusion Matrix Measures for Web Services Ranking. *IEEE Access : Practical Innovations, Open Solutions*, *8*, 90847–90861. DOI: 10.1109/ACCESS.2020.2994222

Hoque, M. J., Razu, M., Jashim, M., & Mostafa, M. (2020). Automation of Traditional Exam Invigilation using CCTV and Bio-Metric. *International Journal of Advanced Computer Science and Applications*, *11*(6), 392–399. DOI: 10.14569/IJACSA.2020.0110651

Hussein, F., Al-Ahmad, A., El-Salhi, S., Alshdaifat, E., & Al-Hami, M. (2022). Advances in Contextual Action Recognition: Automatic Cheating Detection Using Machine Learning Techniques. *Data*, *7*(9), 122. Advance online publication. DOI: 10.3390/data7090122

Hutton, J. S., Dudley, J., Horowitz-Kraus, T., DeWitt, T., & Holland, S. K. (2020). Associations Between Home Literacy Environment, Brain White Matter Integrity and Cognitive Abilities in Preschool-Age Children. *Acta Paediatrica. International Journal of Pediatrics, 109*(7), 1376–1386. DOI: 10.1111/apa.15124 PMID: 31854046

Ifeagwazi, C. M., Chukwuorji, J. C., Egbodo, S. O., & Nwoke, M. B. (2019). Peer pressure, fear of failure and examination cheating behavior in the university: Does gender make the difference? *Cognition, Brain, Behavior: An Interdisciplinary Journal, 23*(1), 43–62. DOI: 10.24193/cbb.2019.23.03

Ijjina, E. P., Joshi, A. S., & Kanahasabai, G. (2020). Detection of Customer Interested Garments in Surveillance Video using Computer Vision. *2020 11th International Conference on Computing, Communication and Networking Technologies (ICCCNT)*, 1–5. DOI: 10.1109/ICCCNT49239.2020.9225571

Ilahi, R., Bimantoro, F., Dwiyansaputra, R., & Farinda, R. (2022). Classification of Pringgasela Typical Songket Using Multi Texton Co-occurrence Descriptor and K-Nearest Neighbor. In *Proceedings of the First Mandalika International Multi-Conference on Science and Engineering 2022, MIMSE 2022 (Informatics and Computer Science)* (pp. 352–366). Atlantis Press International BV. DOI: 10.2991/978-94-6463-084-8_30

Institute of Electrical and Electronics Engineers. Bangalore Section, & Institute of Electrical and Electronics Engineers. (n.d.). *Automated Cheating Detection in Exams using Posture and Emotion Analysis*.

Janke, S., Rudert, S. C., Petersen, Ä., Fritz, T. M., & Daumiller, M. (2021). Cheating in the wake of COVID-19: How dangerous is ad-hoc online testing for academic integrity? *Computers and Education Open, 2*, 100055. DOI: 10.1016/j.caeo.2021.100055

Jiang, P., Ergu, D., Liu, F., Cai, Y., & Ma, B. (2021). A Review of Yolo Algorithm Developments. *Procedia Computer Science, 199*, 1066–1073. DOI: 10.1016/j.procs.2022.01.135

Julianto, A., Sunyoto, A., & Wibowo, F. W. (2022). Optimasi Hyperparameter Convolutional Neural Network Untuk Klasifikasi Penyakit Tanaman Padi. *Teknimedia: Teknologi Informasi Dan Multimedia, 3*(2), 98–105. DOI: 10.46764/teknimedia.v3i2.77

Kaddoura, S., & Gumaei, A. (2022). Towards effective and efficient online exam systems using deep learning-based cheating detection approach. *Intelligent Systems with Applications, 16*, 200153. Advance online publication. DOI: 10.1016/j.iswa.2022.200153

Kim, J., Sung, J.-Y., & Park, S. (2020). Comparison of Faster-RCNN, YOLO, and SSD for Real-Time Vehicle Type Recognition. *2020 IEEE International Conference on Consumer Electronics - Asia (ICCE-Asia)*, 1–4. DOI: 10.1109/ICCE-Asia49877.2020.9277040

Lavanya, G., & Pande, S. D. (2023). Enhancing Real-time Object Detection with YOLO Algorithm. *EAI Endorsed Transactions on Internet of Things*, *10*, 1. DOI: 10.4108/eetiot.4541

Lee, J., & il Hwang, K. (2022). YOLO with adaptive frame control for real-time object detection applications. *Multimedia Tools and Applications*, *81*(25), 36375–36396. DOI: 10.1007/s11042-021-11480-0

Lin, T.-Y., Goyal, P., Girshick, R., He, K., & Dollar, P. (2020). Focal Loss for Dense Object Detection. *IEEE Transactions on Pattern Analysis and Machine Intelligence*, *42*(2), 318–327. DOI: 10.1109/TPAMI.2018.2858826 PMID: 30040631

Liu, C., Zhou, H., Xu, H.-C., Hou, B.-Y., & Yao, L. (2020). Abnormal Behavior Recognition in an Examination Based on Pose Spatio-Temporal Features. *2020 IEEE 5th International Conference on Cloud Computing and Big Data Analytics (ICCCBDA)*, 380–386. DOI: 10.1109/ICCCBDA49378.2020.9095710

Magalhães, S. C., dos Santos, F. N., Machado, P., Moreira, A. P., & Dias, J. (2023). Benchmarking edge computing devices for grape bunches and trunks detection using accelerated object detection single shot multibox deep learning models. *Engineering Applications of Artificial Intelligence*, *117*, 105604. DOI: 10.1016/j.engappai.2022.105604

Mahajan, M., Rajeshwaribajre, R., Sharma, L., & Varma, S. L. (2020). Computer Vision Based Campus Surveillance. *Proceedings - IEEE 2020 2nd International Conference on Advances in Computing, Communication Control and Networking, ICACCCN 2020*, 1013–1018. DOI: 10.1109/ICACCCN51052.2020.9362776

Malhotra, M., & Chhabra, I. (2021). Automatic Invigilation Using Computer Vision. *Proceedings of the 3rd International Conference on Integrated Intelligent Computing Communication & Security (ICIIC 2021)*, *4*, 130–136. https://doi.org/https://doi.org/10.2991/ahis.k.210913.017

Masud, M. M., Hayawi, K., Mathew, S. S., Michael, T., & El Barachi, M. (2022). Smart Online Exam Proctoring Assist for Cheating Detection. *Lecture Notes in Computer Science (Including Subseries Lecture Notes in Artificial Intelligence and Lecture Notes in Bioinformatics), 13087 LNAI*, 118–132. DOI: 10.1007/978-3-030-95405-5_9

Meituan. (2022, June). *YOLOv6 By Meituan Vision AI Department*. Meituan Vision AI Department. https://github.com/meituan/YOLOv6

Nguyen, Q. H., Ly, H.-B., Ho, L. S., Al-Ansari, N., Van Le, H., Tran, V. Q., Prakash, I., & Pham, B. T. (2021). Influence of Data Splitting on Performance of Machine Learning Models in Prediction of Shear Strength of Soil. *Mathematical Problems in Engineering*, *2021*, 1–15. DOI: 10.1155/2021/4832864

Nishchal, J., Reddy, S., & Navya, P. N. (2020). Automated Cheating Detection in Exams using Posture and Emotion Analysis. *2020 IEEE International Conference on Electronics, Computing and Communication Technologies (CONECCT)*, 1–6. DOI: 10.1109/CONECCT50063.2020.9198691

Olorunshola, O. E., Irhebhude, M. E., & Evwiekpaefe, A. E. (2023). A Comparative Study of YOLOv5 and YOLOv7 Object Detection Algorithms. *Journal of Computing and Social Informatics*, *2*(1), 1–12. DOI: 10.33736/jcsi.5070.2023

Ossai, M. C., Ethe, N., Edougha, D. E., & Okeh, O. D. (2023). Academic integrity during examinations, age and gender as predictors of academic performance among high school students. *International Journal of Educational Development*, *100*, 102811. Advance online publication. DOI: 10.1016/j.ijedudev.2023.102811

Park, S., Lee, S., & Park, J. (2020). Data augmentation method for improving the accuracy of human pose estimation with cropped images. *Pattern Recognition Letters*, *136*, 244–250. DOI: 10.1016/j.patrec.2020.06.015

Purwita Sary, I., Ucok Armin, E., Andromeda, S., Engineering, E., & Singaperbangsa Karawang, U. (2023). Performance Comparison of YOLOv5 and YOLOv8 Architectures in Human Detection Using Aerial Images. *Ultima Computing : Jurnal Sistem Komputer*, *15*(1), 8–13. DOI: 10.31937/sk.v15i1.3204

Reddy, K. R., Priya, K. H., & Neelima, N. (2015). Object Detection and Tracking — A Survey. *2015 International Conference on Computational Intelligence and Communication Networks (CICN)*, 418–421. DOI: 10.1109/CICN.2015.317

Senthilkumar, V., Kanagaraj, G., Primya, T., Joycema, J., Joan, M. B., & Vicram, J. A. (2021). Application of AI And Computer Vision To Face Mask And Social Distance Detection in CCTV Video Streams. *2021 International Conference on Advancements in Electrical, Electronics, Communication, Computing and Automation (ICAECA)*, 1–5. DOI: 10.1109/ICAECA52838.2021.9675746

Septiana, D. (2022). *Hubungan Efikasi Diri Dan Stres Akademik Dengan Perilaku Menyontek Siswa* [Universitas Islam Negeri Raden Intan Lampung]. https://repository.radenintan.ac.id/17318/

Setiyono, B., Amini, D. A., & Sulistyaningrum, D. R. (2021). Number plate recognition on vehicle using YOLO - Darknet. *Journal of Physics: Conference Series*, *1821*(1), 012049. Advance online publication. DOI: 10.1088/1742-6596/1821/1/012049

Shandilya, S. K., Srivastav, A., Yemets, K., Datta, A., & Nagar, A. K. (2023). YOLO-based segmented dataset for drone vs. bird detection for deep and machine learning algorithms. *Data in Brief*, *50*, 109355. DOI: 10.1016/j.dib.2023.109355 PMID: 37609648

Sharma, P., Hans, P., & Gupta, S. C. (2020). Classification Of Plant Leaf Diseases Using Machine Learning And Image Preprocessing Techniques. *2020 10th International Conference on Cloud Computing, Data Science & Engineering (Confluence)*, 480–484. DOI: 10.1109/Confluence47617.2020.9057889

Siddhpura, A., & Siddhpura, M. (2020). Plagiarism, contract cheating and other academic misconducts in online engineering education: Analysis, detection and prevention strategies. *Proceedings of 2020 IEEE International Conference on Teaching, Assessment, and Learning for Engineering, TALE 2020*, 112–119. DOI: 10.1109/TALE48869.2020.9368311

Sletten, S. R. (2021). Rethinking Assessment: Replacing Traditional Exams with Paper Reviews. *Journal of Microbiology & Biology Education*, *22*(2), e00109-21. Advance online publication. DOI: 10.1128/jmbe.00109-21 PMID: 34594442

Soleimanitaleb, Z., Keyvanrad, M. A., & Jafari, A. (2019). Object Tracking Methods:A Review. *2019 9th International Conference on Computer and Knowledge Engineering (ICCKE)*, 282–288. DOI: 10.1109/ICCKE48569.2019.8964761

Srinivasan, A., Srikanth, A., Indrajit, H., & Narasimhan, V. (2020). A Novel Approach for Road Accident Detection using DETR Algorithm. *2020 International Conference on Intelligent Data Science Technologies and Applications, IDSTA 2020*, 75–80. DOI: 10.1109/IDSTA50958.2020.9263703

Tripathi, G., & Kumar, R. (2020). Early Prediction of Diabetes Mellitus Using Machine Learning. *2020 8th International Conference on Reliability, Infocom Technologies and Optimization (Trends and Future Directions) (ICRITO)*, 1009–1014. DOI: 10.1109/ICRITO48877.2020.9197832

Varma, G., Subramanian, A., Namboodiri, A., Chandraker, M., & Jawahar, C. V. (2019). IDD: A Dataset for Exploring Problems of Autonomous Navigation in Unconstrained Environments. *2019 IEEE Winter Conference on Applications of Computer Vision (WACV)*, 1743–1751. DOI: 10.1109/WACV.2019.00190

Wu, C., Luo, Q., & Shou, D. (2022). Image Recognition Technology and Development of YOLO Algorithm. *Advances in Transdisciplinary Engineering*, *33*, 32–38. DOI: 10.3233/ATDE221146

Xu, S., Wang, X., Lv, W., Chang, Q., Cui, C., Deng, K., Wang, G., Dang, Q., Wei, S., Du, Y., & Lai, B. (2022). *PP-YOLOE: An evolved version of YOLO*. http://arxiv.org/abs/2203.16250

Zaffar, A., Jawad, M., Shabbir, M., Factory, P. O., & Cantt, W. (2023). A Novel CNN-RNN Model for E-Cheating Detection Based on Video Surveillance. *UW JOURNAL OF CMPUTER SCIENCE, 5*(1). https://uwjcs.org.pk/index.php/ojs/article/view/64

Zhao, Z. Q., Zheng, P., Xu, S. T., & Wu, X. (2019). Object Detection with Deep Learning: A Review. In *IEEE Transactions on Neural Networks and Learning Systems* (Vol. 30, Issue 11, pp. 3212–3232). Institute of Electrical and Electronics Engineers Inc. DOI: 10.1109/TNNLS.2018.2876865

Zhou, Z., Li, Y., Peng, C., Wang, H., & Du, S. (2021). Image Processing: Facilitating Retinanet for Detecting Small Objects. *Journal of Physics: Conference Series*, *1815*(1), 012016. DOI: 10.1088/1742-6596/1815/1/012016

Zoph, B., Cubuk, E. D., Ghiasi, G., Lin, T. Y., Shlens, J., & Le, Q. V. (2020). Learning data augmentation strategies for object detection. In Computer Vision–ECCV 2020: 16th European Conference, Glasgow, UK, August 23–28, 2020 [Springer International Publishing.]. *Proceedings*, *16*(Part XXVII), 566–583.

Zulkifli, N. (2022). Efektivitas Pelaksanaan Ujian Semester Menggunakan Computer Based Test. *Edumaspul: Jurnal Pendidikan*, *6*(1), 186–191. DOI: 10.33487/edumaspul.v6i1.3057

Compilation of References

Aanchal, A., & Prasad, P. W. C. (2021, November). Topic: Scoping review of Blockchain based data storage technique in industrial IoT data management. In 2021 6th International Conference on Innovative Technology in Intelligent System and Industrial Applications (CITISIA) (pp. 1-10). IEEE.

Abbass, H. A. (2002). An evolutionary artificial neural networks approach for breast cancer diagnosis. *Artificial Intelligence in Medicine*, *25*(3), 265–281. DOI: 10.1016/S0933-3657(02)00028-3 PMID: 12069763

Abhilasha, A., Sreenivasulu, A., & Manimozhi, T. KUMAR, P. S., Selvakumar, P., & Singh, P. (2022, April). The Model of Smart Sensing Device For Sensitive Nano-clusters Modification in Sensing Properties. In 2022 2nd International Conference on Advance Computing and Innovative Technologies in Engineering (ICACITE) (pp. 1043-1047). IEEE.

Abhilasha, A., Srccnivasulu, A., Manimozhi, T., Kumar, P. S., Selvakumar, P., & Singh, P. (2022, April). The Model of Smart Sensing Device For Sensitive Nano-clusters Modification in Sensing Properties. In 2022 2nd International Conference on Advance Computing and Innovative Technologies in Engineering (ICACITE) (pp. 1043-1047). IEEE.

Abunasser, B. S., Al-Hiealy, M. R. J., Zaqout, I. S., & Abu-Naser, S. S. (2022). Breast cancer detection and classification using deep learning Xception algorithm. *International Journal of Advanced Computer Science and Applications*, *13*(7). Advance online publication. DOI: 10.14569/IJACSA.2022.0130729

Acemoglu, D., & Autor, D. (2011). "Chapter 12 – Skills, Tasks and Technologies: Implications for Employment and Earnings". In: Handbook of Labor Economics. Ed. by David Card and Orley Ashenfelter. Vol. 4B. Elsevier, pp. 1043–1171.

Acemoglu, D., Anderson, G. W., Beede, D. N., Buffington, C., Childress, E. E., Dinlersoz, E., Foster, L. S., Goldschlag, N., Haltiwanger, J. C., Kroff, Z., Restrepo, P., & Zolas, N. (2022). "Automation and the Workforce: A Firm-Level View from the 2019 Annual Business Survey". In: Technology, Productivity,and Economic Growth. Ed. by Susantu Basu, Lucy Eldridge, John C. Haltiwanger, and Erich Strassner. Studies in Income and Wealth. University of Chicago Press. DOI: 10.3386/w30659

Acemoglu, D. (2024). Harms of AI. In Bullock, J. B., Chen, Y.-C., Himmelreich, J., Hudson, V. M., Korinek, A., Young, M. M., & Zhang, B. (Eds.), *The Oxford Handbook of AI Governance*. Oxford University Press. DOI: 10.1093/oxfordhb/9780197579329.013.65

Adamova, A., Zhukabayeva, T., & Adamov, N. (2024). Machine Learning Algorithms for Intrusion Detection in IoT-enabled Smart Homes. *Procedia Computer Science, 241*, 427–432. https://doi.org/https://doi.org/10.1016/j.procs.2024.08.059. DOI: 10.1016/j.procs.2024.08.059

Adhikari, S. (2021). Method for Intelligent Information Processing Using Software Reuse Library and Blockchain Architecture. In *AIAA Scitech 2021 Forum* (p. 1919). DOI: 10.2514/6.2021-1919

Adibi, M., Sholichin, M., & Prayogo, T. B. (2024). Comparative Planning Of Sprinkler Irrigation and Drip Irrigation for Orange And Avocado Crops in Research Land. *JTRESDA : Jurnal Teknologi dan Rekayasa Sumber Daya Air, 04*(01), 790–803. DOI: 10.21776/ub.jtresda.2024.004.01.067

Adil, M., Simon, R., & Khatri, S. K. (2019). Automated Invigilation System for Detection of Suspicious Activities during Examination. *2019 Amity International Conference on Artificial Intelligence (AICAI)*, 361–366. DOI: 10.1109/AICAI.2019.8701263

Agarwal, R., Kumar, T., Singh, A., & Schutte, D. P. (2025). Empowering Indigenous Economies: Financial and Investment Prediction for Sustainable Development. In Irfan, M., Majeed, M., Kassim, S., & Ofori, K. (Eds.), *The Future of Indigenous Inclusivity: Unlocking Finance for Economic Development* (pp. 63–82). IGI Global Scientific Publishing., DOI: 10.4018/979-8-3693-7924-0.ch004

Aggarwal, M., Khullar, V., Goyal, N., Singh, A., Tolba, A., Thompson, E. B., & Kumar, S. (2023). Pre-trained deep neural network-based features selection supported machine learning for rice leaf disease classification. *Agriculture, 13*(5), 936. DOI: 10.3390/agriculture13050936

Agiollo, A., Conti, M., Kaliyar, P., Lin, T. N., & Pajola, L. (2021). DETONAR: Detection of routing attacks in RPL-based IoT. *IEEE Transactions on Network and Service Management, 18*(2), 1178–1190. DOI: 10.1109/TNSM.2021.3075496

Agrawal, A., Kumar, T., Agarwal, R., & Gupta, A. (2024), "AI-Driven Personalized Risk Management in the Insurance Sector", Taneja, S., Kumar, P., Reepu, Kukreti, M. and Özen, E. (Ed.) Data Alchemy in the Insurance Industry, Emerald Publishing Limited, Leeds, pp. 27-39. https://doi.org/DOI: 10.1108/978-1-83608-582-920241006

Aguerchi, K., Jabrane, Y., Habba, M., & El Hassani, A. H. (2024). A CNN hyper-parameters optimization based on particle swarm optimization for mammography breast cancer classification. *Journal of Imaging, 10*(2), 30. DOI: 10.3390/jimaging10020030 PMID: 38392079

Ahad, M. T., Ameen, M., Islam, M. R., Hasan, M. K., & Islam, M. T. (2023). Comparison of CNN-based deep learning architectures for rice diseases classification. *Artificial Intelligence in Agriculture, 9*, 22–35. DOI: 10.1016/j.aiia.2023.07.001

Ahamed, N. N., Karthikeyan, P., Velliangiri, S., & Gunjan, V. K. (2023). Blockchain Security for the Internet of Things. In *Modern Approaches in IoT and Machine Learning for Cyber Security: Latest Trends in AI* (pp. 75–92). Springer International Publishing.

Ahmad, N., Asif, H. M. S., Saleem, G., Younus, M. U., Anwar, S., & Anjum, M. R. (2021). Leaf image-based plant disease identification using color and texture features. *Wireless Personal Communications, 121*(2), 1139–1168. DOI: 10.1007/s11277-021-09054-2

Ahmad, R., & Alsmadi, I. (2021). Machine learning approaches to IoT security: A systematic literature review. *Internet of Things : Engineering Cyber Physical Human Systems, 14*, 100365. DOI: 10.1016/j.iot.2021.100365

Ahvan, Y. R., & Pour, H. Z. (2016). The correlation of multiple intelligences for the achievements of secondary students. *Educational Research Review, 11*(4), 141–145. DOI: 10.5897/ERR2015.2532

Aiswarya, J. S. S. N. S., Sowbhagya Sree, P. S., Hemanjali, K., Pranavi, P., Babu, V., & Kumar, S. (2022). A review: Various challenges and causes based on cervical cancer. *2022 5th International Conference on Contemporary Computing and Informatics (IC3I)*, 1452–1457. https://doi.org/DOI: 10.1109/IC3I57145.2022.10002142

Ajibade, P., & Mutula, S. M. (2021). Virtual learning: A disruptive service in academic libraries. *Library Hi Tech News, 38*(1), 12–13. DOI: 10.1108/LHTN-07-2020-0067

Ajra, H., Nahar, M. K., Sarkar, L., & Islam, M. S. (2020). Disease detection of plant leaf using image processing and CNN with preventive measures. In *2020 Emerging Technology in Computing, Communication and Electronics (ETCCE)* (pp. 1–6). IEEE. DOI: 10.1109/ETCCE51779.2020.9350890

Akhmad, M. W. (2019). *Pengaruh Kecerdasan Logis-Matematis dan Kecerdasan Spasial-Visual terhadap Kemampuan Pemecahan Masalah Geometri Siswa Kelas IV di MI Darul Ulum Kota Batu*. UIN Maulana Malik Ibrahim.

Akhtar, M. M., Rizvi, D. R., Ahad, M. A., Kanhere, S. S., Amjad, M., & Coviello, G. (2021). Efficient data communication using distributed ledger technology and iota-enabled internet of things for a future machine-to-machine economy. *Sensors (Basel)*, *21*(13), 4354. DOI: 10.3390/s21134354 PMID: 34202173

Ak, M. F. (2020). A comparative analysis of breast cancer detection and diagnosis using data visualization and machine learning applications. *Health Care*, *8*(2), 111. DOI: 10.3390/healthcare8020111PMC PMID: 32357391

Akram, S. V., Malik, P. K., Singh, R., Anita, G., & Tanwar, S. (2020). Adoption of blockchain technology in various realms: Opportunities and challenges. *Security and Privacy*, *3*(5), e109. DOI: 10.1002/spy2.109

Al-Abbadey, M., Fong, M. M. W., Wilde, L. J., Ingham, R., & Ghio, D. (2021). Mobile health apps: An exploration of user-generated reviews in Google Play Store on a physical activity application. *Digital Health*, *7*, 1–12. DOI: 10.1177/20552076211014988 PMID: 34017609

Alahmari, M. (2023). Exploring the Influential Factors Affecting Staff Willingness to Adopt Augmented Reality. *International Journal of Information and Education Technology (IJIET)*, *13*(7), 1078–1084. DOI: 10.18178/ijiet.2023.13.7.1907

Aldabbas, H., & Amin, R. (2021). A novel mechanism to handle address spoofing attacks in SDN based IoT. *Cluster Computing*, *24*(4), 3011–3026. DOI: 10.1007/s10586-021-03309-0

Al-Doulat, A., Nur, N., Karduni, A., Benedict, A., Al-Hossami, E., & Maher, M. Lou, Dou, W., Dorodchi, M., & Niu, X. (2020). Making Sense of Student Success and Risk Through Unsupervised Machine Learning and Interactive Storytelling. *Artificial Intelligence in Education*, 9. https://doi.org/https://doi.org/10.1007/978-3-030-52237-7_1

Ali, G., Mijwil, M. M., Buruga, B. A., & Abotaleb, M. (2024). A Comprehensive review on cybersecurity issues and their mitigation measures in FinTech.

Ali, L. (2024). Influence of abiotic stresses on the production and sustainability of rice crop. *Agrobiological Records*, *16*, 99–108. DOI: 10.47278/journal.abr/2024.017

Al-Kinani, H. K. R., & Al-Omari, A. H. M. (2021). Analytical thinking and its relationship to logical intelligence for scientific fifth-grade students in mathematics. *Turkish Journal of Computer and Mathematics Education*, *12*(10), 3148–3161. https://www.proquest.com/scholarly-journals/analytical-thinking-relationship-logical/docview/2628339997/se-2?accountid=62707

Al-Labadi, L., & Sant, S. (2021). Enhance Learning Experience Using Technology in Class. *Journal of Technology and Science Education*, *11*(1), 44–52. DOI: 10.3926/jotse.1050

Almadani, M. S., & Hussain, F. K. (2023, November). Implementing a Secure Blockchain-Based Wallet System with Multi-Factor Authentication. In 2023 IEEE International Conference on e-Business Engineering (ICEBE) (pp. 23-30). IEEE. DOI: 10.1109/ICEBE59045.2023.00010

Al-Shareeda, M., Ali, M., & Manickam, S. (2023). The blockchain internet of things: Review, opportunities, challenges, and recommendations. *Indonesian Journal of Electrical Engineering and Computer Science*, *31*(3), 1673–1683. DOI: 10.11591/ijeecs.v31.i3.pp1673-1683

Al-Tam, R. M., & Narangale, S. M. (2021). Breast cancer detection and diagnosis using machine learning: A survey. *Journal of Scientific Research*, *65*(5), 265–285. DOI: 10.37398/JSR.2021.650532

Alzoubi, Y. I., Al-Ahmad, A., Kahtan, H., & Jaradat, A. (2022). Internet of things and blockchain integration: Security, privacy, technical, and design challenges. *Future Internet*, *14*(7), 216. DOI: 10.3390/fi14070216

Ambusaidi, M. A., Tan, Z., He, X., Nanda, P., Lu, L. F., & Jamdagni, A. (2014). Intrusion detection method based on nonlinear correlation measure. *International Journal of Internet Protocol Technology*, *8*(2/3), 77. DOI: 10.1504/IJIPT.2014.066377

Amethiya, Y., Pipariya, P., Patel, S., & Shah, M. (2022). Comparative analysis of breast cancer detection using machine learning and biosensors. *Intelligent Medicine*, *2*(2), 69–81. DOI: 10.1016/j.imed.2021.08.004

Amoah, E., & Oh, J. Y. (2020). BLOCKCHAIN IN IoT AND PROJECT MANAGEMENT. *Issues in Information Systems*, *21*(3).

Anami, B. S., Malvade, N. N., & Palaiah, S. (2020). Classification of yield-affecting biotic and abiotic paddy crop stresses using field images. *Information Processing in Agriculture*, *7*(2), 272–285. DOI: 10.1016/j.inpa.2019.08.005

Anas, M., Ul Haq, I., Husnain, G., Ali, S., & Jaffery, F. (2024). Advancing breast cancer detection: Enhancing YOLOv5 network for accurate classification in mammogram images.

Anastasi, G., Franchini, M., Pieroni, S., Buzzi, M., Buzzi, M. C., Leporini, B., & Molinaro, S. (2024). Machine learning techniques in breast cancer preventive diagnosis: A review. *Multimedia Tools and Applications*. https://doi.org/ *(Add DOI if available)*DOI: 10.1007/s11042-024-16809-9

Andrianto, H., Faizal, A., & Armandika, F. (2020). Smartphone application for deep learning-based rice plant disease detection. In *2020 International Conference on Information Technology Systems and Innovation (ICITSI)* (pp. xx–xx). IEEE. DOI: 10.1109/ICITSI50517.2020.9264942

Angara, J. S., & Saripalle, R. S. (2022). Towards a virtual water currency for industrial products using blockchain technology. *Water Policy*, *24*(6), 923–941. DOI: 10.2166/wp.2022.285

Angellina, A., Herwindiati, D. E., & Hendryli, J. (2023). Performa Support Vector Machine Pada Klasifikasi Lahan dan Air Tanah. *JURNAL MEDIA INFORMATIKA BUDIDARMA*, *7*(1), 231. DOI: 10.30865/mib.v7i1.5279

Araka, E., Oboko, R., Maina, E., & Gitonga, R. (2022). Using educational data mining techniques to identify profiles in self-regulated learning: An empirical evaluation. *International Review of Research in Open and Distance Learning*, *23*(1), 131–162. DOI: 10.19173/irrodl.v22i4.5401

Archana, K. S., Srinivasan, S., Bharathi, S. P., Balamurugan, R., Prabakar, T. N., & Britto, A. S. F. (2022). A novel method to improve computational and classification performance of rice plant disease identification. *The Journal of Supercomputing*, *78*(xx), 1–21. https://doi.org/xxx. DOI: 10.1007/s11227-021-04245-x

Arora, Y. K., Sharma, M., Vardhan, M., Sharma, D. K., & Batra, R. (2020). A novel approach for detection and classification of rice leaf diseases. In *Applied Computer Vision and Image Processing:Proceedings of ICCET 2020* (Vol. 1). Springer. DOI: 10.1007/978-981-15-4029-5_10

Arsat, N., Bakar, N. S. A. A., & Yahya, N. (2022, September). Testing in blockchain-based systems: a systematic review. In 2022 10th International Conference on Cyber and IT Service Management (CITSM) (pp. 1-6). IEEE. DOI: 10.1109/CITSM56380.2022.9935846

Ashari, I. F., Banjarnahor, R., Farida, D. R., Aisyah, S. P., Dewi, A. P., & Humaya, N. (2022). Application of Data Mining with the K-Means Clustering Method and Davies Bouldin Index for Grouping IMDB Movies. *Journal of Applied Informatics and Computing, 6*(1), 07–15. DOI: 10.30871/jaic.v6i1.3485

Ashari, I. F., Dwi Nugroho, E., Baraku, R., Novri Yanda, I., & Liwardana, R. (2023). Analysis of Elbow, Silhouette, Davies-Bouldin, Calinski-Harabasz, and Rand-Index Evaluation on K-Means Algorithm for Classifying Flood-Affected Areas in Jakarta. *Journal of Applied Informatics and Computing, 7*(1), 89–97. DOI: 10.30871/jaic.v7i1.4947

Aurangzeb, M., Wang, Y., Iqbal, S., Naveed, A., Ahmed, Z., Alenezi, M., & Shouran, M. (2024). Enhancing cybersecurity in smart grids: Deep black box adversarial attacks and quantum voting ensemble models for blockchain privacy-preserving storage. *Energy Reports, 11*, 2493–2515. DOI: 10.1016/j.egyr.2024.02.010

Awalu, I. L., Kook, P. H., & Lim, J. S. (2019, July). Development of a distributed blockchain evoting system. In Proceedings of the 2019 10th International Conference on E-business, Management and Economics (pp. 207-216). DOI: 10.1145/3345035.3345080

Azbeg, K., Ouchetto, O., Andaloussi, S. J., & Fetjah, L. (2022). A taxonomic review of the use of IoT and blockchain in healthcare applications. *Ingénierie et Recherche Biomédicale : IRBM = Biomedical Engineering and Research, 43*(5), 511–519. DOI: 10.1016/j.irbm.2021.05.003

Azim, M. A., Islam, M. K., Rahman, M. M., & Jahan, F. (2021). An effective feature extraction method for rice leaf disease classification. *Telkomnika, 19*(2), 463–470. DOI: 10.12928/telkomnika.v19i2.16488

Babina, T., Fedyk, A., He, A. X., & Hodson, J. (2024). Firm Investments in Artificial Intelligence Technologies and Changes in Workforce Composition. In Basu, S., Eldridge, L., Haltiwanger, J. C., & Strassner, E. (Eds.), *Technology, Productivity, and Economic Growth.* National Bureau of Economic Research.

Babina, T., Fedyk, A., He, A., & Hodson, J. (2024). Artificial Intelligence, Firm Growth, and Product Innovation. *Journal of Financial Economics, 151*, 151. DOI: 10.1016/j.jfineco.2023.103745

Bahar, A., Prananta, A. W., & Afifudin, M. (2024). *Mobile App Usability, Data Privacy, Brand Loyalty, and Customer Retention in Indonesia ' s Online Transportation Services Industry. 5*(2), 2562–2572.

Bai, Y., ... (2022). Constitutional AI: Harmlessness from AI Feedback. Preprint. arXiv. arXiv: 2212.08073.

Bala, K., & Kaur, P. D. (2022, January). Changing trends of blockchain in IoT: Benefits and challenges. In *2022 12th International Conference on Cloud Computing, Data Science & Engineering (Confluence)* (pp. 324-329). IEEE.

Balaji, V., Krishnamoorthi, R., Saravanan, S., & Poornima, R. (2023). Deep transfer learning technique for multimodal disease classification in plant images. *Contrast Media & Molecular Imaging*, *2023*, 5644727. Advance online publication. DOI: 10.1155/2023/5644727 PMID: 37213211

Balevi, E., & Gitlin, R. D. (2018). Optimizing the number of fog nodes for cloud-fog-thing networks. *IEEE Access : Practical Innovations, Open Solutions*, *6*, 11173–11183. DOI: 10.1109/ACCESS.2018.2808598

Ball, J. E., Anderson, D. T., & Chan, C. S. (2017). Comprehensive Survey of Deep Learning in Remote Sensing: Theories, Tools, And Challenges for the Community. *Journal of Applied Remote Sensing*, *11*(04), 1. DOI: 10.1117/1.JRS.11.042609

Balovsyak, S., Derevyanchuk, O., Kravchenko, H., Ushenko, Y., & Hu, Z. (2023). Clustering students according to their academic achievement using Fuzzy Logic. *International Journal of Modern Education and Computer Science*, *15*(6), 31–43. DOI: 10.5815/ijmecs.2023.06.03

Barba-Guaman, L., Naranjo, J. E., Ortiz, A., & Gonzalez, J. G. P. (2021). Object Detection in Rural Roads Through SSD and YOLO Framework. In H. and D. G. and M. F. and R. C. A. M. Rocha Álvaro and Adeli (Ed.), *Trends and Applications in Information Systems and Technologies* (pp. 176–185). Springer International Publishing. DOI: 10.1007/978-3-030-72657-7_17

Bari, B. S.. (2021). A real-time approach of diagnosing rice leaf disease using deep learning-based faster R-CNN framework. *PeerJ. Computer Science*, *7*, e432. DOI: 10.7717/peerj-cs.432 PMID: 33954231

Bårli, E. M., Yazidi, A., Viedma, E. H., & Haugerud, H. (2021). DoS and DDoS mitigation using variational autoencoders. *Computer Networks*, *199*, 108399. DOI: 10.1016/j.comnet.2021.108399

Barlow, W. E., White, E., Ballard-Barbash, R., Vacek, P. M., Titus-Ernstoff, L., Carney, P. A., Tice, J. A., Buist, D. S. M., Geller, B. M., Rosenberg, R., Yankaskas, B. C., & Kerlikowske, K. (2006). Prospective breast cancer risk prediction model for women undergoing screening mammography. *Journal of the National Cancer Institute*, *98*(17), 1204–1214. DOI: 10.1093/jnci/djj331 PMID: 16954473

Bashir, F., & Warraich, N. F. (2023). Future libraries' blockchain opportunities and challenges: a systematic literature review and research agenda. Digital Library Perspectives, (ahead-of-print).

Bashir, F., & Warraich, N. F. (2022). Prospects of semantic web and blockchain technologies in libraries. In *Blockchain and Deep Learning: Future Trends and Enabling Technologies* (pp. 31–45). Springer International Publishing. DOI: 10.1007/978-3-030-95419-2_2

Bathalapalli, V. K., Mohanty, S. P., Kougianos, E., Yanambaka, V. P., Baniya, B. K., & Rout, B. (2021, December). A puf-based approach for sustainable cybersecurity in smart agriculture. In *2021 19th OITS International Conference on Information Technology (OCIT)* (pp. 375-380). IEEE. DOI: 10.1109/OCIT53463.2021.00080

Batool, S., Rashid, J., Nisar, M. W., Kim, J., Kwon, H.-Y., & Hussain, A. (2023). Educational data mining to predict students' academic performance: A survey study. *Education and Information Technologies*, *28*(1), 905–971. DOI: 10.1007/s10639-022-11152-y

Bayih, A. Z., Morales, J., Assabie, Y., & de By, R. A. (2022). Utilization of internet of things and wireless sensor networks for sustainable smallholder agriculture. *Sensors (Basel)*, *22*(9), 3273. DOI: 10.3390/s22093273 PMID: 35590963

Bera, S., Dey, T., Mukherjee, A., & Buyya, R. (2023). E-CropReco: A dew-edge-based multi-parametric crop recommendation framework for internet of agricultural things. *The Journal of Supercomputing*, *79*(11), 1–35. DOI: 10.1007/s11227-023-05131-4

Berutu, S. S., Budiati, H., & Gulo, F. (2023). Data preprocessing approach for machine learning-based sentiment classification. *INFOTEL*, *15*(4), 317–325. DOI: 10.20895/infotel.v15i4.1030

Beyene, M., Toussaint, P. A., Thiebes, S., Schlesner, M., Brors, B., & Sunyaev, A. (2022). A scoping review of distributed ledger technology in genomics: Thematic analysis and directions for future research. *Journal of the American Medical Informatics Association : JAMIA*, *29*(8), 1433–1444. DOI: 10.1093/jamia/ocac077 PMID: 35595301

Bhabay, Y., & Mane, A. (2022). Rice diseases pests and nutritional deficiency classification using convolutional neural network. *Grenze International Journal of Engineering & Technology (GIJET)*, *8*(2). [No page numbers available]

Bhanu, K. N., Mahadevaswamy, H. S., & Jasmine, H. J. (2020). IoT based Smart System for Enhanced Irrigation in Agriculture. *2020 International Conference on Electronics and Sustainable Communication Systems (ICESC)*, 760–765. DOI: 10.1109/ICESC48915.2020.9156026

Bharanidharan, N.. (2023). *Multiclass paddy disease detection using filter-based feature transformation technique*. IEEE., DOI: 10.1109/ACCESS.2023.3322587

Bhartiya, V. P., Janghel, R. R., & Rathore, Y. K. (2022). Rice leaf disease prediction using machine learning. In *2022 Second International Conference on Power, Control and Computing Technologies (ICPC2T)* (pp. 1–6). IEEE. DOI: 10.1109/ICPC2T53885.2022.9776692

Bhattacharya, S., Mukherjee, A., & Phadikar, S. (2020). A deep learning approach for the classification of rice leaf diseases. In *Intelligence Enabled Research: DoSIER 2019* (pp. 61–69). DOI: 10.1007/978-981-15-2021-1_8

Bhimavarapu, U. (2023). Prediction and classification of rice leaves using the improved PSO clustering and improved CNN. *Multimedia Tools and Applications*, *82*(14), 21701–21714. DOI: 10.1007/s11042-023-14631-7

Bhugra, S., Chaudhury, S., & Lall, B. (2015). Use of leaf colour for drought stress analysis in rice. In *2015 Fifth National Conference on Computer Vision, Pattern Recognition, Image Processing and Graphics (NCVPRIPG)* (pp. 1–6). IEEE. DOI: 10.1109/NCVPRIPG.2015.7490060

Bhushan, B., Sahoo, C., Sinha, P., & Khamparia, A. (2021). Unification of Blockchain and Internet of Things (BIoT): Requirements, working model, challenges and future directions. *Wireless Networks*, *27*(1), 55–90. DOI: 10.1007/s11276-020-02445-6

Bhuyan, M. H., Bhattacharyya, D. K., & Kalita, J. K. (2014). Network Anomaly Detection: Methods, Systems and Tools. *IEEE Communications Surveys and Tutorials. IEEE Communications Surveys and Tutorials*, *16*(1), 303–336. DOI: 10.1109/SURV.2013.052213.00046

Bhuyan, P., Saikia, P. A., Barman, S. D., & Deka, P. C. (2023). SE_SPnet: Rice leaf disease prediction using stacked parallel convolutional neural network with squeeze-and-excitation. *Expert Systems: International Journal of Knowledge Engineering and Neural Networks*, *40*(7), e13304. DOI: 10.1111/exsy.13304

Bijoy, M. H.. (2024). *Towards sustainable agriculture: A novel approach for rice leaf disease detection using dCNN and enhanced dataset*. IEEE., DOI: 10.1109/ACCESS.2024.3344556

Bodkhe, U., Tanwar, S., Parekh, K., Khanpara, P., Tyagi, S., Kumar, N., & Alazab, M. (2020). Blockchain for industry 4.0: A comprehensive review. *IEEE Access : Practical Innovations, Open Solutions, 8*, 79764–79800. DOI: 10.1109/ACCESS.2020.2988579

Bonett, D. G., & Wright, T. A. (2015). Cronbach's alpha reliability: Interval estimation, hypothesis testing, and sample size planning. *Journal of Organizational Behavior, 36*(1), 3–15. DOI: 10.1002/job.1960

Bouzar-Benlabiod, L., Harrar, K., Yamoun, L., Khodja, M. Y., & Akhloufi, M. A. (2023). A novel breast cancer detection architecture based on a CNN-CBR system for mammogram classification. *Computers in Biology and Medicine, 163*, 107133. DOI: 10.1016/j.compbiomed.2023.107133 PMID: 37327756

Brandão, J. G., Castro Junior, A. P., Pacheco, V. M. G., Rodrigues, C. G., Belo, O. M. O., Coimbra, A. P., & Calixto, W. P. (2025). Optimization of machine learning models for sentiment analysis in social media. *Information Sciences, 694*(October 2024), 121704. DOI: 10.1016/j.ins.2024.121704

Breiman, L. (2001). No Title. *Machine Learning, 45*(1), 5–32. DOI: 10.1023/A:1010933404324

Brotsis, S., Limniotis, K., Bendiab, G., Kolokotronis, N., & Shiaeles, S. (2021). On the suitability of blockchain platforms for IoT applications: Architectures, security, privacy, and performance. *Computer Networks, 191*(September 2020), 108005. DOI: 10.1016/j.comnet.2021.108005

Budiyanto, R., Yohanie, D. D., & Widodo, S. (2018). Pengaruh Kecerdasan Logis-Matematis terhadap Kemampuan Pemecahan Masalah Matematis Siswa pada Penyelesaian Soal Cerita Materi SPLDV. *Simki-Techsain, 2*(10), 1–8.

Bujang, M. A., Omar, E. D., & Baharum, N. A. (2018). A review on sample size determination for Cronbach's alpha test: A simple guide for researchers. *The Malaysian Journal of Medical Sciences : MJMS, 25*(6), 85–99. DOI: 10.21315/mjms2018.25.6.9 PMID: 30914882

Cai, H., & King, I. (2020). Education Technology for Online Learning in Times of Crisis. *Proceedings of 2020 IEEE International Conference on Teaching, Assessment, and Learning for Engineering, TALE 2020*, 758–763. DOI: 10.1109/TALE48869.2020.9368387

Cai, J., Luo, J., Wang, S., & Yang, S. (2018). Feature selection in machine learning: A new perspective. *Neurocomputing, 300*, 70–79. DOI: 10.1016/j.neucom.2017.11.077

Caldarelli, G., & Ellul, J. (2021). Trusted academic transcripts on the blockchain: A systematic literature review. *Applied Sciences (Basel, Switzerland), 11*(4), 1842. DOI: 10.3390/app11041842

Caliskan, A., Bryson, J. J., & Narayanan, A. (2017). Semantics Derived Automatically from Language Corpora Contain Human-Like Biases. *Science, 356*(6334), 183–186. DOI: 10.1126/science.aal4230 PMID: 28408601

Cambron-Goulet, É., Dumas, J. P., Bergeron, É., Bergeron, L., & St-Onge, C. (2019). Guidelines for Creating Written Clinical Reasoning Exams: Insight from a Delphi Study. *Health Profession Education, 5*(3), 237–247. DOI: 10.1016/j.hpe.2018.09.001

Cao, L., & Yang, H. (2019, August). Building virtual digital library based on P2P and Blockchain. In 2019 11th International Conference on Intelligent Human-Machine Systems and Cybernetics (IHMSC) (Vol. 2, pp. 341-345). IEEE.

Cao, F., Xing, B., Luo, J., Li, D., Qian, Y., Zhang, C., Bai, H., & Zhang, H. (2023). An Efficient Object Detection Algorithm Based on Improved YOLOv5 for High-Spatial-Resolution Remote Sensing Images. *Remote Sensing (Basel), 15*(15), 3755. DOI: 10.3390/rs15153755

Cárcamo, J., Cuevas, A., & Rodríguez, L. A. (2024). On uniqueness of the set of k-means. arXiv preprint arXiv:2410.13495.

Cavallo, A. (2018). More Amazon Effects: Online Competition and Pricing Behaviors. Working Paper 25138. National Bureau of Economic Research.

Cazzaniga, M., Jaumotte, F., Li, L., Melina, G., Panton, A. J., Pizzinelli, C., Rockall, E., & Tavares, M. M. (2024). *Gen-AI: Artificial Intelligence and the Future of Work. Staff Discussion Note 2024/001*. International Monetary Fund. DOI: 10.5089/9798400262548.006

Chalkias, K. K., Lindstrøm, J., Maram, D., Riva, B., Roy, A., Sonnino, A., & Wang, J. (2024, May). Fastcrypto: Pioneering Cryptography Via Continuous Benchmarking. In Companion of the 15th ACM/SPEC International Conference on Performance Engineering (pp. 227-234).

Chandola, V., Banerjee, A., & Kumar, V. (2009). Anomaly detection. *ACM Computing Surveys, 41*(3), 1–58. DOI: 10.1145/1541880.1541882

Chattopadhyay, P., Patel, H. P., & Parmar, V. (2022, August). Internet of Things (IoT) in Smart Agriculture. In *2022 3rd International Conference on Electronics and Sustainable Communication Systems (ICESC)* (pp. 536-540). IEEE.

Chaudhari, D. J., & Malathi, K. (2023). Detection and prediction of rice leaf disease using a hybrid CNN-SVM model. *Optical Memory and Neural Networks (Information Optics)*, *32*(1), 39–57. DOI: 10.3103/S1060992X2301006X

Chauhan, N., & Dwivedi, R. K. (2022, April). Designing a secure smart healthcare system with blockchain: a review. In *2022 1st International Conference on Informatics (ICI)* (pp. 214-219). IEEE. DOI: 10.1109/ICI53355.2022.9786895

Chembakottu, B., Li, H., & Khomh, F. (2023). A large-scale exploratory study of android sports apps in the google play store. *Information and Software Technology*, *164*(June), 107321. DOI: 10.1016/j.infsof.2023.107321

Chemodanov, D., Calyam, P., & Palaniappan, K. (2020). "Fog computing to enable geospatial video analytics for disaster-incident situational awareness" in Fog Comput. *Theory into Practice*, •••, 473–503.

Chen, B., Azad, S., Fowler, M., West, M., & Zilles, C. (2020). Learning to Cheat: Quantifying Changes in Score Advantage of Unproctored Assessments Over Time. *L@S 2020 - Proceedings of the 7th ACM Conference on Learning @ Scale*, 197–206. DOI: 10.1145/3386527.3405925

Chen, F., Xiao, Z., Cui, L., Lin, Q., Li, J., & Yu, S. (2020). Blockchain for Internet of Things applications: A review and open issues. *Journal of Network and Computer Applications*, *172*, 102839. DOI: 10.1016/j.jnca.2020.102839

Chen, J., Chen, Y., Du, X., Li, C., Lu, J., Zhao, S., & Zhou, X. (2013). Big data challenge: A data management perspective. *Frontiers of Computer Science*, *7*(2), 157–164. DOI: 10.1007/s11704-013-3903-7

Chen, S. (2022). Improved Fuzzy Algorithm for College Students' Academic Early Warning. *Mathematical Problems in Engineering*, *2022*, 1–9. Advance online publication. DOI: 10.1155/2022/5764800

Chhadikar, N., Bhamare, P., Patil, K., & Kumari, S. (2019). Image Processing Based Tracking and Counting Vehicles. *2019 3rd International Conference on Electronics, Communication and Aerospace Technology (ICECA)*, 335–339. DOI: 10.1109/ICECA.2019.8822070

Chiu, W. Y., Meng, W., & Li, W. (2021, December). LibBlock-towards decentralized library system based on blockchain and IPFS. In 2021 18th International Conference on Privacy, Security and Trust (PST) (pp. 1-9). IEEE. DOI: 10.1109/PST52912.2021.9647821

Choudhari, G., Desai, R., Dagale, P., & Dashetwar, I. (2023). Breast cancer prediction using various machine learning algorithms and their comparative analysis. In *Proceedings of the 2023 IEEE Pune Section International Conference (PuneCon)* (pp. 1–6). DOI: 10.1109/PuneCon58714.2023.10450082

Chowdhury, R. R., & Abas, P. E. (2022). A survey on device fingerprinting approach for resource-constraint IoT devices: Comparative study and research challenges. *Internet of Things : Engineering Cyber Physical Human Systems, 20*, 100632. DOI: 10.1016/j.iot.2022.100632

Chowhan, P., & Chakraborty, A. P. (2022). Remote sensing technology—a new dimension in detection, quantification and tracking of abiotic and biotic stresses. In *Plant Stress: Challenges and Management in the New Decade* (pp. 445–457). Springer. https://doi.org/DOI: 10.1007/978-3-030-92336-1_20

Cohen, L., Manion, L., & Morrison, K. (2017). Validity and reliability. In *Research methods in education* (pp. 245–284). Routledge. DOI: 10.4324/9781315456539-14

Comas-Forgas, R., Lancaster, T., Calvo-Sastre, A., & Sureda-Negre, J. (2021). Exam Cheating and Academic Integrity Breaches During The COVID-19 Pandemic: An analysis of internet search activity in Spain. *Heliyon, 7*(10), e08233. Advance online publication. DOI: 10.1016/j.heliyon.2021.e08233 PMID: 34722942

Cong, X., Li, S., Chen, F., Liu, C., & Meng, Y. (2023). A Review of YOLO Object Detection Algorithms based on Deep Learning. *Frontiers in Computing and Intelligent Systems, 4*(2), 17–20. DOI: 10.54097/fcis.v4i2.9730

Cristianini, N., & Shawe-Taylor, J. (2000). *An Introduction to Support Vector Machines and Other Kernel-based Learning Methods.* DOI: 10.1017/CBO9780511801389

Danielis, P., Beckmann, M., & Skodzik, J. (2020). "An ISO-compliant test procedure for technical risk analyses of IoT systems based on STRIDE" in 2020 IEEE 44th Annual Computers,Software, and Applications Conference (COMPSAC) (IEEE), 499–504.

Danielsson, J., & Uthemann, A. (2024). On the Use of Artificial Intelligence in Financial Regulations and the Impact on Financial Stability. Working Paper 4604628. SSRN.

Daniya, T., & Vigneshwari, S. (2023). Rice plant leaf disease detection and classification using optimization enabled deep learning. *Journal of Environmental Informatics, 42*(1). Advance online publication. DOI: 10.3808/jei.202300492

Dar, C. R., Ruey, C. F., & Yu, H. L. (2000). Breast cancer diagnosis using self-organizing map for sonography. *Ultrasound in Medicine & Biology*, *26*(3), 405–411. DOI: 10.1016/S0301-5629(99)00156-8 PMID: 10773370

Dar, R. A., Rasool, M., & Assad, A. (2022). Breast cancer detection using deep learning: Datasets, methods, and challenges ahead. *Computers in Biology and Medicine*, *149*, 106073. DOI: 10.1016/j.compbiomed.2022.106073 PMID: 36103745

Das, S., Dutta, D., & Ghosh, P. (2023). Deep learning analysis of rice blast disease using remote sensing images. *IEEE Geoscience and Remote Sensing Letters*, *20*, 1–5. DOI: 10.1109/LGRS.2023.3244324

Dean, J., & Ghemawat, S. (2008). MapReduce. *Communications of the ACM*, *51*(1), 107–113. DOI: 10.1145/1327452.1327492

Dewan Perwakilan Rakyat Republik Indonesia, P. R. I. (2022, October 17). *UU No. 27 Tahun 2022 Perlindungan Data Pribadi*. BPK RI. https://peraturan.bpk.go.id/Details/229798/uu-no-27-tahun-2022

Dewi, A. L., Pardede, H. F., Suryawati, E., Pratiwi, H., Heryana, A., Yuliani, A. R., & Ramdan, A. (2024). Two-Stage Object Detection for Autonomous Vehicles With VGG-16 Based Faster R-CNN. *Jurnal Elektronika dan Telekomunikasi (JET)*, *24*(1), 25. DOI: 10.55981/jet.551

Dewi, D. A. I. C., & Pramita, D. A. K. (2019). Analisis Perbandingan Metode Elbow dan Silhouette pada Algoritma Clustering K-Medoids dalam Pengelompokan Produksi Kerajinan Bali. *Matrix : Jurnal Manajemen Teknologi Dan Informatika*, *9*(3), 102–109. DOI: 10.31940/matrix.v9i3.1662

Dinc, E., Kuscu, M., Bilgin, B. A., & Akan, O. B. (2019). Internet of everything: A unifying framework beyond internet of things. In *Harnessing the Internet of Everything (IoE) for Accelerated Innovation Opportunities* (pp. 1–30). IGI Global. DOI: 10.4018/978-1-5225-7332-6.ch001

Diwan, T., Anirudh, G., & Tembhurne, J. V. (2023). Object Detection Using YOLO: Challenges, Architectural Successors, Datasets and Applications. *Multimedia Tools and Applications*, *82*(6), 9243–9275. DOI: 10.1007/s11042-022-13644-y PMID: 35968414

Doan, T. V. T., Messai, M. L., Gavin, G., & Darmont, J. (2023). A survey on implementations of homomorphic encryption schemes. *The Journal of Supercomputing*, *79*(13), 1–42. DOI: 10.1007/s11227-023-05233-z

Dogra, R., Rani, S., Singh, A., Albahar, M. A., Barrera, A. E., & Alkhayyat, A. (2023). Deep learning model for detection of brown spot rice leaf disease with smart agriculture. *Computers & Electrical Engineering, 109*, 108659. DOI: 10.1016/j.compeleceng.2023.108659

Dubes, R. C. (1987). How many clusters are best? - An experiment. *Pattern Recognition, 20*(6), 645–663. DOI: 10.1016/0031-3203(87)90034-3

Dubey, R. K., & Choubey, D. K. (2024). Reliable detection of blast disease in rice plant using optimized artificial neural network. *Agronomy Journal, 116*(3), 1099–1111. DOI: 10.1002/agj2.21449

Ducrée, J., Gravitt, M., Walshe, R., Bartling, S., Etzrodt, M., & Harrington, T. (2020). Open platform concept for blockchain-enabled crowdsourcing of technology development and supply chains. *Frontiers in Blockchain, 3*, 586525. DOI: 10.3389/fbloc.2020.586525

Dulce, E., & Hurtado, J. (2020, October). The role of the blockchain technology in the elderly care solutions: A systematic mapping study. In *International Workshop on Gerontechnology* (pp. 23-34). Cham: Springer International Publishing.

Durve, M., Bonaccorso, F., Montessori, A., Lauricella, M., Tiribocchi, A., & Succi, S. (2021). Tracking Droplets in Soft Granular Flows with Deep Learning Techniques. *The European Physical Journal Plus, 136*(8), 864. DOI: 10.1140/epjp/s13360-021-01849-3 PMID: 34458055

Dutta, M., Bhadra, K., Naskar, M., & Chatterjee, P. (2024). Plant signaling and response to abiotic stress. In *Molecular Dynamics of Plant Stress and Its Management* (pp. 211–241). Springer., DOI: 10.1007/978-981-97-1699-9_9

Dwivedi, S. K., Roy, P., Karda, C., Agrawal, S., & Amin, R. (2021). Blockchain-based internet of things and industrial IoT: A comprehensive survey. *Security and Communication Networks, 2021*, 1–21. DOI: 10.1155/2021/7142048

Eddine, M. S., Ferrag, M. A., Friha, O., & Maglaras, L. (2021). EASBF: An efficient authentication scheme over blockchain for fog computing-enabled internet of vehicles. *Journal of Information Security and Applications, 59*(March), 102802. DOI: 10.1016/j.jisa.2021.102802

Eisen, N., Lee, N. T., Galliher, C., & Katz, J. (2023). AI Can Strengthen U.s.Democracy—and Weaken It. Brookings. url: https://www.brookings.edu/articles/ai-can-strengthen-u-s- democracy-and-weaken-it/

Eisfeldt, A. L., Schubert, G., & Miao, B. Z. (2023). Generative AI and Firm Values. Working Paper 31222. National Bureau of Economic Research.

Ekpenyong, J. A., & Owan, V. J. (2022). Application of Blockchain Technology for Distributed Management of Digital Research Library Holdings. In *Innovative Technologies for Enhancing Knowledge Access in Academic Libraries* (pp. 196–212). IGI Global. DOI: 10.4018/978-1-6684-3364-5.ch012

Ellingrud, K., Sanghvi, S., Dandona, G. S., Madgavkar, A., Chui, M., White, O., & Hasebe, P. (2023). *Generative AI and the Future of Work in America*. McKinsey Global Institute.

Ellis, R., Cleland, J., Scrimgeour, D. S. G., Lee, A. J., Hines, J., & Brennan, P. A. (2023). Establishing The Predictive Validity of the Intercollegiate Membership of the Royal Colleges of Surgeons Written Examination: MRCS Part A. *The Surgeon*, *21*(6), 323–330. Advance online publication. DOI: 10.1016/j.surge.2023.07.004 PMID: 37544852

Enjellina, D. (2021). Effect of Duration and Interval of Drip Irrigation System on Growth and Yield of Pakcoy Mustard (Brassica rapa L. Ssp. Chinensis). *Agrifor*, *20*(2), 311. DOI: 10.31293/agrifor.v20i2.5757

Esmaeili, F., Behnam, B., & Esmaeili, K. (2014, September). A Study of Relationship between Multiple Intelligences and Writing Ability of Iranian Female and Male Students. *Mediterranean Journal of Social Sciences*. Advance online publication. DOI: 10.5901/mjss.2014.v5n20p2663

European Union Lex. (2016, May 4). *Regulation - 2016/679 - EN - GDPR- EUR-Lex*. European Union. https://eur-lex.europa.eu/eli/reg/2016/679/oj

Fadillah, A. (2019). Hubungan Antara Efikasi Diri Dengan Perilaku Menyontek Pada Mahasiswa. *Psikoborneo: Jurnal Ilmiah Psikologi*, *7*(4), 657–664. DOI: 10.30872/psikoborneo.v7i4.4846

Fang, Y., Ye, J., & Wang, H. (2020). Realization of Intelligent Invigilation System Based on Adaptive Threshold. *2020 5th International Conference on Computer and Communication Systems (ICCCS)*, 201–205. DOI: 10.1109/ICCCS49078.2020.9118581

Fan, Z., & Yan, S. (2016). Clustering of College Students Based on Improved K-means Algorithm. *International Computer Symposium*. DOI: 10.1109/ICS.2016.0139

Farokhah, L. (2019). Implementasi K-Nearest Neighbor Untuk Klasifikasi Bunga Dengan Ekstraksi Fitur Warna RGB. [JTIIK]. *Jurnal Teknologi Informasi Dan Ilmu Komputer*, *7*(6), 1129–1135. DOI: 10.25126/jtiik.2020722608

Farooq, M. S., Javid, R., Riaz, S., & Atal, Z. (2022). Internet of Things (IoT) Technologies in Greenhouse Farming: A Systematic Literature Review. *IEEE Access : Practical Innovations, Open Solutions*.

Fatorachian, H., & Kazemi, H. (2021). Impact of industry 4.0 on supply chain performance. *Production Planning and Control*, *32*(1), 63–81. DOI: 10.1080/09537287.2020.1712487

Fauzi, F., & Monawati, M. (2018). Hubungan Antara Kecerdasan Logis Matematika dan Kedisiplinan Belajar Siswa Kelas V SD Negeri 1 Pagar Air Kabupaten Aceh Besar. Pesona Dasar: Jurnal Pendidikan Dasar dan Humaniora, 6(1).

Fauziah, K. R., Nurhayati, N., & Arsyad, M. (2015). Analisis Hubungan Antara Kecerdasan Logis-Matematis dengan Hasil Belajar Fisika Peserta Didik Kelas XI IPA SMA Negeri Di Kabupaten Jeneponto. Jurnal Sains dan Pendidikan Fisika, 11(3), 239-244.

Febrianti, A. F., Cabral, A. H., & Anuraga, G. (2018). K-Means Clustering Dengan Metode Elbow Untuk Pengelompokan Kabupaten Dan Kota Di Jawa Timur. *Seminar Nasional Hasil Riset Dan Pengabdian -SNHRP*, 863–870. http://karyailmiah.unipasby.ac.id/wp-content/uploads/2019/04/K-Means-Artikel.pdf

Feng, X., Liu, W., Zhuang, Y., Wei, Y., Wu, Y., & He, Y. (2020). Hyperspectral imaging combined with machine learning as a tool to obtain high-throughput plant salt-stress phenotyping. *The Plant Journal*, *101*(6), 1448–1461. DOI: 10.1111/tpj.14597 PMID: 31680357

Ferdig, R. E., Baumgartner, E., Hartshorne, R., Kaplan-Rakowski, R., & Mouza, C. (2020). Teaching, Technology, and Teacher Education During the COVID-19 Pandemic: Stories from the Field. *AACE-Association for the Advancement of Computing in Education*, 125–128.

Fernández-Caramés, T. M., & Fraga-Lamas, P. (2018). A Review on the Use of Blockchain for the Internet of Things. *IEEE Access : Practical Innovations, Open Solutions*, *6*, 32979–33001. DOI: 10.1109/ACCESS.2018.2842685

Ferrag, M. A., & Shu, L. (2021). The performance evaluation of blockchain-based security and privacy systems for the Internet of Things: A tutorial. *IEEE Internet of Things Journal*, *8*(24), 17236–17260. DOI: 10.1109/JIOT.2021.3078072

Ferrara, P., Mandal, A. K., Cortesi, A., & Spoto, F. (2021). Static analysis for discovering IoT vulnerabilities. *International Journal of Software Tools for Technology Transfer*, *23*(1), 71–88. DOI: 10.1007/s10009-020-00592-x

Firnando, F. M., Muslikh, A. R., & Iriananda, S. W. (2024). Analyzing InceptionV3 and InceptionResNetV2 with data augmentation for rice leaf disease classification. *Journal of Future Artificial Intelligence and Technologies*, *1*(1), 1–11. DOI: 10.62411/faith.2024-4

Flores, A., Morales, A., Campos, G., & Gelso, J. (2021, December). Energy Efficiency Using IOTA Tangle for Greenhouse Agriculture. In *Annual International Conference on Information Management and Big Data* (pp. 122-138). Cham: Springer International Publishing.

Frey, C. B., & Osborne, M. A. (2017). The Future of Employment: How Susceptible AreJobs to Computerisation? *Technological Forecasting and Social Change, 114*, 254–280. DOI: 10.1016/j.techfore.2016.08.019

Fukagawa, N. K., & Ziska, L. H. (2019). Rice: Importance for global nutrition. *Journal of Nutritional Science and Vitaminology, 65*(Supplement), S2–S3. DOI: 10.3177/jnsv.65.S2 PMID: 31619630

Gaba, S., Khan, H., Almalki, K. J., Jabbari, A., Budhiraja, I., Kumar, V., Singh, A., Singh, K. K., Askar, S. S., & Abouhawwash, M. (2023). Holochain: An agent-centric distributed hash table security in smart iot applications. *IEEE Access : Practical Innovations, Open Solutions, 11*, 81205–81223. DOI: 10.1109/ACCESS.2023.3300220

Ganatra, N., & Patel, A. (2020). A multiclass plant leaf disease detection using image processing and machine learning techniques. *International Journal on Emerging Technologies, 11*(2), 1082–1086.

Ganguli, D.. (2022). "Predictability and Surprise in Large Generative Models". In: *Proceedings of the 2022 ACM Conference on Fairness, Accountability, and Transparency*. Association for Computing Machinery, pp. 1747–1764. DOI: 10.1145/3531146.3533229

Gao, C., Wu, X., Huang, Y., & Zhang, W. (2024). The impact of water-saving irrigation on rice growth and comprehensive evaluation of irrigation strategies. *Agronomy (Basel), 14*(7), 1363. DOI: 10.3390/agronomy14071363

Gao, Z., Yu, K., Zhu, Q., Liu, Y., & He, Y. (2020). Deep learning application in plant stress imaging: A review. *AgriEngineering, 2*(3), 368–392. DOI: 10.3390/agriengineering2030029

Gardner, H. E. (2011). *Frames of mind: The theory of multiple intelligences*. Basic books.

Garg, D., Goel, P., Pandya, S., Ganatra, A., & Kotecha, K. (2018). A Deep Learning Approach for Face Detection using YOLO. *2018 IEEE Punecon*, 1–4. DOI: 10.1109/PUNECON.2018.8745376

Gatti, R. R., Singh, C., Agrawal, R., & Serrao, F. J. (Eds.). (2023). *Self-Powered Cyber Physical Systems*. John Wiley & Sons. DOI: 10.1002/9781119842026

Ge, M., Syed, N. F., Fu, X., Baig, Z., & Robles-Kelly, A. (2021). Towards a deep learning-driven intrusion detection approach for Internet of Things. *Computer Networks, 186*(December 2020), 107784. DOI: 10.1016/j.comnet.2020.107784

George, G. M., & Jayashree, L. S. (2022). Fusion of Blockchain-IoT network to improve supply chain traceability using Ethermint Smart chain: A Review. *KSII Transactions on Internet and Information Systems*, *16*, 3694–3722.

Ghosal, S., & Sarkar, K. (2020). Rice leaf diseases classification using CNN with transfer learning. In *2020 IEEE Calcutta Conference (Calcon)* (pp. xx–xx). IEEE. DOI: 10.1109/CALCON49167.2020.9106423

Ghose, P., Uddin, M. A., Islam, M. M., Islam, M., & Acharjee, U. K. (2022). A breast cancer detection model using a tuned SVM classifier. *2022 25th International Conference on Computer and Information Technology (ICCIT)*, 102–107. https://doi.org/DOI: 10.1109/ICCIT57492.2022.10003819

Ghosh, S., Sayyed, S., Wani, K., Mhatre, M., & Hyder Ali, H. (2016). *Smart irrigation: A smart drip irrigation system using cloud, android and data mining* (hlm. 239). DOI: 10.1109/ICAECCT.2016.7942589

Giattino, C., Mathieu, E., Samborska, V., & Roser, M. (2023). Artificial Intelligence. Our World in Data. url: https://ourworldindata.org/artificial-intelligence. Horowitz, Michael C. (2018). "Artificial Intelligence, International Competition, and the Balance of Power". Texas National Security Review 1(3), pp. 36–57.

Glória, A., Dionisio, C., Simões, G., Cardoso, J., & Sebastião, P. (2020). Water Management for Sustainable Irrigation Systems Using Internet-of-Things. *Sensors (Basel)*, *20*(5), 5. Advance online publication. DOI: 10.3390/s20051402 PMID: 32143482

Gogoi, M., Barman, S. D., Bhuyan, P., & Saikia, P. A. (2023). Classification and detection of rice diseases using a 3-stage CNN architecture with transfer learning approach. *Agriculture*, *13*(8), 1505. DOI: 10.3390/agriculture13081505

Goodfellow, I., Bengio, Y., & Courville, A. (2016). *Deep Learning*. MIT Press.

Goodman, B., & Flaxman, S. R. (2017). European Union Regulations on Algorithmic Decision Making and a "Right to Explanation." *˜the œAI Magazine. AI Magazine*, *38*(3), 50–57. DOI: 10.1609/aimag.v38i3.2741

Gou, C., Li, S., Zou, D., Xu, X., & Zhang, M. (2024). Machine and deep learning: Artificial intelligence application in biotic and abiotic stress management in plants. *Frontiers in Bioscience (Landmark Edition)*, *29*(1), 20–45. DOI: 10.31083/j.fbl2901020 PMID: 38287813

Goudarzi, S., Anisi, M. H., Soleymani, S. A., Ayob, M., & Zeadally, S. (2021). An IoT-based prediction technique for efficient energy consumption in buildings. *IEEE Transactions on Green Communications and Networking*, *5*(4), 2076–2088. DOI: 10.1109/TGCN.2021.3091388

Goyal, P., Bhat, M., Yadav, S. K., Gupta, R., & Bansal, R. (2024). A deep learning approach for early detection of drought stress in maize using proximal scale digital images. *Neural Computing & Applications*, *36*(4), 1899–1913. DOI: 10.1007/s00521-023-09219-z

Gros, D., Habermann, T., Kirstein, G., Meschede, C., Ruhrberg, S. D., Schmidt, A., & Siebenlist, T. (2018). Anaphora Resolution. *International Journal of Information Retrieval Research*, *8*(3), 33–45. DOI: 10.4018/IJIRR.2018070103

Guo, Y., Yin, X., Zhao, X., Yang, D., & Bai, Y. (2019). Hyperspectral image classification with SVM and guided filter. *EURASIP Journal on Wireless Communications and Networking*, *2019*(1), 56. DOI: 10.1186/s13638-019-1346-z

Gupta, A., & Kamthania, D. (2021). Study of Sentiment on Google Play Store Applications. SSRN *Electronic Journal*. DOI: 10.2139/ssrn.3833926

Gupta, A., Shridhar, K., & Dhillon, P. K. (2015). A review of breast cancer awareness among women in India: Cancer literate or awareness deficit? [Research Online]. *European Journal of Cancer*, *51*(14), 2058–2066. DOI: 10.1016/j.ejca.2015.07.008 PMID: 26232859

Gupta, D., Bhatt, P., & Bhatt, S. (2020, December). A game theoretic analysis for cooperative smart farming. In *2020 IEEE International Conference on Big Data (Big Data)* (pp. 2350-2359). IEEE. DOI: 10.1109/BigData50022.2020.9377935

Gupta, S., Garg, R., Gupta, N., Alnumay, W. S., Ghosh, U., & Sharma, P. K. (2021). Energy-efficient dynamic homomorphic security scheme for fog computing in IoT networks. *Journal of Information Security and Applications*, *58*(February), 102768. DOI: 10.1016/j.jisa.2021.102768

Gupta, V. (2022). State of Art of Blockchain Technology in the Educational Sector. *Technology, Business, Innovation, and Entrepreneurship in Industry*, *4*(0), 299–314.

Guven, I., & Parlak, M. (2022, November). Blockchain, AI and IoT Empowered Swarm Drones for Precision Agriculture Applications. In *2022 IEEE 1st Global Emerging Technology Blockchain Forum: Blockchain & Beyond (iGETblockchain)* (pp. 1-6). IEEE.

Halim, A. A. A., Mohamad, R., Rahman, F. Y. A., Harun, H., & Anas, N. M. (2023). IoT based smart irrigation, control, and monitoring system for chilli plants using NodeMCU-ESP8266. *Bulletin of Electrical Engineering and Informatics*, *12*(5), 3053–3060. DOI: 10.11591/eei.v12i5.5266

Hana, Z., Nujumuddin, N., & Sucilestari, R. (2020). Memahami dan Mengukur Kecerdasan Logis Matematis: Sebuah Telaah Pustaka. *DIKDAS MATAPPA: Jurnal Ilmu Pendidikan Dasar*, *3*(2), 197–206.

Haq, T. U., Shah, T., Siddiqui, G. F., Iqbal, M. Z., Hameed, I. A., & Jamil, H. (2021). Improved twofish algorithm: A digital image enciphering application. *IEEE Access : Practical Innovations, Open Solutions*, *9*, 76518–76530. DOI: 10.1109/ACCESS.2021.3081792

Haque, M. A., Haque, S., Zeba, S., Kumar, K., Ahmad, S., Rahman, M., & Ahmed, L. (2023). Sustainable and efficient E-learning internet of things system through blockchain technology. *E-Learning and Digital Media*, •••, 20427530231156711.

Haridasan, A., Thomas, J., & Raj, E. D. (2023). Deep learning system for paddy plant disease detection and classification. *Environmental Monitoring and Assessment*, *195*(1), 120. DOI: 10.1007/s10661-022-10656-x PMID: 36399232

Hartanti, N. T. (2020). Metode Elbow dan K-Means Guna Mengukur Kesiapan Siswa SMK Dalam Ujian Nasional. *Jurnal Nasional Teknologi Dan Sistem Informasi*, *6*(2), 82–89. DOI: 10.25077/TEKNOSI.v6i2.2020.82-89

Haruna, Y., Qin, S., & Mbyamm Kiki, M. J. (2023). An improved approach to detection of rice leaf disease with GAN-based data augmentation pipeline. *Applied Sciences (Basel, Switzerland)*, *13*(3), 1346. DOI: 10.3390/app13031346

Hasnain, M., Albogamy, F. R., Alamri, S. S., Ghani, I., & Mehboob, B. (2023). The Hyperledger fabric as a Blockchain framework preserves the security of electronic health records. *Frontiers in Public Health*, *11*, 1272787. DOI: 10.3389/fpubh.2023.1272787 PMID: 38089022

Hasnain, M., Pasha, M. F., Ghani, I., Imran, M., Alzahrani, M. Y., & Budiarto, R. (2020). Evaluating Trust Prediction and Confusion Matrix Measures for Web Services Ranking. *IEEE Access : Practical Innovations, Open Solutions*, *8*, 90847–90861. DOI: 10.1109/ACCESS.2020.2994222

Hassan, A., Afrouzi, H. N., Siang, C. H., Ahmed, J., Mehranzamir, K., & Wooi, C.-L. (2022). A survey and bibliometric analysis of different communication technologies available for smart meters. *Cleaner Engineering and Technology*, *7*, 100424. https://doi.org/https://doi.org/10.1016/j.clet.2022.100424. DOI: 10.1016/j.clet.2022.100424

Hassan, E. E., Chung, L. L., Sulaima, M. F., Bahaman, N., & Kadir, A. F. A. (2022). Smart irrigation system with photovoltaic supply. *Bulletin of Electrical Engineering and Informatics*, *11*(1), 29–41. DOI: 10.11591/eei.v11i1.3338

Hassani, H. L., Bahnasse, A., Martin, E., Roland, C., Bouattane, O., & Diouri, M. E. M. (2021). Vulnerability and security risk assessment in a IIoT environment in compliance with standard IEC 62443. *Procedia Computer Science*, *191*, 33–40. DOI: 10.1016/j.procs.2021.07.008

Hassija, V., Zeadally, S., Jain, I., Tahiliani, A., Chamola, V., & Gupta, S. (2021). Framework for determining the suitability of blockchain: Criteria and issues to consider. *Transactions on Emerging Telecommunications Technologies*, *32*(10), e4334. DOI: 10.1002/ett.4334

He, W., Huang, J., Gao, N., & Chen, H. (2022, January). Data Distribution Scheme for Information Service Based on Blockchain. In 2022 2nd International Conference on Consumer Electronics and Computer Engineering (ICCECE) (pp. 21-25). IEEE. DOI: 10.1109/ICCECE54139.2022.9712811

Heale, R., & Twycross, A. (2015). Validity and reliability in quantitative studies. *Evidence-Based Nursing*, *18*(3), 66–67. DOI: 10.1136/eb-2015-102129 PMID: 25979629

Hernanda, R. A. P., Doi, R., & Okamoto, M. (2024). Current trends in the use of thermal imagery in assessing plant stresses: A review. *Computers and Electronics in Agriculture*, *224*, 109227. DOI: 10.1016/j.compag.2024.109227

Hernández-López, J. A., Andrade, H. J., & Barrios, M. (2024). Agricultural drought assessment in dry zones of Tolima, Colombia, using an approach based on water balance and vegetation water stress. *The Science of the Total Environment*, *921*, 171144. DOI: 10.1016/j.scitotenv.2024.171144 PMID: 38401721

Hidayat, S., Rismayati, R., Tajuddin, M., & Merawati, N. L. P. (2020). Segmentation of university customers loyalty based on RFM analysis using fuzzy c-means clustering. *Jurnal Teknologi Dan Sistem Komputer*, *8*(2), 133–139. DOI: 10.14710/jtsiskom.8.2.2020.133-139

Hoque, M. J., Razu, M., Jashim, M., & Mostafa, M. (2020). Automation of Traditional Exam Invigilation using CCTV and Bio-Metric. *International Journal of Advanced Computer Science and Applications*, *11*(6), 392–399. DOI: 10.14569/IJACSA.2020.0110651

Hota, P. K., Hota, L., & Dash, P. K. (2022). Blockchain in Digital Libraries: State of the Art, Trends, and Challenges. Machine Learning Adoption in Blockchain-Based Intelligent Manufacturing, 17-32.

https://academicintegrity.org/. (2021). *THE FUNDAMENTAL VALUES OF ACADEMIC INTEGRITY*. International Center for Academic Integrity. https://academicintegrity.org/images/pdfs/20019_ICAI-Fundamental-Values_R12.pdf

Huang, Y. (2024). *"Is the Impact of AI Different from That of IT?" New Technologies, Digitalization, and AI: The Future Is Here. IMF Research Perspectives 26*. Klare, Brendan F., Mark J. Burge, Joshua C. Klontz, Richard W. Vorder Bruegge, and Anil.

Hu, L. (2022, February). Intelligent value added service platform of smart library based on blockchain technology. In *2022 IEEE International Conference on Electrical Engineering, Big Data and Algorithms (EEBDA)* (pp. 1391-1394). IEEE. DOI: 10.1109/EEBDA53927.2022.9744776

Huo, J., & Pan, B. (2022). Study the path planning of intelligent robots and the application of blockchain technology. *Energy Reports*, *8*, 5235–5245. DOI: 10.1016/j.egyr.2022.03.204

Hussein, F., Al-Ahmad, A., El-Salhi, S., Alshdaifat, E., & Al-Hami, M. (2022). Advances in Contextual Action Recognition: Automatic Cheating Detection Using Machine Learning Techniques. *Data*, *7*(9), 122. Advance online publication. DOI: 10.3390/data7090122

Hutabarat, B. F., Peslinof, M., Afrianto, M. F., & Fendriani, Y. (t.t.). Water Parameter Monitoring Database System Based On Internet of Things (IoT) Using Thingspeak Platform. *Journal Online of Physics, 8*(2), 42–50.

Hutton, J. S., Dudley, J., Horowitz-Kraus, T., DeWitt, T., & Holland, S. K. (2020). Associations Between Home Literacy Environment, Brain White Matter Integrity and Cognitive Abilities in Preschool-Age Children. *Acta Paediatrica. International Journal of Pediatrics*, *109*(7), 1376–1386. DOI: 10.1111/apa.15124 PMID: 31854046

Ibrahim, S., Zulkifli, N. A., Sabri, N., Shari, A. A., & Noordin, M. R. M. (2019). Rice grain classification using multi-class support vector machine (SVM). *IAES International Journal of Artificial Intelligence (IJ-AI), 8*(3), Article 3. DOI: 10.11591/ijai.v8.i3.pp215-220

Ibrahim, A.-W. S., & Atya, D. B. (2022). Detection of diseases in rice leaf using deep learning and machine learning techniques. *Webology*, *19*(1), 1493–1503. DOI: 10.14704/WEB/V19I1/WEB19100

Idrus, A., Tarihoran, N., Supriatna, U., Tohir, A., Suwarni, S., & Rahim, R. (2022). Distance Analysis Measuring for Clustering using K-Means and Davies Bouldin Index Algorithm. *TEM Journal, 11*(4), 1871–1876. https://doi.org/DOI: 10.18421/TEM114-55

Ifeagwazi, C. M., Chukwuorji, J. C., Egbodo, S. O., & Nwoke, M. B. (2019). Peer pressure, fear of failure and examination cheating behavior in the university: Does gender make the difference? *Cognition, Brain, Behavior: An Interdisciplinary Journal, 23*(1), 43–62. DOI: 10.24193/cbb.2019.23.03

Ijjina, E. P., Joshi, A. S., & Kanahasabai, G. (2020). Detection of Customer Interested Garments in Surveillance Video using Computer Vision. *2020 11th International Conference on Computing, Communication and Networking Technologies (ICCCNT)*, 1–5. DOI: 10.1109/ICCCNT49239.2020.9225571

Ilahi, R., Bimantoro, F., Dwiyansaputra, R., & Farinda, R. (2022). Classification of Pringgasela Typical Songket Using Multi Texton Co-occurrence Descriptor and K-Nearest Neighbor. In *Proceedings of the First Mandalika International Multi-Conference on Science and Engineering 2022, MIMSE 2022 (Informatics and Computer Science)* (pp. 352–366). Atlantis Press International BV. DOI: 10.2991/978-94-6463-084-8_30

Imamah, Y., U. L., Djunaidy, A., Pardede, E., & Purnomo, M. H. (2024). Clustering Approach for Modeling Course Difficulty Level in Adaptive Learning. *2024 IEEE International Conference on Computational Intelligence and Virtual Environments for Measurement Systems and Applications (CIVEMSA)*, 1–6. DOI: 10.1109/CIVEMSA58715.2024.10639956

Imteaj, A., Thakker, U., Wang, S., Li, J., & Amini, M. H. (2021). A survey on federated learning for resource-constrained IoT devices. *IEEE Internet of Things Journal, 9*(1), 1–24. DOI: 10.1109/JIOT.2021.3095077

Institute of Electrical and Electronics Engineers. Bangalore Section, & Institute of Electrical and Electronics Engineers. (n.d.). *Automated Cheating Detection in Exams using Posture and Emotion Analysis.*

Iqbal, J.. (2023). Early detection and classification of rice brown spot and bacterial blight diseases using digital image processing. *Journal of Computing & Biomedical Informatics, 4*(2), 98–109.

Islam, F., Ali, B., Wang, Y., & Xia, X. (2024). Transcriptomic reprogramming of rice cultivars in response to herbicide, salt and their combined stresses. *Plant Stress, 12*, 100504. DOI: 10.1016/j.stress.2024.100504

Islam, M. J., Datta, R., & Iqbal, A. (2023). Actual rating calculation of the zoom cloud meetings app using user reviews on google play store with sentiment annotation of BERT and hybridization of RNN and LSTM. *Expert Systems with Applications*, *223*(March), 119919. DOI: 10.1016/j.eswa.2023.119919 PMID: 36969371

Ismaya, N., & Zyen, A. K. (2022). Penentuan siswa berprestasi menggunakan metode k-means clustering di smp takhassus al qur'an sadamiyyah. Jtinfo. *Jurnal Teknik Informatika*, *1*(1), 64–48.

Jadhawar, A. A. (2024). Library as a Service (LaaS): Innovating Resource Sharing Among Universities. *Library Progress International*, *44*(1), 87–104.

Jain, K. (2012). Face Recognition Performance: Role of Demographic Information. *IEEE Transactions on Information Forensics and Security*, *7*(6), 1789–1801. DOI: 10.1109/TIFS.2012.2214212

Jain, R. K. (2023). Experimental performance of smart IoT-enabled drip irrigation system using and controlled through web-based applications. *Smart Agricultural Technology*, *4*, 100215. DOI: 10.1016/j.atech.2023.100215

Janke, S., Rudert, S. C., Petersen, Ä., Fritz, T. M., & Daumiller, M. (2021). Cheating in the wake of COVID-19: How dangerous is ad-hoc online testing for academic integrity? *Computers and Education Open*, *2*, 100055. DOI: 10.1016/j.caeo.2021.100055

Jayashri, N., Rampur, V., Gangodkar, D., Abirami, M., Balarengadurai, C., & Kumar, A. (2023). Improved block chain system for high secured IoT integrated supply chain. *Measurement. Sensors*, *25*, 100633.

Jazzar, M., & Hamad, M. (2022). "An analysis study of IoT and dos attack perspective" in Proceedings of international conference on intelligent cyber-physical systems: ICPS 2021 (Singapore: Springer Nature Singapore), 127–142.

Jesie, R. S., Godwin Premi, M. S., & Jarin, T. (2024). Comparative analysis of paddy leaf diseases sensing with a hybrid convolutional neural network model. *Measurement. Sensors*, *31*, 100966. DOI: 10.1016/j.measen.2023.100966

Jha, S. K. (2023). Application of blockchain technology in libraries and information centers services. *Library Hi Tech News*. Advance online publication. DOI: 10.1108/LHTN-02-2023-0020

Jiang, F., Lu, Y., Chen, Y., Cai, D., & Li, G. (2020). Image recognition of four rice leaf diseases based on deep learning and support vector machine. *Computers and Electronics in Agriculture*, *179*, 105824. DOI: 10.1016/j.compag.2020.105824

Jiang, P., Ergu, D., Liu, F., Cai, Y., & Ma, B. (2021). A Review of Yolo Algorithm Developments. *Procedia Computer Science, 199*, 1066–1073. DOI: 10.1016/j.procs.2022.01.135

Jiang, Z., Dong, Z., Jiang, W., & Yang, Y. (2021). Recognition of rice leaf diseases and wheat leaf diseases based on multi-task deep transfer learning. *Computers and Electronics in Agriculture, 186*, 106184. DOI: 10.1016/j.compag.2021.106184

Jiménez-Macías, A., Muñoz-Merino, P. J., & Kloos, C. D. (2024). Student Clustering Through KMeans to Enhance Teaching. *2024 XVI Congreso de Tecnología, Aprendizaje y Enseñanza de La Electrónica (TAEE)*, 1–5. DOI: 10.1109/TAEE59541.2024.10604956

Jin, S., Tang, Q., Li, Y., & Hu, X. (2024). Insights into plant sensory mechanisms under abiotic stresses. *Plants, 13*(14), 1907. DOI: 10.3390/plants13141907 PMID: 39065434

JNSS, J., & Ganesh, E. N. (2023). A novel blockchain-based lightweight encryption technique in fog based iot for personal healthcare data application. *International Journal of Intelligent Systems and Applications in Engineering, 11*(3s), 119–128.

Jony, A. I., & Arnob, A. K. B. (2024). Deep learning paradigms for breast cancer diagnosis: A comparative study on Wisconsin diagnostic dataset. *Malaysian Journal of Science and Advanced Technology, 4*(1), 109–117. DOI: 10.56532/mjsat.v4i2.245

Julianto, A., Sunyoto, A., & Wibowo, F. W. (2022). Optimasi Hyperparameter Convolutional Neural Network Untuk Klasifikasi Penyakit Tanaman Padi. *Teknimedia: Teknologi Informasi Dan Multimedia, 3*(2), 98–105. DOI: 10.46764/teknimedia.v3i2.77

Kaddoura, S., & Gumaei, A. (2022). Towards effective and efficient online exam systems using deep learning-based cheating detection approach. *Intelligent Systems with Applications, 16*, 200153. Advance online publication. DOI: 10.1016/j.iswa.2022.200153

Kalaichelavan, K., Malik, H., Husnu, N., & Sreenath, S. (2020). What Do People Complain about Drone Apps? A Large-Scale Empirical Study of Google Play Store Reviews. *Procedia Computer Science, 170*, 547–554. DOI: 10.1016/j.procs.2020.03.124

Kambayo, A. N., Berutu, S. S., & Nshimiyimana, A. (2025). *Aspect-Based Sentiment Analysis Using Latent Dirichlet Allocation (LDA) and DistilBERT on Threads App Reviews. 9*(01), 25–34.

Kanipriya, M., Sugumaran, D., Pandiyarajan, V., Yogananda, B. S., Selvakumar, P., & Mohan, S. (2024, January). Analysis of Electromyography Signals for Control of Mechanical Prosthesis using Machine Learning Techniques. In 2024 2nd International Conference on Intelligent Data Communication Technologies and Internet of Things (IDCIoT) (pp. 1059-1067). IEEE.

Karlsson, K., Jiang, W., Wicker, S., Adams, D., Ma, E., van Renesse, R., & Weatherspoon, H. (2018, July). Vegvisir: A partition-tolerant blockchain for the internet-of-things. In *2018 IEEE 38th International Conference on Distributed Computing Systems (ICDCS)* (pp. 1150-1158). IEEE. DOI: 10.1109/ICDCS.2018.00114

Karmakar, R., Chatterjee, S., Das, A. K., & Mandal, A. (2023). BCPUML: Breast cancer prediction using machine learning approach—A performance analysis. *SN Computer Science*, *4*(4), 377. Advance online publication. DOI: 10.1007/s42979-023-01825-x

Karsenti, T. (2019). *Artificial intelligence in education : The urgent need to prepare teachers for tomorrow ' s schools. 27*(1), 105–111. DOI: 10.18162/fp.2019.a166

Kartikeyan, P., & Shrivastava, G. (2021). Review on emerging trends in detection of plant diseases using image processing with machine learning. *International Journal of Computer Applications*, *975*(8887), 39–48. DOI: 10.5120/ijca2021920990

Kasahara, A., Mitchell, J., Yang, J., Cuomo, R. E., McMann, T. J., & Mackey, T. K. (2024). Digital technologies used in clinical trial recruitment and enrollment including application to trial diversity and inclusion: A systematic review. *Digital Health*, *10*, 20552076241242390. DOI: 10.1177/20552076241242390 PMID: 38559578

Kathar, S., Hardel, H., Alam, Z., Phansalkar, S., & Sajjan, R. (2021, December). Auction System for Agricultural Trade Using Blockchain Technology: A Survey with Proposed Framework. In *2021 IEEE Pune Section International Conference (PuneCon)* (pp. 1-10). IEEE. DOI: 10.1109/PuneCon52575.2021.9686543

Kathole, A. B., Katti, J., Lonare, S., & Dharmale, G. (2023). Identify and classify pests in the agricultural sector using metaheuristics deep learning approach. *Franklin Open*, *3*, 100024. https://doi.org/https://doi.org/10.1016/j.fraope.2023.100024. DOI: 10.1016/j.fraope.2023.100024

Kaur, U., Kaur, M., & Khang, A. (2023). Futuristic Technologies in Agriculture Challenges and Future Prospects. *Handbook of Research on AI-Equipped IoT Applications in High-Tech Agriculture*, 347-364.

Kavitha, S. (2024, March). Identification of nutrient deficiency based on leaf image data using machine learning. In *2024 International Conference on Emerging Smart Computing and Informatics (ESCI)* (pp. [insert pages if available]). IEEE. https:// doi.org/ [Replace DOI if necessary]DOI: 10.1109/ESCI59138.2024.10565432

Khan, A. U., Ma, Z., Li, M., Zhi, L., Hu, W., & Yang, X. (2023). From traditional to emerging technologies in supporting smart libraries. A bibliometric and thematic approach from 2013 to 2022. Library Hi Tech, (ahead-of-print).

Khan, M. Q., Malik, F., Rahman, N., Pakhtunkhwa, K., & Arabia, S. (2024). *Optimized Sentiment Classification of Google Play Store App Ratings Using Advanced Machine Learning Models*. 252–266. DOI: 10.21015/vtse.v12i4.1997

Khan, A. R., Akbar, M., Alam, T. M., & Rauf, A. (2023). Automatic diagnosis of rice leaves diseases using hybrid deep learning model. *Journal of Advances in Information Technology, 14*(3), 418–425. DOI: 10.12720/jait.14.3.418-425

Khan, A. U., Zhang, Z., Taleby Ahvanooey, M., & Rafique, W. (2022). Opinion mining towards blockchain technology adoption for accessing digital library resources. *Aslib Journal of Information Management, 74*(1), 135–157. DOI: 10.1108/ AJIM-01-2021-0016

Khan, M. A., & Salah, K. (2018). IoT security: Review, blockchain solutions, and open challenges. *Future Generation Computer Systems, 82*, 395–411. DOI: 10.1016/j. future.2017.11.022

Khashan, O. A., & Khafajah, N. M. (2023). Efficient hybrid centralized and blockchain-based authentication architecture for heterogeneous IoT systems. *Journal of King Saud University. Computer and Information Sciences, 35*(2), 726–739. DOI: 10.1016/j.jksuci.2023.01.011

Khodadady, E., & Dastgahian, B. S. (2013). Logical-Mathematical Intelligence and Its Relationship with English Language Proficiency. *American Journal of Scientific Research, 89*, 57–68.

Khurshid, A., Alsaaidi, R., Aslam, M., & Raza, S. (2022). EU cybersecurity act and IoT certification: Landscape, perspective and a proposed template scheme. *IEEE Access : Practical Innovations, Open Solutions, 10*, 129932–129948. DOI: 10.1109/ ACCESS.2022.3225973

Kim, J., Sung, J.-Y., & Park, S. (2020). Comparison of Faster-RCNN, YOLO, and SSD for Real-Time Vehicle Type Recognition. *2020 IEEE International Conference on Consumer Electronics - Asia (ICCE-Asia)*, 1–4. DOI: 10.1109/ICCE-Asia49877.2020.9277040

Kiratiratanapruk, K., Kositchaiwat, P., Suwanvajokkasikit, K., & Lursinsap, C. (2022). Automatic detection of rice disease in images of various leaf sizes. *arXiv Preprint*, arXiv:2206.07344. https://doi.org//arXiv.2206.07344DOI: 10.48550

Kogan, L., Papanikolaou, D., Schmidt, L. D. W., & Seegmiller, B. (2023). Technology and Labor Displacement: Evidence from Linking Patents with Worker-Level Data. Working Paper 31846. National Bureau of Economic Research.

Kombate, Y., & Houngue, P. (2022). A bibliometric method for analyzing IoT hacking solutions in network services. *Procedia Computer Science*, *203*, 141–148.

Korkmaz, O. (2012). The Impact of Critical Thinking and Logico-Mathematical Intelligence on Algorithmic Design Skills *. *Journal of Educational Computing Research*, *46*(2), 173–193. https://doi.org/http://dx.doi.org/10.2190/EC.46.2.d. DOI: 10.2190/EC.46.2.d

Kösemen, C., Dalkiliç, G., & Öksüzer, Ş. (2022). Performance analysis of lightweight Internet of things devices on blockchainnetworks. *Turkish Journal of Electrical Engineering and Computer Sciences*, *30*(2), 328–343. DOI: 10.3906/elk-2103-110

Kose, U., Prasath, V. S., Mondal, M. R. H., Podder, P., & Bharati, S. (Eds.). (2022). *Artificial Intelligence and Smart Agriculture Technology*. CRC Press. DOI: 10.1201/9781003299059

Krishnamoorthy, N.. (2021). Rice leaf diseases prediction using deep neural networks with transfer learning. *Environmental Research*, *198*, 111275. DOI: 10.1016/j.envres.2021.111275 PMID: 33989629

Krishnan, R. S., Julie, E. G., Robinson, Y. H., Raja, S., Kumar, R., Thong, P. H., & Son, L. H. (2020). Fuzzy Logic based Smart Irrigation System using Internet of Things. *Journal of Cleaner Production*, *252*, 119902. DOI: 10.1016/j.jclepro.2019.119902

Kukreja, V., Sharma, R., & Vats, S. (2023). Revolutionizing rice farming: Automated identification and classification of rice leaf blight disease using deep learning. In *2023 Third International Conference on Secure Cyber Computing and Communication (ICSCCC)* (pp. 173–177). IEEE. DOI: 10.1109/ICSCCC58608.2023.10176408

Kulkarni, P., & Shastri, S. (2024). Rice leaf diseases detection using machine learning. *Journal of Scientific Research and Technology*, *2024*, 17–22. DOI: 10.61808/jsrt81

Kumar, T., & Kaur, S. (2023). Ethical Aspects of Cybersecurity in E-Commerce. In Cybersecurity for Decision Makers (1st ed.). CRC Press. DOI: 10.1201/9781003319887

Kumar, A., Bhardwaj, A., Sharma, A., & Gupta, A. (2023). Multi-scale feature fusion-based lightweight dual stream transformer for detection of paddy leaf disease. *Environmental Monitoring and Assessment, 195*(9), 1020. DOI: 10.1007/s10661-023-11628-5 PMID: 37548778

Kumar, A., Sharma, S., Goyal, N., Singh, A., Cheng, X., & Singh, P. (2021). Secure and energy-efficient smart building architecture with emerging technology IoT. *Computer Communications, 176*(June), 207–217. DOI: 10.1016/j.comcom.2021.06.003

Kumari, A., & Sharma, I. (2023, October). Augmentation of Internet of Things Security and Privacy with IOTA Tangle Network. In *2023 4th IEEE Global Conference for Advancement in Technology (GCAT)* (pp. 1-5). IEEE. DOI: 10.1109/GCAT59970.2023.10353394

Kumari, S., & Muthulakshmi, P. (2023). Artificial Intelligence—Blockchain Enabled Technology for Internet of Things: Research Statements, Open Issues, and Possible Applications in the Near Future. *Privacy Preservation of Genomic and Medical Data*, 433-480.

Kumar, K., & Kannan, E. (2022). Detection of rice plant disease using AdaBoostSVM classifier. *Agronomy Journal, 114*(4), 2213–2229. DOI: 10.1002/agj2.21070

Kumar, P. M., Jeswanth, A. R., & Manani, J. (2022). Sentimental Analysis on Quick Commerce Platform Based on Customer Reviews in Google Play Store. [IJRISE]. *International Journal of Research in Science & Engineering*, *2*(06), 29–42.

Kumar, R., & Sharma, R. (2022). Leveraging blockchain for ensuring trust in IoT: A survey. *Journal of King Saud University. Computer and Information Sciences, 34*(10), 8599–8622. DOI: 10.1016/j.jksuci.2021.09.004

Kumar, V. S., Jaganathan, M., Viswanathan, A., Umamaheswari, M., & Vignesh, J. (2023). Rice leaf disease detection based on bidirectional feature attention pyramid network with YOLO v5 model. *Environmental Research Communications*, *5*(6), 065014. DOI: 10.1088/2515-7620/acdece

Kundu, R., Chauhan, U., & Chauhan, S. P. S. (2022). Plant leaf disease detection using image processing. In *2022 2nd International Conference on Innovative Practices in Technology and Management (ICIPTM)* (Vol. 2, pp. 78–83). IEEE. DOI: 10.1109/ICIPTM54933.2022.9754170

Lamanna, C., & La Manna, M. (2018). BitView: using blockchain technology to validate and diffuse global usage data for academic publications. In Maturity and Innovation in Digital Libraries: 20th International Conference on Asia-Pacific Digital Libraries, ICADL 2018, Hamilton, New Zealand, November 19-22, 2018 [Springer International Publishing.]. *Proceedings*, *20*, 267–277.

Lambrecht, A., & Tucker, C. E. (2018). Algorithmic Bias? An Empirical Study into Apparent Gender-Based Discrimination in the Display of STEM Career Ads. Working Paper 2852260. SSRN. Landemore, Hélène (2023). "Fostering More Inclusive Democracy with AI". Finance &amp. *Development*, *60*(4).

Lamtzidis, O., & Gialelis, J. (2018, December). An IOTA based distributed sensor node system. In *2018 IEEE Globecom Workshops (GC Wkshps)* (pp. 1-6). IEEE. DOI: 10.1109/GLOCOMW.2018.8644153

Lata, N., & Owan, V. J. (2022). Contemporary trends and technologies in research libraries: An overview. Innovative technologies for enhancing knowledge access in academic libraries, 40-56.

Latif, S., Idrees, Z., Huma, Z., & Ahmad, J. (2021). Blockchain technology for the industrial Internet of Things: A comprehensive survey on security challenges, architectures, applications, and future research directions. *Transactions on Emerging Telecommunications Technologies*, *32*(11), e4337. DOI: 10.1002/ett.4337

Lavanya, G., & Pande, S. D. (2023). Enhancing Real-time Object Detection with YOLO Algorithm. *EAI Endorsed Transactions on Internet of Things*, *10*, 1. DOI: 10.4108/eetiot.4541

LeCun, Y., Bengio, Y., & Hinton, G. (2015). Deep learning. *Nature*, *521*(7553), 436–444. DOI: 10.1038/nature14539 PMID: 26017442

Lee, J., & il Hwang, K. (2022). YOLO with adaptive frame control for real-time object detection applications. *Multimedia Tools and Applications*, *81*(25), 36375–36396. DOI: 10.1007/s11042-021-11480-0

Leelavathy, B., & Kovvur, R. M. R. (2021). Prediction of biotic stress in paddy crop using deep convolutional neural networks. In *Proceedings of International Conference on Computational Intelligence and Data Engineering: ICCIDE 2020* (pp. 361–368). Springer. DOI: 10.1007/978-981-15-8767-2_29

Lee, T. F., & Chen, W. Y. (2021). Lightweight fog computing-based authentication protocols using physically unclonable functions for internet of medical things. *Journal of Information Security and Applications*, *59*(701), 102817. DOI: 10.1016/j.jisa.2021.102817

Lengoatha, L., & Seymour, F. L. (2020, September). Determinant factors of intention to adopt blockchain technology across academic libraries. In Conference of the South African Institute of Computer Scientists and Information Technologists 2020 (pp. 244-250).

Li, F., & Liao, K. (2022, April). Research on multi service collaboration mode of university smart library based on blockchain. In 2022 7th International Conference on Cloud Computing and Big Data Analytics (ICCCBDA) (pp. 217-220). IEEE. DOI: 10.1109/ICCCBDA55098.2022.9778878

Liang, X., & Kim, Y. (2021). "A survey on security attacks and solutions in the IoT network"in 2021 IEEE 11th Annual Computing and Communication Workshop and Conference (CCWC) (IEEE), 853–859.

Lin, T.-Y., Goyal, P., Girshick, R., He, K., & Dollar, P. (2020). Focal Loss for Dense Object Detection. *IEEE Transactions on Pattern Analysis and Machine Intelligence*, *42*(2), 318–327. DOI: 10.1109/TPAMI.2018.2858826 PMID: 30040631

Liu, C., Zhou, H., Xu, H.-C., Hou, B.-Y., & Yao, L. (2020). Abnormal Behavior Recognition in an Examination Based on Pose Spatio-Temporal Features. *2020 IEEE 5th International Conference on Cloud Computing and Big Data Analytics (ICCCBDA)*, 380–386. DOI: 10.1109/ICCCBDA49378.2020.9095710

Liu, X., Yu, W., Liang, F., Griffith, D., & Golmie, N. (2021). On deep reinforcement learning security for Industrial Internet of Things. *Computer Communications, 168*(December 2020), 20–32. DOI: 10.1016/j.comcom.2020.12.013

Liu, J., & Ma, W. (2007). An effective recognition method of breast cancer based on PCA and SVM algorithm. In *Proceedings of the International Conference on Management of e-Commerce and e-Government* (pp. 57–64). Springer. DOI: 10.1007/978-3-540-77413-6_8

Lombardi, M., Pascale, F., & Santaniello, D. (2021). Internet of things: A general overview between architectures, protocols and applications. *Information (Basel), 12*(2), 1–21. DOI: 10.3390/info12020087

López-Oriona, Á., Vilar, J. A., & D'Urso, P. (2023). Hard and soft clustering of categorical time series based on two novel distances with an application to biological sequences. *Information Sciences*, *624*, 467–492. DOI: 10.1016/j.ins.2022.12.065

Lund, E. H., Jaccheri, L., Li, J., Cico, O., & Bai, X. (2019, May). Blockchain and sustainability: A systematic mapping study. In 2019 IEEE/ACM 2nd International Workshop on Emerging Trends in Software Engineering for Blockchain (WETSEB) (pp. 16-23). IEEE. DOI: 10.1109/WETSEB.2019.00009

Lu, Y., Tao, X., Jiang, F., Du, J., Li, G., & Liu, Y. (2024). Image recognition of rice leaf diseases using atrous convolutional neural network and improved transfer learning algorithm. *Multimedia Tools and Applications*, *83*(5), 12799–12817. DOI: 10.1007/s11042-023-16047-9

Lv, Z., Yang, X., Yu, J., Li, L., Liu, A., & Li, C. (2023, September). Network Traffic Forensics and Traceability Technology Based on Blockchain. In 2023 6th International Conference on Information Communication and Signal Processing (ICICSP) (pp. 717-723). IEEE. DOI: 10.1109/ICICSP59554.2023.10390572

Madgule, M., Vinayaka, N., Sonkhaskar, Y. M., Patel, D. R., Karthikeyan, R., & Selvakumar, P. (2023). Mechanical properties and microstructure of activated TIG welded similar joints of Inconel alloys by desirability approaches. *Materials Today: Proceedings*, *77*, 528–533.

Ma, F., & Li, Z. (2020). Future prospect of library and information science in China in the context of new liberal arts. *Journal of Library Science in China*, *46*(06), 4–15.

Magalhães, S. C., dos Santos, F. N., Machado, P., Moreira, A. P., & Dias, J. (2023). Benchmarking edge computing devices for grape bunches and trunks detection using accelerated object detection single shot multibox deep learning models. *Engineering Applications of Artificial Intelligence*, *117*, 105604. DOI: 10.1016/j.engappai.2022.105604

Mahadev Madgule, N. (2023). Mechanical properties and microstructure of activated TIG welded similar joints of Inconel alloys by desirability approaches. *Materials Today: Proceedings*, *77*, 528–533. DOI: 10.1016/j.matpr.2022.12.250

Mahadevan, K., Punitha, A., & Suresh, J. (2024). Automatic recognition of rice plant leaf diseases detection using deep neural network with improved threshold neural network. *e-Prime - Advances in Electrical Engineering. Electronics and Energy*, *8*, 100534. DOI: 10.1016/j.prime.2024.100534

Mahajan, M., Rajeshwaribajre, R., Sharma, L., & Varma, S. L. (2020). Computer Vision Based Campus Surveillance. *Proceedings - IEEE 2020 2nd International Conference on Advances in Computing, Communication Control and Networking, ICACCCN 2020*, 1013–1018. DOI: 10.1109/ICACCCN51052.2020.9362776

Mahmoud, R., Yousuf, T., Aloul, F., & Zualkernan, I. (2015). "Internet of things (IoT) security: current status, challenges and prospective measures" in 2015 10th International Conference for Internet Technology and Secured Transactions (ICITST) (IEEE), 336–341.

Mahoney, M. V., & Chan, P. K. (2003). An Analysis of the 1999 DARPA/Lincoln Laboratory Evaluation Data for Network Anomaly Detection. In *Lecture notes in computer science* (pp. 220–237). DOI: 10.1007/978-3-540-45248-5_13

Maidment-Otlet, R. (2021). Digital-first approaches and the library brand in a post-pandemic world. In Libraries, Digital Information, and COVID (pp. 103-110). Chandos Publishing. DOI: 10.1016/B978-0-323-88493-8.00016-1

Maiti, S., Juneja, S., Goyal, R., & Chaudhry, N. (2023). Reviewing the landscape of deep learning approaches for breast cancer detection. In *Proceedings of the 2023 International Conference on Computing, Communication, and Intelligent Systems (ICCCIS)* (pp. 193–199). DOI: 10.1109/ICCCIS60361.2023.10425219

Makarenko, I., Semushin, S., Suhai, S., Kazmi, S. A., Oracevic, A., & Hussain, R. (2020). "A comparative analysis of cryptographic algorithms in the internet of things" in 2020 InternationalScientific and Technical Conference Modern Computer Network Technologies(MoNeTeC) (IEEE), 1–8.

Malhotra, M., & Chhabra, I. (2021). Automatic Invigilation Using Computer Vision. *Proceedings of the 3rd International Conference on Integrated Intelligent Computing Communication & Security (ICIIC 2021), 4*, 130–136. https://doi.org/https://doi .org/10.2991/ahis.k.210913.017

Mallick, D. K., Ray, R., & Dash, S. R. (2020). Detection and classification of crop diseases from its leaves using image processing. In *Smart Intelligent Computing and Applications:Proceedings of the Third International Conference on Smart Computing and Informatics* (Vol. 1, pp. 539–547). Springer. DOI: 10.1007/978-981-13-9282-5_20

Malvade, N. N., Anami, B. S., & Palaiah, S. (2022). A comparative analysis of paddy crop biotic stress classification using pre-trained deep neural networks. *Artificial Intelligence in Agriculture, 6*, 167–175. DOI: 10.1016/j.aiia.2022.09.001

Mamat, H. S., Zaini, N. A., & Halim, S. A. (2020). Paddy leaf disease recognition system using image processing techniques and support vector machine. *ESTEEM Academic Journal, 16*, 41–50.

Mani, M., & Agarwal, R. (2022). Payment Banks in India: Insights From the Customers' Perception. [IJABIM]. *International Journal of Asian Business and Information Management, 13*(1), 1–15. DOI: 10.4018/IJABIM.297851

Manohar, N., & Gowda, K. J. (2020). Image processing system based identification and classification of leaf disease: A case study on paddy leaf. In *2020 International Conference on Electronics and Sustainable Communication Systems (ICESC)* (pp. xx–xx). IEEE.

Masenya, T. M., & Chisita, C. T. (2022). Futurizing library services in a technology-driven dispensation: Reflections on selected academic libraries in Zimbabwe and South Africa. In *Innovative technologies for enhancing knowledge access in academic libraries* (pp. 1–21). IGI Global. DOI: 10.4018/978-1-6684-3364-5.ch001

Masud, M. M., Hayawi, K., Mathew, S. S., Michael, T., & El Barachi, M. (2022). Smart Online Exam Proctoring Assist for Cheating Detection. *Lecture Notes in Computer Science (Including Subseries Lecture Notes in Artificial Intelligence and Lecture Notes in Bioinformatics), 13087 LNAI*, 118–132. DOI: 10.1007/978-3-030-95405-5_9

Matin, M. M. H., Islam, M. M., Rahman, M. M., & Ahmed, M. (2020). An efficient disease detection technique of rice leaf using AlexNet. *Journal of Computer and Communications*, *8*(12), 49–57. DOI: 10.4236/jcc.2020.812005

Maylawati, D. S., Priatna, T., Sugilar, H., & Ramdhani, M. A. (2020). *Data science for digital culture improvement in higher education using K-means clustering and text analytics. 10*(5), 4569–4580. DOI: 10.11591/ijece.v10i5.pp4569-4580

McElheran, K., Li, J. F., Brynjolfsson, E., Kroff, Z., Dinlersoz, E., Foster, L., & Zolas, N. (2024). AI adoption in America: Who, what, and where. *Journal of Economics & Management Strategy*, *33*(2), 375–415.

Meenalochinia, G., & Ramkumar, S. (2021). Survey of machine learning algorithms for breast cancer detection using mammogram images. *Materials Today: Proceedings*, *37*(2), 2738–2743. DOI: 10.1016/j.matpr.2020.08.543

Meituan. (2022, June). *YOLOv6 By Meituan Vision AI Department*. Meituan Vision AI Department. https://github.com/meituan/YOLOv6

Mekha, P., & Teeyasuksaet, N. (2021). Image classification of rice leaf diseases using random forest algorithm. In *2021 Joint International Conference on Digital Arts, Media and Technology with ECTI Northern Section Conference on Electrical, Electronics, Computer and Telecommunication Engineering* (pp. xx–xx). IEEE. DOI: 10.1109/ECTIDAMTNCON51128.2021.9425696

Milsan, A. L., & Wewe, M. (2018). Hubungan Antara Kecerdasan Logis Matematis dengan Hasil Belajar Matematika. *Journal of Education Technology*, *2*(2), 65–69. DOI: 10.23887/jet.v2i2.16183

Mistry, I., Tanwar, S., Tyagi, S., & Kumar, N. (2020). Blockchain for 5G-enabled IoT for industrial automation: A systematic review, solutions, and challenges. *Mechanical Systems and Signal Processing*, *135*, 106382. DOI: 10.1016/j.ymssp.2019.106382

Mitra, A., Singhal, A., Mohanty, S. P., Kougianos, E., & Ray, C. (2022). eCrop: A novel framework for automatic crop damage estimation in smart agriculture. *SN Computer Science*, *3*(4), 319. DOI: 10.1007/s42979-022-01216-8

Mohamed Nafuri, A. F., Sani, N. S., Zainudin, N. F. A., Rahman, A. H. A., & Aliff, M. (2022). Clustering Analysis for Classifying Student Academic Performance in Higher Education. *Applied Sciences (Basel, Switzerland)*, *12*(19), 9467. Advance online publication. DOI: 10.3390/app12199467

Mohammed, L., & Yusoff, Y. (2023). Detection and classification of plant leaf diseases using digital image processing methods: A review. *ASEAN Engineering Journal*, *13*(1), 1–9. DOI: 10.11113/aej.v13.17460

Mohanta, B. K., Jena, D., Panda, S. S., & Sobhanayak, S. (2019). Blockchain technology: A survey on applications and security privacy challenges. *Internet of Things : Engineering Cyber Physical Human Systems*, *8*, 100107. DOI: 10.1016/j.iot.2019.100107

Mohapatra, D., & Das, N. (2023). A precise model for accurate rice disease diagnosis: A transfer learning approach. *Proceedings of the Indian National Science Academy. Part A, Physical Sciences*, *89*(1), 162–171. DOI: 10.1007/s43538-022-00149-3

Mohd Pozi, M. S., Muruti, G., Abu Bakar, A., Jatowt, A., & Kawai, Y. (2018, May). Preserving author editing history using blockchain technology. In *Proceedings of the 18th ACM/IEEE on Joint Conference on Digital Libraries* (pp. 165-168). DOI: 10.1145/3197026.3197068

Mohd Talib, N. I., Abd Majid, N. A., & Sahran, S. (2023). Identification of student behavioral patterns in higher education using k-means clustering and support vector machine. *Applied Sciences (Basel, Switzerland)*, *13*(5), 3267. DOI: 10.3390/app13053267

Mohri, M., Rostamizadeh, A., & Talwalkar, A. (2018). *Foundations of Machine Learning* (2nd ed.). MIT Press.

Mohyuddin, G., Khan, M. A., Haseeb, A., Mahpara, S., Waseem, M., & Saleh, A. M. (2024). Evaluation of Machine Learning approaches for precision farming in Smart Agriculture System-A comprehensive Review. *IEEE Access : Practical Innovations, Open Solutions*.

Monteiro, E. S., da Rosa Righi, R., Barbosa, J. L. V., & Alberti, A. M. (2021). APTM: A model for pervasive traceability of agrochemicals. *Applied Sciences (Basel, Switzerland), 11*(17), 8149. DOI: 10.3390/app11178149

Morchid, A., Muhammad Alblushi, I. G., Khalid, H. M., El Alami, R., Sitaramanan, S. R., & Muyeen, S. M. (2024). High-technology agriculture system to enhance food security: A concept of smart irrigation system using Internet of Things and cloud computing. *Journal of the Saudi Society of Agricultural Sciences*. Advance online publication. DOI: 10.1016/j.jssas.2024.02.001

Mousavi, S. K., & Ghaffari, A. (2021). Data cryptography in the Internet of Things using the artificial bee colony algorithm in a smart irrigation system. *Journal of Information Security and Applications, 61*, 102945. DOI: 10.1016/j.jisa.2021.102945

Mridha, K., Shaw, R. N., Kumar, D., & Ghosh, A. (2022). Driver Drowsiness Alert System Using Real-Time Detection. AI and IoT for Smart City Applications, 65-81.

Mudaliar, N. K., Redekar, S., Kadam, N. M., & Ramachandran, J. (2023). Machine learning-based prediction model for cancer detection. *2023 2nd International Conference on Applied Artificial Intelligence and Computing (ICAAIC)*, 383–388. https://doi.org/DOI: 10.1109/ICAAIC57855.2023.10193819

Mughnyanti, M., Efendi, S., & Zarlis, M. (2020). Analysis of determining centroid clustering x-means algorithm with davies-bouldin index evaluation. *IOP Conference Series. Materials Science and Engineering, 725*(1), 012128. Advance online publication. DOI: 10.1088/1757-899X/725/1/012128

Murphy, K. M., Singh, D., Zhang, H., & Crisp, P. A. (2024). Deep learning in image-based plant phenotyping. *Annual Review of Plant Biology, 75*(1), 1–26. DOI: 10.1146/annurev-arplant-070523-042828 PMID: 38382904

Mustopa, A., Pratama, E. B., Hendini, A., & Risdiansyah, D. (2020, November). Analysis of user reviews for the pedulilindungi application on google play using the support vector machine and naive bayes algorithm based on particle swarm optimization. In *2020 Fifth International Conference on Informatics and Computing (ICIC)* (pp. 1-7). IEEE.

Muthmainnah, R. N., Eminita, V., & Ramadhan, A. I. (2020). CLUSTER ANALYSIS OF STUDENTS ' MULTIPLE INTELLIGENCES IN FIRST GRADE AT MADRASAH IBTIDAIYAH AL-INAYAH IN INDONESIA. *Journal of Critical Reviews, 7*(13), 929–933.

Nagasubramanian, K., Jones, S., Singh, A. K., Sarkar, S., Singh, A., & Ganapathysubramanian, B. (2020). Usefulness of interpretability methods to explain deep learning based plant stress phenotyping. *arXiv Preprint*, arXiv:2007.05729. https://arxiv.org/abs/2007.05729

Naji, M. A., El Filali, S., Aarika, K., Benlahmar, E. L. H., Abdelouhahid, R. A., & Debauche, O. (2021). Machine learning algorithms for breast cancer prediction and diagnosis. *Procedia Computer Science*, *191*, 487–492. DOI: 10.1016/j.procs.2021.07.062

Nalini, S., Krishnaraj, N., Jayasankar, T., Vinothkumar, K., Sagai Francis Britto, A., Subramaniam, K., & Bharatiraja, C. (2021). Paddy leaf disease detection using an optimized deep neural network. *Computers, Materials & Continua*, *68*(1), 1117–1128. DOI: 10.32604/cmc.2021.012431

Nallamala, S. H., Mishra, P., & Koneru, S. V. (2019). Breast cancer detection using machine learning way. *International Journal of Recent Technology and Engineering*, *8*(2–3), 1402–1405.

Namdech, P., Rotjanawasuthorn, S., Singh, B., & Kaunert, C. (2025). Harnessing Artificial Intelligence For Lifelong Education: Opportunities And Challenges. *Asian Education and Learning Review*, *3*(2), 4–4.

Nazari, M., Hussain, A., & Musilek, P. (2023). Applications of Clustering Methods for Different Aspects of Electric Vehicles. *Electronics (Basel)*, *12*(4), 790. Advance online publication. DOI: 10.3390/electronics12040790

Ndunagu, J. N., Ukhurebor, K. E., Akaaza, M., & Onyancha, R. B. (2022). Development of a Wireless Sensor Network and IoT-based Smart Irrigation System. *Applied and Environmental Soil Science*, *2022*(1), 7678570. DOI: 10.1155/2022/7678570

Nguyen, Q. H., Ly, H.-B., Ho, L. S., Al-Ansari, N., Van Le, H., Tran, V. Q., Prakash, I., & Pham, B. T. (2021). Influence of Data Splitting on Performance of Machine Learning Models in Prediction of Shear Strength of Soil. *Mathematical Problems in Engineering*, *2021*, 1–15. DOI: 10.1155/2021/4832864

Nigam, A., Tiwari, A. K., & Pandey, A. (2020). Paddy leaf diseases recognition and classification using PCA and BFO-DNN algorithm by image processing. *Materials Today: Proceedings*, *33*, 4856–4862. DOI: 10.1016/j.matpr.2020.08.397

Ni, J., Zhang, K., Lin, X., & Shen, X. (2017). Securing fog computing for internet of things applications: Challenges and solutions. *IEEE Communications Surveys and Tutorials*, *20*(1), 601–628. DOI: 10.1109/COMST.2017.2762345

Nishchal, J., Reddy, S., & Navya, P. N. (2020). Automated Cheating Detection in Exams using Posture and Emotion Analysis. *2020 IEEE International Conference on Electronics, Computing and Communication Technologies (CONECCT)*, 1–6. DOI: 10.1109/CONECCT50063.2020.9198691

Nurzaelani, M. M., & Wibowo, S. (2015). Hubungan Antara Kecerdasan Logis-Matematis dan Komunikasi Interpersonal dengan Hasil Belajar Mata Pelajaran Matematika. *Jurnal Teknologi Pendidikan*, *4*(1), 56–72.

O'Shaughnessy, M. (2022). One of the Biggest Problems in Regulating AI Is Agreeing on a Definition. Carnegie Endowment. url: https://carnegieendowment.org/2022/10/06/one-of-biggest- problems-in-regulating-ai-is-agreeing-on-definition-pub-88100.

Obermeyer, Z., Nissan, R., Stern, M., Eanef, S., Bembeneck, E. J., & Mullainathan, S. (2021). *Algorithmic Bias Playbook*. The Center for Applied Artificial Intelligence at Chicago Booth.

Ojobor, R. C. (2023). 1 Blockchain Technology for Library Services: Challenges and Opportunities for Libraries from a Nigerian Perspective. *Information Services for a Sustainable Society: Current Developments in an Era of Information Disorder*, *183*, 7–23. DOI: 10.1515/9783110772753-003

Ökten, İ., & Yüzgeç, U. (2022). Rice plant disease detection using image processing and probabilistic neural network. In *International Congress of Electrical and Computer Engineering* (pp. xx–xx). Springer International Publishing., https://doi.org/xxx DOI: 10.1007/978-3-031-01984-5_7

Olorunshola, O. E., Irhebhude, M. E., & Evwiekpaefe, A. E. (2023). A Comparative Study of YOLOv5 and YOLOv7 Object Detection Algorithms. *Journal of Computing and Social Informatics*, *2*(1), 1–12. DOI: 10.33736/jcsi.5070.2023

Omame, I. M., & Alex-Nmecha, J. C. (2021). Application of blockchain in libraries and information centers. In *Handbook of Research on Knowledge and Organization Systems in Library and Information Science* (pp. 384–397). IGI Global. DOI: 10.4018/978-1-7998-7258-0.ch020

Ordóñez, J., Alexopoulos, A., Koutras, K., Kalogeras, A., Stefanidis, K., & Martos, V. (2023). Blockchain in Agriculture: A PESTELS Analysis. *IEEE Access : Practical Innovations, Open Solutions*.

Ossai, M. C., Ethe, N., Edougha, D. E., & Okeh, O. D. (2023). Academic integrity during examinations, age and gender as predictors of academic performance among high school students. *International Journal of Educational Development*, *100*, 102811. Advance online publication. DOI: 10.1016/j.ijedudev.2023.102811

Oyelude, A. A. (2022). Trending issues in advancing blockchain technology in libraries, archives and museums. *Library Hi Tech News*, *39*(6), 6–7. DOI: 10.1108/LHTN-06-2021-0040

Pacheco, J., Benitez, V. H., Felix-Herran, L. C., & Satam, P. (2020). Artificial neural networks- based intrusion detection system for internet of things fog nodes. *IEEE Access : Practical Innovations, Open Solutions*, *8*, 73907–73918. DOI: 10.1109/ACCESS.2020.2988055

Padilla, D. A., Pajes, R. A. I., & De Guzman, J. T. (2020). Detection of corn leaf diseases using convolutional neural network with OpenMP implementation. In *2020 IEEE 12th International Conference on Humanoid, Nanotechnology, Information Technology, Communication and Control, Environment, and Management (HNICEM)* (pp. xx–xx). IEEE. DOI: 10.1109/HNICEM51456.2020.9400004

Padol, P. B., & Yadav, A. A. (2016). SVM classifier based grape leaf disease detection. *2016 Conference on Advances in Signal Processing (CASP)*, 175–179. DOI: 10.1109/CASP.2016.7746160

Paganelli, A. L., & Paganelli, A. L. (2023). Blockchain and the Research Libraries: Expanding Interlibrary Loan and Protecting Privacy. In Research Anthology on Convergence of Blockchain, Internet of Things, and Security (pp. 1152-1166). IGI Global.

Pahlawanto, R. D. S., Salsabila, H., & Pratiwi, K. R. (2024). Detection and prediction of rice plant diseases using convolutional neural network (CNN) method. *Journal of Student Research Exploration*, *2*(1), 22–33. DOI: 10.52465/josre.v2i1.254

Pallathadka, H., Chowdhary, C. L., & Ahmad, S. (2022). Application of machine learning techniques in rice leaf disease detection. *Materials Today: Proceedings*, *51*, 2277–2280. DOI: 10.1016/j.matpr.2021.11.398

Pambudi, A., Purnama, S., Ayuninggati, T., Santoso, N. P. L., & Oktariyani, A. (2021, November). Legality on digital document using blockchain technology: an exhaustive study. In *2021 Sixth International Conference on Informatics and Computing (ICIC)* (pp. 1-6). IEEE. DOI: 10.1109/ICIC54025.2021.9632860

Panda, S. S., Mohanta, B. K., Dey, M. R., Satapathy, U., & Jena, D. (2020, July). Distributed ledger technology for securing IoT. In *2020 11th international conference on computing, communication and networking technologies (ICCCNT)* (pp. 1-6). IEEE. DOI: 10.1109/ICCCNT49239.2020.9225333

Panda, S., & Kaur, N. (2023). Blockchain: A New Technology in Library System and Management. In Handbook of Research on Advancements of Contactless Technology and Service Innovation in Library and Information Science (pp. 211-230). IGI Global. DOI: 10.4018/978-1-6684-7693-2.ch011

Pan, J., Wang, T., & Wu, Q. (2023). RiceNet: A two stage machine learning method for rice disease identification. *Biosystems Engineering*, *225*, 25–40. DOI: 10.1016/j.biosystemseng.2022.11.007

Paramasivam, S., Palanisamy, S. K., Cinthaikinian, S., Palanisamy, V., Mariappan, R., & Selvakumar, P. K. (2025). Biosensors and its diverse applications in healthcare systems. *Zeitschrift für Physikalische Chemie*, *239*(1), 37–43.

Paricherla, M., Ritonga, M., Shinde, S. R., Chaudhari, S. M., Linur, R., & Raghuvanshi, A. (2023). Machine learning techniques for accurate classification and detection of intrusions in computer network. *Bulletin of Electrical Engineering and Informatics*, *12*(4), 2340–2347. DOI: 10.11591/eei.v12i4.4708

Park, S., Lee, S., & Park, J. (2020). Data augmentation method for improving the accuracy of human pose estimation with cropped images. *Pattern Recognition Letters*, *136*, 244–250. DOI: 10.1016/j.patrec.2020.06.015

Parvathi Sangeetha, B., Kumar, N., Ambalgi, A. P., Abdul Haleem, S. L., Thilagam, K., & Vijayakumar, P. (2022). IOT based smart irrigation management system for environmental sustainability in India. *Sustainable Energy Technologies and Assessments*, *52*, 101973. DOI: 10.1016/j.seta.2022.101973

Parven, N., Rahman, M. M., Nahar, N., & Rahman, M. M. (2020). Detection and recognition of paddy plant leaf diseases using machine learning technique. *International Journal of Innovative Technology and Exploring Engineering*, *9*(1), 546–551. DOI: 10.35940/ijitee.E2509.039520

Patel, H., & Shrimali, B. (2023). AgriOnBlock: Secured data harvesting for agriculture sector using blockchain technology. *ICT Express*, *9*(2), 150–159. https://doi.org/https://doi.org/10.1016/j.icte.2021.07.003. DOI: 10.1016/j.icte.2021.07.003

Patidar, S., ... (2020). Rice plant disease detection and classification using deep residual learning. In *Machine Learning, Image Processing, Network Security and Data Sciences:Second International Conference, MIND 2020,Silchar, India,July 30–31, 2020, Proceedings, Part I* (Vol. 2, pp. xx–xx). Springer Singapore. DOI: 10.1007/978-981-15-6315-7_23

Patil, P. C., Patil, D. D., & Subramanium, P. (2020). Detection and identification of rice leaf diseases using multiclass SVM. *System*, *7*(06), 1–5.

Paulavičius, R., Grigaitis, S., Igumenov, A., & Filatovas, E. (2019). A decade of blockchain: Review of the current status, challenges, and future directions. *Informatica (Vilnius), 30*(4), 729–748. DOI: 10.15388/Informatica.2019.227

Paul, R., Ghosh, N., Sau, S., Chakrabarti, A., & Mohapatra, P. (2021). Blockchain based secure smart city architecture using low resource IoTs. *Computer Networks, 196*(February), 108234. DOI: 10.1016/j.comnet.2021.108234

Pehlivan, A., & Durgut, M. (2017). The effect of logical-mathematical intelligence on financial accounting achievement according to multiple intelligence theory. *Journal of Education & Social Policy, 4*(3), 132–139.

Peng, J., Gao, H., Zhang, Z., & Chen, Y. (2023). RiceDRA-net: Precise identification of rice leaf diseases with complex backgrounds using a res-attention mechanism. *Applied Sciences (Basel, Switzerland), 13*(8), 4928. DOI: 10.3390/app13084928

Pereira, G. P., Chaari, M. Z., & Daroge, F. (2023). IoT-enabled smart drip irrigation system using ESP32. *IoT, 4*(3), 221–243.

Petchiammal, M., Rajarajeswari, S., & Bhuvaneswari, T. (2023). Paddy doctor: A visual image dataset for automated paddy disease classification and benchmarking. In *Proceedings of the 6th Joint International Conference on Data Science & Management of Data (10th ACM IKDD CODS and 28th COMAD)* (pp. 17–24). ACM. DOI: 10.1145/3570991.3570994

Petrovic, S. (2006). A comparison between the silhouette index and the davies-bouldin index in labelling ids clusters. *Proceedings of the 11th Nordic Workshop of Secure IT Systems*, 2006, 53–64.

Phang, S. K., Chiang, T. H. A., Happonen, A., & Chang, M. M. L. (2023). From Satellite to UAV-Based Remote Sensing: A Review on Precision Agriculture. *IEEE Access : Practical Innovations, Open Solutions, 11*, 127057–127076. DOI: 10.1109/ACCESS.2023.3330886

Pikatza, A., Sainz, N., González, I., & Emaldi, M. (2023, July). Development of an Open Source IoT-Blockchain Platform for Traceability of Fresh Products from Farm to Fork. In *International Congress on Blockchain and Applications* (pp. 487-497). Cham: Springer Nature Switzerland. DOI: 10.1007/978-3-031-45155-3_46

Pizzinelli, C., Panton, A. J., Ms, M. M. T., Cazzaniga, M., & Li, L. (2023). Labor Market Exposure to AI: Cross-country Differences and Distributional Implications. Working Paper 2023/216. International Monetary Fund.

Podder, A. K., Al Bukhari, A., Islam, S., Mia, S., Mohammed, M. A., Kumar, N. M., & Abdulkareem, K. H. (2021). IoT based smart agrotech system for verification of Urban farming parameters. *Microprocessors and Microsystems*, *82*, 104025. DOI: 10.1016/j.micpro.2021.104025

Poornappriya, T. S., & Gopinath, R. (2022). Rice plant disease identification using artificial intelligence approaches. *International Journal of Electrical Engineering and Technology*, *11*(10), 392–402.

Pothen, M. E., & Pai, M. L. (2020). Detection of rice leaf diseases using image processing. In *2020 Fourth International Conference on Computing Methodologies and Communication (ICCMC)* (pp. xx–xx). IEEE. DOI: 10.1109/ICCMC48092.2020.ICCMC-00080

Prakash, C., Singh, L. P., Gupta, A., & Lohan, S. K. (2023). Advancements in smart farming: A comprehensive review of IoT, wireless communication, sensors, and hardware for agricultural automation. *Sensors and Actuators. A, Physical*, *362*, 114605. DOI: 10.1016/j.sna.2023.114605

Prakash, R., Anoop, V. S., & Asharaf, S. (2022). Blockchain technology for cybersecurity: A text mining literature analysis. *International Journal of Information Management Data Insights*, *2*(2), 100112. DOI: 10.1016/j.jjimei.2022.100112

Pratmanto, D., Rousyati, R., Wati, F. F., Widodo, A. E., Suleman, S., & Wijianto, R. (2020). App Review Sentiment Analysis Shopee Application in Google Play Store Using Naive Bayes Algorithm. *Journal of Physics: Conference Series*, *1641*(1), 012043. Advance online publication. DOI: 10.1088/1742-6596/1641/1/012043

Prodan, M., Paraschiv, E., & Stanciu, A. (2023). Applying deep learning methods for mammography analysis and breast cancer detection. *Applied Sciences (Basel, Switzerland)*, *13*(7), 4272. DOI: 10.3390/app13074272

Pullo, S., Pareschi, R., Piantadosi, V., Salzano, F., & Carlini, R. (2023, December). Integrating IOTA's Tangle with the Internet of Things for Sustainable Agriculture: A Proof-of-Concept Study on Rice Cultivation. []. MDPI.]. *Informatics (MDPI)*, *11*(1), 3. DOI: 10.3390/informatics11010003

Purnamaningsih, C., & Aziz, A. (2014). Pemanfaatan Metode K-Means Clustering dalam Penentuan Penjurusan Siswa SMA. *JURNAL ITSMART*, *3*(1), 27–33. DOI: 10.20961/its.v3i1.644

Purwita Sary, I., Ucok Armin, E., Andromeda, S., Engineering, E., & Singaperbangsa Karawang, U. (2023). Performance Comparison of YOLOv5 and YOLOv8 Architectures in Human Detection Using Aerial Images. *Ultima Computing : Jurnal Sistem Komputer*, *15*(1), 8–13. DOI: 10.31937/sk.v15i1.3204

Puspitasari, A., Rapani, & Ambarita, A. (2018). Hubungan Kecerdasan Verbal dan Logika-Matematika dengan Meyelesaikan Soal Cerita Matematika. *Jurnal Pedagogi*, *7*(2), 430–439.

Putra, R. E., Tjandrasa, H., & Suciati, N. (2020). Severity classification of non-proliferative diabetic retinopathy using convolutional support vector machine. *International Journal of Intelligent Engineering and Systems*, *13*(4), 156–170. DOI: 10.22266/ijies2020.0831.14

Qian, C. (2021, April). Digital copyright management model of university library based on blockchain technology. In *2021 International Conference on Computer, Blockchain and Financial Development (CBFD)* (pp. 508-513). IEEE. DOI: 10.1109/CBFD52659.2021.00108

Quan, R. (2021). National Smart Library System: Opening a New Chapter for the Transformation toward Smart Library. *Journal of Library Science in China*, *47*, 4–14.

Rachman, R. K., Mahmud, R., Susilo, E., & Widodo, T. (2024). Enhanced vision transformer and transfer learning approach to improve rice disease recognition. *Journal of Computing Theories and Applications*, *1*(4), 446–460. DOI: 10.62411/jcta.10459

Radeva, I., & Popchev, I. (2022). Blockchain-Enabled Supply-Chain in Crop Production Framework. *Cybernetics and Information Technologies*, *22*(1), 151–170. DOI: 10.2478/cait-2022-0010

Radovancovici, M., Galiş, D., & Pungilă, C. (2022, September). A Practical Analysis of Techniques for Minting Genetic Information as NFTs in Blockchain Technology. In 2022 24th International Symposium on Symbolic and Numeric Algorithms for Scientific Computing (SYNASC) (pp. 331-337). IEEE. DOI: 10.1109/SYNASC57785.2022.00060

Raehanah, R., & Apriani, R. (2019). Pengaruh kecerdasan logis matematis terhadap hasil belajar matematika dasar. *Jurnal Pijar MIPA*, *14*(3), 112–117.

Rahman, C. R., Arko, P. S., Ali, M. E., Khan, M. A. I., Apon, S. H., Nowrin, F., & Hasan, K. (2020). Identification and recognition of rice diseases and pests using convolutional neural networks. *Biosystems Engineering*, *194*, 112–120. DOI: 10.1016/j.biosystemseng.2020.03.020

Rahman, K. S., Rakib, M. R. I., Salehin, M. M., Ali, M. R., & Rahman, A. (2024). Assessment of paddy leaves disease severity level using image processing technique. *Smart Agricultural Technology*, *7*, 100410. DOI: 10.1016/j.atech.2024.100410

Rahman, M. A., Zaman, N., Asyhari, A. T., Sadat, S. M. N., Pillai, P., & Arshah, R. A. (2021). SPY-BOT: Machine learning-enabled post filtering for Social Network-Integrated Industrial Internet of Things. *Ad Hoc Networks*, *121*(June), 102588. DOI: 10.1016/j.adhoc.2021.102588

Rahman, M. M., Ahmed, S., Akter, S., & Islam, M. R. (2024). Green vanguards: Harnessing the power of plant antioxidants, signal catalysts, and genetic engineering to combat reactive oxygen species under multiple abiotic stresses. *Plant Stress*, *100547*, 100547. Advance online publication. DOI: 10.1016/j.stress.2024.100547

Rahman, S. I., Ahmed, S., Fariha, T. A., Mohammad, A., Haque, M. N. M., Chellappan, S., & Noor, J. (2024). Unsupervised machine learning approach for tailoring educational content to individual student weaknesses. *High-Confidence Computing*, *4*(4), 100228. DOI: 10.1016/j.hcc.2024.100228

Raiaan, M. A. K., Fahad, N. M., Mukta, M. S. H., & Shatabda, S. (2024). MammoLight: A lightweight convolutional neural network for diagnosing breast cancer from mammography images. *Biomedical Signal Processing and Control*, *94*, 106279. DOI: 10.1016/j.bspc.2024.106279

Rajpoot, V., Tiwari, A., & Jalal, A. S. (2023). Automatic early detection of rice leaf diseases using hybrid deep learning and machine learning methods. *Multimedia Tools and Applications*, *82*(23), 36091–36117. DOI: 10.1007/s11042-023-14969-y

Rakhee, M., & Elayidom, S. (2023, January). Blockchain Technology as a Solution for Vulnerabilities in Internet of Things-A Survey. In *2023 Advanced Computing and Communication Technologies for High Performance Applications (ACCTHPA)* (pp. 1-6). IEEE. DOI: 10.1109/ACCTHPA57160.2023.10083337

Ramakrishnan, T., Mohan Gift, M. D., Chitradevi, S., Jegan, R., & Subha Hency Jose, P. Nagaraja, Rajneesh Sharma H.N., Selvakumar P., Sintayehu Mekuria Hailegiorgis, (2022). "Study of Numerous Resins Used in Polymer Matrix Composite Materials" Advances in Materials Science and Engineering, vol. 2022, Article ID 1088926, 8 pages, 2022. DOI: 10.1155/2022/1088926

Ramakrishnan, T., Mohan Gift, M. D., Chitradevi, S., Jegan, R., Hency Jose, P. S., Nagaraja, H. N., & Hailegiorgis, S. M. (2022). Study of numerous resins used in polymer matrix composite materials. *Advances in Materials Science and Engineering*, *2022*(1), 1088926.

Ramesh, S., & Vydeki, D. (2020). Recognition and classification of paddy leaf diseases using optimized deep neural network with Jaya algorithm. *Information Processing in Agriculture*, *7*(2), 249–260. DOI: 10.1016/j.inpa.2019.09.002

Rammer, C., Fernández, G. P., & Czarnitzki, D. (2022). Artificial Intelligence and Industrial Innovation: Evidence from German Firm-Level Data. *Research Policy*, *51*(7), 104555. DOI: 10.1016/j.respol.2022.104555

Ramos-Fernández, L., Cáceres-Jara, M., Aquino, M. A., & Espinoza, G. (2024). Water stress index and stomatal conductance under different irrigation regimes with thermal sensors in rice fields on the northern coast of Peru. *Remote Sensing (Basel)*, *16*(5), 796. DOI: 10.3390/rs16050796

Ramya, R. S., Ramesh, B. D., Kumudavalli, M. V., Joshi, P., & Venugopal, K. R. (2023, December). Blockchain-Integrated Metaverse for Academia: A Solution for Virtual Classes, Lectures, and Project Fundings. In International Conference on Information and Communication Technology for Competitive Strategies (pp. 387-396). Singapore: Springer Nature Singapore.

Rana, A., Gupta, S., & Gupta, B. (2024). A comprehensive framework for quantitative risk assessment of organizational networks using FAIR-modified attack trees. *Frontiers of Computer Science*, *6*, 1304288. DOI: 10.3389/fcomp.2024.1304288

Ravindran, M. (2022). Energy Efficient Clustering for Precision Agriculture using LEACH Protocol. *NeuroQuantology : An Interdisciplinary Journal of Neuroscience and Quantum Physics*, *20*(7), 2702.

Rawat, R., Chakrawarti, R. K., Sarangi, S. K., Patel, J., Bhardwaj, V., Rawat, A., & Rawat, H. (Eds.). (2023). *Quantum Computing in Cybersecurity*. John Wiley & Sons. DOI: 10.1002/9781394167401

Reddy, K. R., Priya, K. H., & Neelima, N. (2015). Object Detection and Tracking — A Survey. *2015 International Conference on Computational Intelligence and Communication Networks (CICN)*, 418–421. DOI: 10.1109/CICN.2015.317

Rejeb, A., Rejeb, K., Abdollahi, A., Al-Turjman, F., & Treiblmaier, H. (2022). The Interplay between the Internet of Things and agriculture: A bibliometric analysis and research agenda. *Internet of Things : Engineering Cyber Physical Human Systems*, *19*(July), 100580. DOI: 10.1016/j.iot.2022.100580

Restrepo, P. (2023). Automation: Theory, Evidence, and Outlook. Working Paper 31910. National Bureau of Economic Research.

Rhaman, M. S., Islam, M. R., Jahan, M., & Akter, M. (2024). Phytohormone strigolactone: Involvement in guard cell signaling and abiotic stress tolerance in plants. *Journal of Plant Growth Regulation*, *43*(12), 1–14. DOI: 10.1007/s00344-024-11421-9

Riahi, A., Challal, Y., Natalizio, E., Chtourou, Z., & Bouabdallah, A. (2013). "A systemic approach for IoT security" in 2013 IEEE International Conference on Distributed Computing in Sensor Systems (IEEE), 351–355.

Ritharson, P. I., Raimond, K., Mary, X. A., Robert, J. E., & J, A. (2024). DeepRice: A deep learning and deep feature based classification of rice leaf disease subtypes. *Artificial Intelligence in Agriculture*, *11*, 34–49. DOI: 10.1016/j.aiia.2023.11.001

Roikha, M. (2016). *Pengaruh Kecerdasan Mtematis-Logis dan Kecerdasan Visual-Spasial terhadap Prestasi Belajar Matematika Materi Bangun Ruang Kelas V di SD Tamansiswa Turen*. UIN Maulana Malik Ibrahim, Malang.

Rokach, L. (2009). Ensemble-based classifiers. *Artificial Intelligence Review*, *33*(1–2), 1–39. DOI: 10.1007/s10462-009-9124-7

Roscher, R., Bohn, B., Duarte, M. F., & Garcke, J. (2020). Explainable machine learning for scientific insights and discoveries. *IEEE Access : Practical Innovations, Open Solutions*, *8*, 42200–42216. DOI: 10.1109/ACCESS.2020.2976199

Roski, M., Sebastian, R., Ewerth, R., Hoppe, A., & Nehring, A. (2024). Learning analytics and the Universal Design for Learning (UDL): A clustering approach. *Computers \& Education, 214*, 105028.

Roth, B., Becker, N., Romeyke, S., Schäfer, S., Domnick, F., & Spinath, F. M. (2015). Intelligence and school grades : A meta-analysis. *Intelligence*, *53*, 118–137. DOI: 10.1016/j.intell.2015.09.002

Rouhani, S., Belchior, R., Cruz, R. S., & Deters, R. (2021). Distributed attribute-based access control system using permissioned blockchain. *World Wide Web (Bussum)*, *24*(5), 1–28. DOI: 10.1007/s11280-021-00874-7

Rovshenov, A., & Peker, S. (2022). Performance comparison of different machine learning techniques for early prediction of breast cancer using Wisconsin Breast Cancer Dataset. *2022 3rd International Informatics and Software Engineering Conference (IISEC)*, 1–6. https://doi.org/DOI: 10.1109/IISEC56805.2022.10009384

Rudin, C. (2019). Stop explaining black box machine learning models for high stakes decisions and use interpretable models instead. *Nature Machine Intelligence*, *1*(5), 206–215. DOI: 10.1038/s42256-019-0048-x PMID: 35603010

Rudrakar, S., & Rughani, P. (2022). *IoT based Agriculture (IoTA)*. Architecture, Cyber Attack, Cyber Crime and Digital Forensics Challenges.

Ryu, J.-H., Jeong, H., & Cho, J. (2020). Performances of vegetation indices on paddy rice at elevated air temperature, heat stress, and herbicide damage. *Remote Sensing (Basel)*, *12*(16), 2654. DOI: 10.3390/rs12162654

Sadiq, S., Umer, M., Ullah, S., Mirjalili, S., Rupapara, V., & Nappi, M. (2021). Discrepancy detection between actual user reviews and numeric ratings of Google App store using deep learning. *Expert Systems with Applications*, *181*(June), 115111. Advance online publication. DOI: 10.1016/j.eswa.2021.115111

Saeed, N., Wen, F., & Afzal, M. Z. (2024). Decentralized peer-to-peer energy trading in microgrids: Leveraging blockchain technology and smart contracts. *Energy Reports*, *12*, 1753–1764. DOI: 10.1016/j.egyr.2024.07.053

Sagarika, G. K., Krishna Prasad, S. J., & Mohana Kumar, S. (2020). Paddy plant disease classification and prediction using convolutional neural network. In *2020 International Conference on Recent Trends on Electronics, Information, Communication & Technology (RTEICT)* (pp. 829–833). IEEE. DOI: 10.1109/RTEICT49044.2020.9315634

Sahu, A. K., Sharma, S., Tanveer, M., & Raja, R. (2021). Internet of Things attack detection using hybrid Deep Learning Model. *Computer Communications*, *176*(December 2020), 146–154. DOI: 10.1016/j.comcom.2021.05.024

Saifullah, M., Iqbal, M., Hussain, M., & Ahmed, S. (2019). Estimation of water stress on rice crop using ecological parameters. *International Journal of Agricultural Sustainable Development*, *1*(1), 17–29. DOI: 10.33411/IJASD/20190103

Sakhamuri, S., & Kompalli, V. S. (2020). RETRACTED: An overview on prediction of plant leaves disease using image processing techniques. *IOP Conference Series. Materials Science and Engineering*, *981*(2), 022006. DOI: 10.1088/1757-899X/981/2/022024

Sakthipriya, S., & Naresh, R. (2024). Precision agriculture based on convolutional neural network in rice production nutrient management using machine learning genetic algorithm. *Engineering Applications of Artificial Intelligence*, *130*, 107682. DOI: 10.1016/j.engappai.2023.107682

Salama, R., Al-Turjman, F., Altrjman, C., Kumar, S., & Chaudhary, P. (2023, April). A comprehensive survey of blockchain-powered cybersecurity-a survey. In *2023 International Conference on Computational Intelligence, Communication Technology and Networking (CICTN)* (pp. 774-777). IEEE. DOI: 10.1109/CICTN57981.2023.10141282

Salem, O., Alsubhi, K., Shaafi, A., Gheryani, M., Mehaoua, A., & Boutaba, R. (2021). Man-in- the-middle attack mitigation in internet of medical things. *IEEE Transactions on Industrial Informatics, 18*(3), 2053–2062. DOI: 10.1109/TII.2021.3089462

Sandhu, J. K., Kaur, A., & Kaushal, C. (2022). Analysis of breast cancer in early stage by using machine learning algorithms: A review. *2022 IEEE International Conference on Current Development in Engineering and Technology (CCET)*, 1–7. https://doi.org/DOI: 10.1109/CCET56127.2022.00010

Sankareshwaran, S. P., Hariharan, M., & Maheswari, T. (2023). Optimizing rice plant disease detection with crossover boosted artificial hummingbird algorithm based AX-RetinaNet. *Environmental Monitoring and Assessment, 195*(9), 1070. DOI: 10.1007/s10661-023-11612-z PMID: 37610473

Saputra, R. A., Suharyanto, , Wasiyanti, S., Saefudin, D. F., Supriyatna, A., & Wibowo, A. (2020). Rice leaf disease image classifications using KNN based on GLCM feature extraction. *Journal of Physics: Conference Series, 1641*(1), 012065. DOI: 10.1088/1742-6596/1641/1/012080

Sathyavani, R., JaganMohan, K., & Kalaavathi, B. (2022). Classification of nutrient deficiencies in rice crop using DenseNet-BC. *Materials Today: Proceedings, 56*, 1783–1789. DOI: 10.1016/j.matpr.2021.10.466

Saxe, J., & Berlin, K. (2015). Deep Neural Network Based Malware Detection Using Two Dimensional Binary Program Features. *arXiv (Cornell University)*. https://doi.org//arxiv.1508.03096DOI: 10.1109/MALWARE.2015.7413680

Sazzad, T. M. S., Rahman, M. M., Islam, M. M., & Hossain, M. S. (2020). An image processing framework to identify rice blast. In *2020 International Congress on Human-Computer Interaction, Optimization and Robotic Applications (HORA)* (pp. 1–5). IEEE. DOI: 10.1109/HORA49412.2020.9152912

Schneider, C., Weinmann, M., Mohr, P. N. C., & vom Brocke, J. (2021). When the stars shine too bright: The influence of multidimensional ratings on online consumer ratings. *Management Science, 67*(6), 3871–3898. DOI: 10.1287/mnsc.2020.3654

Scholar, Q., & Zamir, S. (2018). A Comparative Study of Multiple Intelligence of Science and Arts Group Students. *Journal of Research in Social Sciences*, *6*(2), 2305–6533.

Selvakumar, P. (2023). Biomedical applications of polymers in biosensors, cancer vaccines and drug delivery systems. *Biopolimery i Kletka*, *39*(4), 245–256. DOI: 10.7124/bc.000AA0

Selvakumar, P., Mariappan, P., Shunmugam, S., Paul, R., & Jabez, S. (2024).. . *Zeitschrift für Physikalische Chemie*, *2024*. Advance online publication. DOI: 10.1515/zpch-2023-0450

Sendros, A., Drosatos, G., Efraimidis, P. S., & Tsirliganis, N. C. (2022). Blockchain applications in agriculture: A scoping review. *Applied Sciences (Basel, Switzerland)*, *12*(16), 8061. DOI: 10.3390/app12168061

Sengupta, A., Gill, S. S., Das, A., & De, D. (2021). Mobile edge computing based internet of agricultural things: a systematic review and future directions. *Mobile Edge Computing*, 415-441.

Senhadji, A., Tieman, A., Gemayel, E., & Benedek, D. (2021). *A Post-Pandemic Assessment of the Sustainable Development Goals. Staff Discussion Note 2021/003*. International Monetary Fund.

Senthilkumar, V., Kanagaraj, G., Primya, T., Joycema, J., Joan, M. B., & Vicram, J. A. (2021). Application of AI And Computer Vision To Face Mask And Social Distance Detection in CCTV Video Streams. *2021 International Conference on Advancements in Electrical, Electronics, Communication, Computing and Automation (ICAECA)*, 1–5. DOI: 10.1109/ICAECA52838.2021.9675746

Septiana, D. (2022). *Hubungan Efikasi Diri Dan Stres Akademik Dengan Perilaku Menyontek Siswa* [Universitas Islam Negeri Raden Intan Lampung]. https://repository.radenintan.ac.id/17318/

Setiawan, R., Zein, H., Azdy, R. A., & Sulistyowati, S. (2023). Rice leaf disease classification with machine learning: An approach using Nu-SVM. *Indonesian Journal of Data and Science*, *4*(3), 136–144. DOI: 10.56705/ijodas.v4i3.114

Setiyono, B., Amini, D. A., & Sulistyaningrum, D. R. (2021). Number plate recognition on vehicle using YOLO - Darknet. *Journal of Physics: Conference Series*, *1821*(1), 012049. Advance online publication. DOI: 10.1088/1742-6596/1821/1/012049

Shabbir, A., Ali, R. H., Shabbir, M. Z., Abideen, Z. U., Khan, T. A., Ijaz, A. Z., Ali, N., & Abu Bakar, M. (2023). Genetic algorithm-based feature selection for accurate breast cancer classification. In *Proceedings of the 2023 International Conference on IT and Industrial Technologies (ICIT)* (pp. 1–6). DOI: 10.1109/ICIT59216.2023.10335827

Shabsigh, G., & Boukherouaa, E. B. (2023). *Generative Artificial Intelligence in Finance: Risk Considerations. Fintech Note 2023/006*. International Monetary Fund.

Shahzad, K., Khan, S. A., & Iqbal, A. (2024). Effects of blockchain technology (BT) on the university librarians and libraries: a systematic literature review (SLR). Library Hi Tech, (ahead-of-print).

Shandilya, S. K., Srivastav, A., Yemets, K., Datta, A., & Nagar, A. K. (2023). YOLO-based segmented dataset for drone vs. bird detection for deep and machine learning algorithms. *Data in Brief*, *50*, 109355. DOI: 10.1016/j.dib.2023.109355 PMID: 37609648

Shanmugavadivel, K., Sampath, S. H., Nandhakumar, P., Mahalingam, P., Subramanian, M., Kumaresan, P. K., & Priyadharshini, R. (2022). An analysis of machine learning models for sentiment analysis of Tamil code-mixed data. *Computer Speech & Language*, *76*(January), 101407. DOI: 10.1016/j.csl.2022.101407

Sharma, P., Hans, P., & Gupta, S. C. (2020). Classification Of Plant Leaf Diseases Using Machine Learning And Image Preprocessing Techniques. *2020 10th International Conference on Cloud Computing, Data Science & Engineering (Confluence)*, 480–484. DOI: 10.1109/Confluence47617.2020.9057889

Sharma, S., & Singh, B. (2025). Blockchain-Based Secure Digital Library System. In *Blockchain Technology for Cyber Defense, Cybersecurity, and Countermeasures: Techniques, Solutions, and Applications* (pp. 195–210). IGI Global.

Sharma, A., & Singh, B. (2022). Measuring Impact of E-commerce on Small Scale Business: A Systematic Review. *Journal of Corporate Governance and International Business Law*, *5*(1).

Sharma, A., Singh, S. K., Badwal, E., Kumar, S., Gupta, B. B., Arya, V., Chui, K. T., & Santaniello, D. (2023). Fuzzy Based Clustering of Consumers' Big Data in Industrial Applications. *2023 IEEE International Conference on Consumer Electronics (ICCE)*, 1–3. DOI: 10.1109/ICCE56470.2023.10043451

Sharma, M., Kumar, C. J., & Bhattacharyya, D. K. (2024). Machine/deep learning techniques for disease and nutrient deficiency disorder diagnosis in rice crops: A systematic review. *Biosystems Engineering*, *244*, 77–92. DOI: 10.1016/j.biosystemseng.2024.05.014

Sharma, M., Kumar, C. J., & Deka, A. (2022). Early diagnosis of rice plant disease using machine learning techniques. *Archiv für Phytopathologie und Pflanzenschutz*, *55*(3), 259–283. DOI: 10.1080/03235408.2021.2015866

Sharma, S. S., & Dutta, G. (2021). SentiDraw: Using star ratings of reviews to develop domain specific sentiment lexicon for polarity determination. *Information Processing & Management*, *58*(1), 102412. Advance online publication. DOI: 10.1016/j.ipm.2020.102412

Sharma, S., Aggarwal, A., & Choudhury, T. (2018). Breast cancer detection using machine learning algorithms. In *Proceedings of the 2018 International Conference on Computational Techniques, Electronics and Mechanical Systems (CTEMS)* (pp. 114–118). IEEE. DOI: 10.1109/CTEMS.2018.8769187

Sharma, S., & Batth, R. S. (2020, June). Blockchain technology for higher education sytem: A mirror review. In *2020 International Conference on Intelligent Engineering and Management (ICIEM)* (pp. 348-353). IEEE. DOI: 10.1109/ICIEM48762.2020.9160274

Sharma, V., Mir, A. A., & Sarwr, A. (2020). Detection of rice disease using Bayes' classifier and minimum distance classifier. *Journal of Multimedia Information System*, *7*(1), 17–24. DOI: 10.33851/JMIS.2020.7.1.17

Shivsharan, N. M., Kambli, S. G., Nikharge, O. V., & Kudatarkar, K. U. (2023, June). Facilitating Cryptocurrency Analysis and Watchlist Management with a Web-Based Platform. In 2023 3rd International Conference on Intelligent Technologies (CONIT) (pp. 1-6). IEEE. DOI: 10.1109/CONIT59222.2023.10205550

Shi, W., Cao, J., Zhang, Q., Li, Y., & Xu, L. (2016). Edge Computing: Vision and Challenges. *IEEE Internet of Things Journal*, *3*(5), 637–646. DOI: 10.1109/JIOT.2016.2579198

Shi, X., An, X., Zhao, Q., Liu, H., Xia, L., Sun, X., & Guo, Y. (2019). State-of-the-art internet of things in protected agriculture. *Sensors (Basel)*, *19*(8), 1833. DOI: 10.3390/s19081833 PMID: 30999637

Shkembi, K., Kochovski, P., Papaioannou, T. G., Barelle, C., & Stankovski, V. (2023). Semantic Web and blockchain technologies: Convergence, challenges and research trends. *Journal of Web Semantics*, *79*, 100809. DOI: 10.1016/j.websem.2023.100809

Shrivastava, G., & Patidar, H. (2022). Rice plant disease identification decision support model using machine learning. *ICTACT Journal on Soft Computing, 12*(3), 2676–2681. DOI: 10.21917/ijsc.2022.0375

Shrivastava, V. K., & Pradhan, M. K. (2021). Rice plant disease classification using color features: A machine learning paradigm. *Journal of Plant Pathology, 103*(1), 17–26. DOI: 10.1007/s42161-020-00683-3

Shuiqing, H. U. A. N. G., Xiaoguang, W. A. N. G., Cuijuan, X. I. A., & Ouyang, J. (2022). Advancing the work on ancient classics in the new era and accelerating innovative and intelligent development. Journal of library and information sciences in agriculture, 34(5), 4.

Siddhpura, A., & Siddhpura, M. (2020). Plagiarism, contract cheating and other academic misconducts in online engineering education: Analysis, detection and prevention strategies. *Proceedings of 2020 IEEE International Conference on Teaching, Assessment, and Learning for Engineering, TALE 2020*, 112–119. DOI: 10.1109/TALE48869.2020.9368311

Simhadri, C. G., Kondaveeti, H. K., Vatsavayi, V. K., Mitra, A., & Ananthachari, P. (2024). Deep learning for rice leaf disease detection: A systematic literature review on emerging trends, methodologies and techniques. *Information Processing in Agriculture*. Advance online publication. DOI: 10.1016/j.inpa.2024.04.006

Singh, B. (2023). Blockchain Technology in Renovating Healthcare: Legal and Future Perspectives. In *Revolutionizing Healthcare Through Artificial Intelligence and Internet of Things Applications* (pp. 177-186). IGI Global.

Singh, B. (2023). Federated Learning for Envision Future Trajectory Smart Transport System for Climate Preservation and Smart Green Planet: Insights into Global Governance and SDG-9 (Industry, Innovation and Infrastructure). *National Journal of Environmental Law, 6*(2), 6–17.

Singh, B., & Kaunert, C. (2024). Integration of Cutting-Edge Technologies such as Internet of Things (IoT) and 5G in Health Monitoring Systems: A Comprehensive Legal Analysis and Futuristic Outcomes. *GLS Law Journal, 6*(1), 13–20. DOI: 10.69974/glslawjournal.v6i1.123

Singh, J. P., Pradhan, C., & Das, S. C. (2020). Image processing and machine learning techniques to detect and classify paddy leaf diseases. *RE:view*.

Singh, M., Verma, P., Thakur, R., & Singh, A. (2024). Signaling for abiotic stress tolerance in plants: A review. *Agricultural Reviews (Karnal), 45*(2), 1–10. DOI: 10.18805/ag.R-2375

Singh, S. P., Das, B., Pandey, R. P., Singh, H., & Chauhan, D. S. (2023). Custom convolutional neural network for detection and classification of rice plant diseases. *Procedia Computer Science, 218*, 2026–2040. DOI: 10.1016/j.procs.2023.01.179

Siregar, M. U. Comparative Study of K-Means Clustering Algorithm and K-Medoids Clustering in Student Data Clustering. JISKA (Jurnal Informatika Sunan Kalijaga), 7(2), 91-99.

Siwach, V., Sehrawat, H., & Dahiya, O. (2022, October). Hybrid Blockchain Security Mechanism for protection of data related to Precision Agriculture. In *2022 10th International Conference on Reliability, Infocom Technologies and Optimization (Trends and Future Directions)(ICRITO)* (pp. 1-7). IEEE.

Sletten, S. R. (2021). Rethinking Assessment: Replacing Traditional Exams with Paper Reviews. *Journal of Microbiology & Biology Education*, *22*(2), e00109-21. Advance online publication. DOI: 10.1128/jmbe.00109-21 PMID: 34594442

Sneha, S., Sharma, B. S., Ruchitha, H. K., Kodipalli, A., Rao, T., & Rohini, B. R. (2023). Analysis of various machine learning classifiers for prediction and classification of breast cancer. In *Proceedings of the 2023 International Conference on Computational Intelligence for Information, Security and Communication Applications (CIISCA)* (pp. 285–290). DOI: 10.1109/CIISCA59740.2023.00062

Soleimanitaleb, Z., Keyvanrad, M. A., & Jafari, A. (2019). Object Tracking Methods:A Review. *2019 9th International Conference on Computer and Knowledge Engineering (ICCKE)*, 282–288. DOI: 10.1109/ICCKE48569.2019.8964761

Sommer, R., & Paxson, V. (2010). *Outside the Closed World: On Using Machine Learning for Network Intrusion Detection*. https://doi.org/DOI: 10.1109/SP.2010.25

Soydan, O., & Taner, A. (2022). Identification of rice leaf blight disease by using image processing techniques. *Anadolu Tarım Bilimleri Dergisi*, *37*(2), 341–360. DOI: 10.7161/omuanajas.1004187

Srinidhi, B., & Bhargavi, M. S. (2023). An XAI approach to predictive analytics of pancreatic cancer. In *Proceedings of the 2023 International Conference on Information Technology (ICIT)*.

Srinivasan, A., Srikanth, A., Indrajit, H., & Narasimhan, V. (2020). A Novel Approach for Road Accident Detection using DETR Algorithm. *2020 International Conference on Intelligent Data Science Technologies and Applications, IDSTA 2020*, 75–80. DOI: 10.1109/IDSTA50958.2020.9263703

Stana, M. C., Goga, N., Marian, C. V., Popa, R., Vulpe, C. M., & Taslitchi, C. (2021, May). G-Cloud Briefcase-Electronic Archive for Academic Certificates and General Certificates of Education Documents Using Public Private Hyperspace for E-Government Library Services Based on NOSQL Databases. In *2021 IEEE International Black Sea Conference on Communications and Networking (BlackSeaCom)* (pp. 1-5). IEEE. DOI: 10.1109/BlackSeaCom52164.2021.9527826

Starczewski, A., & Krzyżak, A. (2015). Performance evaluation of the silhouette index. *International Conference on Artificial Intelligence and Soft Computing*, 49–58.

Stephen, A., Punitha, A., & Chandrasekar, A. (2024). Optimal deep generative adversarial network and convolutional neural network for rice leaf disease prediction. *The Visual Computer*, *40*(2), 919–936. DOI: 10.1007/s00371-023-02823-z

Subariyanto, M. I., Muncarno, & Supriyadi. (2017). Hubungan Kecerdasan Verbal dan Kecerdasan Logika-Matematika dengan Kemampuan Menyelesaikan Soal Cerita Matematika. *Jurnal P*, *6*(7), 1–13.

Sudeep, U. S., Naidu, K. N., Girish, P. S., Nikesh, T. N., & Sunanda, C. (2022). Brain Tumor Classification using a Support Vector Machine. *International Journal of Computer Applications*, *184*(28), 15–17. DOI: 10.5120/ijca2022922347

Sudhesh, K. M.. (2023). AI based rice leaf disease identification enhanced by dynamic mode decomposition. *Engineering Applications of Artificial Intelligence*, *120*, 105836. https://doi.org/xxx. DOI: 10.1016/j.engappai.2023.105836

Sulistiyawati, A., & Supriyanto, E. (2020). Implementasi Algoritma K-means Clustring dalam Penetuan Siswa Kelas Unggulan. *Jurnal Tekno Kompak*, *15*(2), 25–36. DOI: 10.33365/jtk.v15i2.1162

Sulistyaningrum, D. R., Rasyida, A., & Setiyono, B. (2020). Rice disease classification based on leaf image using multilevel support vector machine (SVM). *Journal of Physics: Conference Series*, *1490*(1), 012017. DOI: 10.1088/1742-6596/1490/1/012053

Sun, J., & Wang, Y. (2023, November). Data Security Mechanism of Digital Library Based on Blockchain. In 2023 International Conference on Ambient Intelligence, Knowledge Informatics and Industrial Electronics (AIKIIE) (pp. 01-06). IEEE. DOI: 10.1109/AIKIIE60097.2023.10390497

Sutton, R. S., & Barto, A. G. (2018). *Reinforcement Learning* (2nd ed.). MIT Press.

Sya'iyah, K., Yuliansyah, H., & Arfiani, I. (2019). Clustering Student Data Based On K - Means Algorithms. *INTERNATIONAL JOURNAL OF SCIENTIFIC & TECHNOLOGY RESEARCH*, *8*(08), 1014–1018.

Syahputra, H., & Wibowo, A. (2023). Comparison of Support Vector Machine (SVM) and Random Forest Algorithm for Detection of Negative Content on Websites. *Jurnal Ilmiah Teknik Elektro Komputer Dan Informatika*, *9*(1), 1. Advance online publication. DOI: 10.26555/jiteki.v9i1.25861

Taber, K. S. (2018). The use of Cronbach's alpha when developing and reporting research instruments in science education. *Research in Science Education*, *48*(6), 1273–1296. DOI: 10.1007/s11165-016-9602-2

Taha, M. F., Saeed, A., Ahmed, H., & El-Sayed, A. (2022). Using deep convolutional neural network for image-based diagnosis of nutrient deficiencies in plants grown in aquaponics. *Chemosensors (Basel, Switzerland)*, *10*(2), 45. DOI: 10.3390/chemosensors10020045

Taherdoost, H. (2016). Validity and reliability of the research instrument; how to test the validation of a questionnaire/survey in a research. *International Journal of Academic Research in Management (IJARM), 5*.

Tanwar, V., Anand, V., Upadhyay, D., & Singh, M. (2024). Classification of Cauliflower Leaf Diseases was Accomplished with a High Degree of Accuracy using the CNN-SVM Approach. *2024 4th International Conference on Intelligent Technologies (CONIT)*, 1–5. DOI: 10.1109/CONIT61985.2024.10626016

Tanwar, R., Chhabra, Y., Rattan, P., & Rani, S. (2022, September). Blockchain in IoT Networks for Precision Agriculture. In *International Conference on Innovative Computing and Communications* [Singapore: Springer Nature Singapore.]. *Proceedings of ICICC*, *2*, 137–147.

Tawornittayakun, J., & Leelasantitham, A. (2020, June). A comparative study of library information storage system through Blockchain technology: A case study of undergraduate researches. In 2020 17th International Conference on Electrical Engineering/Electronics, Computer, Telecommunications and Information Technology (ECTI-CON) (pp. 5-8). IEEE. DOI: 10.1109/ECTI-CON49241.2020.9158239

Tejaswini, P., Pooja, G., & Rupa, G. (2022). Rice leaf disease classification using CNN. In *IOP Conference Series: Earth and Environmental Science* (Vol. 1032, No. 1, p. 012037). IOP Publishing. DOI: 10.1088/1755-1315/1032/1/012017

Tholkapiyan, M., Aruna Devi, B., Bhatt, D., Saravana Kumar, E., Kirubakaran, S., & Kumar, R. (2023). Performance analysis of rice plant diseases identification and classification methodology. *Wireless Personal Communications*, *130*(2), 1317–1341. DOI: 10.1007/s11277-023-10333-3

Thomas, T., Joseph, D., Philip, D., & Nair, S. (2020). Identification of rice genotypes for seedling stage multiple abiotic stress tolerance. *Plant Physiology Reports*, *25*(4), 697–706. DOI: 10.1007/s40502-020-00549-x

Tirri, K., Nokelainen, P., & Komulainen, E. (2013). Multiple intelligences : Can they be measured? *Psychological Test and Assessment Modeling*, *55*(4), 438–461.

Toledo-Pérez, D. C., Rodríguez-Reséndiz, J., Gómez-Loenzo, R. A., & Jauregui-Correa, J. C. (2019). Support Vector Machine-Based EMG Signal Classification Techniques: A Review. *Applied Sciences (Basel, Switzerland)*, *9*(20), 20. Advance online publication. DOI: 10.3390/app9204402

Trammell, P., & Korinek, A. (2023). Economic Growth under Transformative AI. Working Paper 31815. National Bureau of Economic Research.

Tripathi, G., & Kumar, R. (2020). Early Prediction of Diabetes Mellitus Using Machine Learning. *2020 8th International Conference on Reliability, Infocom Technologies and Optimization (Trends and Future Directions) (ICRITO)*, 1009–1014. DOI: 10.1109/ICRITO48877.2020.9197832

Tripathy, R., Mandala, J., Pappu, S. R., & Gopisetty, G. K. D. (2024). Optimization based rice leaf disease classification in federated learning. *Multimedia Tools and Applications*, *83*(29), 1–27. DOI: 10.1007/s11042-023-18085-9

Tusher, A. N.. (2022). Automatic recognition of plant leaf diseases using deep learning (multilayer CNN) and image processing. In *International Conference on Image Processing and Capsule Networks* (pp. xx–xx). Springer International Publishing., https://doi.org/xxx DOI: 10.1007/978-3-031-12413-6_11

Tyagi, S., Reddy, S. R. N., Anand, R., & Sabharwal, A. (2024). Enhancing rice crop health: A light weighted CNN-based disease detection system with mobile application integration. *Multimedia Tools and Applications*, *83*(16), 48799–48829. DOI: 10.1007/s11042-023-17449-5

Udutalapally, V., Mohanty, S. P., Pallagani, V., & Khandelwal, V. (2020). sCrop: A novel device for sustainable automatic disease prediction, crop selection, and irrigation in Internet-of-Agro-Things for smart agriculture. *IEEE Sensors Journal*, *21*(16), 17525–17538. DOI: 10.1109/JSEN.2020.3032438

Umaeza, A., & Budhi, W. (2016). Hubungan Antara Kemampuan Numerik, Verbal dan Menyelesaikan Soal Cerita Fisika Dengan Prestasi Belajar Fisika. *Jurnal Ilmiah Pendidikan Fisika*, *3*(1), 45–46.

Umagapi, I. T., & Umaternate, B. (2023). Uji Kinerja K-Means Clustering Menggunakan Davies-Bouldin Index Pada Pengelompokan Data Prestasi Siswa. *Prosiding SISFOTEK*, 303–308. https://seminar.iaii.or.id/index.php/SISFOTEK/article/view/411%0A

Unal, D., Hammoudeh, M., Khan, M. A., Abuarqoub, A., Epiphaniou, G., & Hamila, R. (2021). Integration of federated machine learning and blockchain for the provision of secure big data analytics for Internet of Things. *Computers & Security*, *109*, 102393. DOI: 10.1016/j.cose.2021.102393

Upadhyaya, A., Singh, Y., & Anand, P. (2022, May). Blockchain for IoT: A Comprehensive Review for Precision Agricultural Networks. In *The International Conference on Recent Innovations in Computing* (pp. 787-802). Singapore: Springer Nature Singapore.

Upadhyay, S. K., & Kumar, A. (2022). A novel approach for rice plant diseases classification with deep convolutional neural network. *International Journal of Information Technology : an Official Journal of Bharati Vidyapeeth's Institute of Computer Applications and Management*, *14*(1), 185–199. DOI: 10.1007/s41870-021-00817-5

Van Noorden, R., & Perkel, J. M. (2023). AI and Science: What 1,600 Researchers Think. *Nature*, *621*(7980), 672–675. DOI: 10.1038/d41586-023-02980-0 PMID: 37758894

Vangipuram, S. L., Mohanty, S. P., & Kougianos, E. (2023, October). CroPAiD: Protection of Information in Agriculture Cyber-Physical Systems Using Distributed Storage and Ledger. In *IFIP International Internet of Things Conference* (pp. 375-394). Cham: Springer Nature Switzerland.

Vangipuram, S. L., Mohanty, S. P., Kougianos, E., & Ray, C. (2022). G-DaM: A Distributed Data Storage with Blockchain Framework for Management of Groundwater Quality Data. *Sensors (Basel)*, *22*(22), 8725. DOI: 10.3390/s22228725 PMID: 36433322

Varian, H. (2019). Artificial Intelligence, Economics, and Industrial Organization. In Agrawal, A., Gans, J., & Goldfarb, A. (Eds.), *The Economics of Artificial Intelligence: An Agenda*. University of Chicago Press. DOI: 10.7208/chicago/9780226613475.003.0016

Varma, G., Subramanian, A., Namboodiri, A., Chandraker, M., & Jawahar, C. V. (2019). IDD: A Dataset for Exploring Problems of Autonomous Navigation in Unconstrained Environments. *2019 IEEE Winter Conference on Applications of Computer Vision (WACV)*, 1743–1751. DOI: 10.1109/WACV.2019.00190

Vaske, J. J., Beaman, J., & Sponarski, C. C. (2017). Rethinking internal consistency in Cronbach's alpha. *Leisure Sciences*, *39*(2), 163–173. DOI: 10.1080/01490400.2015.1127189

Vaz, A., Tendulkar, A., & Palaniappan, S. (2023). Tracing the Evolution of Blockchain and Its Implications to Higher Education. In Advancements in Artificial Intelligence, Blockchain Technology, and IoT in Higher Education (pp. 203-225). Apple Academic Press. DOI: 10.1201/9781003300458-11

Verma, A., Bhattacharya, P., Madhani, N., Trivedi, C., Bhushan, B., Tanwar, S., Sharma, G., Bokoro, P. N., & Sharma, R. (2022). Blockchain for industry 5.0: Vision, opportunities, key enablers, and future directions. *IEEE Access : Practical Innovations, Open Solutions*, *10*, 69160–69199. DOI: 10.1109/ACCESS.2022.3186892

Verma, A., & Ranga, V. (2020). Machine learning based intrusion detection systems for IoT applications. *Wireless Personal Communications*, *111*(4), 2287–2310. DOI: 10.1007/s11277-019-06986-8

Verma, R. (2025). The future of academic libraries: Blockchain for secure and transparent information management. In *Driving Socio-Economic Growth With AI and Blockchain* (pp. 401–432). IGI Global. DOI: 10.4018/979-8-3693-8664-4.ch017

Verma, S., Kumar, P., & Singh, J. P. (2023). A unified lightweight CNN-based model for disease detection and identification in corn, rice, and wheat. *Journal of the Institution of Electronics and Telecommunication Engineers*, •••, 1–12. DOI: 10.1080/03772063.2023.2169025

Verma, S., Nizam, S., & Verma, P. K. (2013). Biotic and abiotic stress signaling in plants. In Sarwat, M., Ahmad, A., & Abdin, M. (Eds.), *Stress signaling in plants: Genomics and proteomics perspective* (Vol. 1). Springer., DOI: 10.1007/978-1-4614-6372-6_2

Vilas-Boas, J. L., Rodrigues, J. J., & Alberti, A. M. (2023). Convergence of Distributed Ledger Technologies with Digital Twins, IoT, and AI for fresh food logistics: Challenges and opportunities. *Journal of Industrial Information Integration*, *31*, 100393. DOI: 10.1016/j.jii.2022.100393

Violino, S., Figorilli, S., Ferrigno, M., Manganiello, V., Pallottino, F., Costa, C., & Menesatti, P. (2023). A data-driven bibliometric review on precision irrigation. *Smart Agricultural Technology*, *5*, 100320. https://doi.org/https://doi.org/10.1016/j.atech.2023.100320. DOI: 10.1016/j.atech.2023.100320

Virgil, P., & Diodato, P. G. (1994). *Dictionary of Bibliometrics* (1st Editio). Routledge. https://doi.org/https://doi.org/10.4324/9780203714133

Vishnoi, V. K., Kumar, K., & Kumar, B. (2021). Plant disease detection using computational intelligence and image processing. *Journal of Plant Diseases and Protection, 128*(1), 19–53. DOI: 10.1007/s41348-020-00368-0

Voulgaris, S., Fotiou, N., Siris, V. A., Polyzos, G. C., Jaatinen, M., & Oikonomidis, Y. (2019). Blockchain technology for intelligent environments. *Future Internet, 11*(10), 213. DOI: 10.3390/fi11100213

Vubangsi, M., & Al-Turjman, F. (2022, August). Design and implementation of a conference attendance monitoring system using blockchain and AI technologies. In *2022 International Conference on Artificial Intelligence in Everything (AIE)* (pp. 197-202). IEEE. DOI: 10.1109/AIE57029.2022.00044

Wadhawan, R., Garg, M., & Sahani, A. K. (2020). Rice plant leaf disease detection and severity estimation. In *2020 IEEE 15th International Conference on Industrial and Information Systems (ICIIS)* (pp. xx–xx). IEEE. DOI: 10.1109/ICIIS51140.2020.9342653

Wahid, A., Juliady, I., Zain, S. G., & Parenreng, J. M. (2022). Secure Wireless Sensor Network using Cryptography for Smart Farming Systems. *Internet of Things and Artificial Intelligence Journal, 2*(4), 248–262. DOI: 10.31763/iota.v2i4.554

Wang, F., Zhao, H., & Ma, J. Y. (2019). Research and practice progress of data provenance from the perspective of data science. *Journal of Library Science in China, 45*(5), 79–100.

Wang, Q., Liu, L., Liu, Z., Qiao, W., & Liu, F. (2022). Application of Fuzzy Clustering in Higher Education General Management Based on Internet Environment. *Mathematical Problems in Engineering, 2022*, 1–11. Advance online publication. DOI: 10.1155/2022/3438666

Wang, Y. (2024). Exploring the Application of Early Childhood Psychological Education Based on K-means Cluster Analysis. *Journal of Electrical Systems, 20*(9s), 64–71.

Watchareeruetai, U., Jaroonchon, N., Jongmool, T., & Limthanaporn, N. (2018, March). Identification of plant nutrient deficiencies using convolutional neural networks. In *2018 International Electrical Engineering Congress (iEECON)* (pp. 1–4). IEEE. DOI: 10.1109/IEECON.2018.8712217

Webb, J., & Hume, D. (2018). "Campus IoT collaboration and governance using the NIST cybersecurity framework" in Living in the internet of things: cybersecurity of the IoT- 2018 (IET), 1–7.

Webb, M. (2019). The Impact of Artificial Intelligence on the Labor Market. Working Paper 3482150. SSRN.

Wedashwara, W., Jatmika, A. H., Zubaidi, A., & Arimbawa, I. W. A. (2021). *Smart solar powered hydroponics system using internet of things and fuzzy association rule mining*. IOP Publishing. DOI: 10.1088/1755-1315/712/1/012007

Wedashwara, W., Wayan Agus Arimbawa, I., Hidayat Jatmika, A., Zubaidi, A., & Mulyana, T. (2020). IoT based Smart Small Scale Solar Energy Planning using Evolutionary Fuzzy Association Rule Mining. *2020 International Conference on Advancement in Data Science, E-learning and Information Systems (ICADEIS)*, 1–6. DOI: 10.1109/ICADEIS49811.2020.9276905

Wedashwara, W., Yadnya, M. S., Sudiarta, I. W., Arimbawa, I. W. A., & Mulyana, T. (2023). Solar Powered Vibration Propagation Analysis System using nRF24l01 based WSN and FRBR. *JOIV: International Journal on Informatics Visualization, 7*(1), 15–21. DOI: 10.30630/joiv.7.1.1592

Widiarina, W., & Wahono, R. S. (2015). Algoritma cluster dinamik untuk optimasi cluster pada algoritma k-means dalam pemetaan nasabah potensial. *Journal of Intelligent Systems, 1*(1), 33–36.

Wiguna, B. J. P. K., Suwarma, I. R., & Liliawati, W. (2018). STEM-based science learning implementation to identify student's personal intelligences profiles. *Journal of Physics: Conference Series, 1013*(1), 012082. Advance online publication. DOI: 10.1088/1742-6596/1013/1/012082

Wijaya, Y. A., Kurniady, D. A., Setyanto, E., Tarihoran, W. S., Rusmana, D., & Rahim, R. (2021). Davies bouldin index algorithm for optimizing clustering case studies mapping school facilities. *TEM J, 10*(3), 1099–1103. DOI: 10.18421/TEM103-13

Wijesekara, D. S. N., & Arachchige, P. (2024). Blockchain and Artificial Intelligence for Big Data Analytics in Networking: Leadingedge Frameworks. *Journal of Engineering Science & Technology Review, 17*(3).

Wu, C., Luo, Q., & Shou, D. (2022). Image Recognition Technology and Development of YOLO Algorithm. *Advances in Transdisciplinary Engineering, 33*, 32–38. DOI: 10.3233/ATDE221146

Wu, M., Li, Q., Bilal, M., Xu, X., Zhang, J., & Hou, J. (2021). Multi-label active learning from crowds for secure IIoT. *Ad Hoc Networks, 121*(February), 102594. DOI: 10.1016/j.adhoc.2021.102594

Wu, Y., Wang, Z., Ma, Y., & Leung, V. C. M. (2021). Deep reinforcement learning for blockchain in industrial IoT: A survey. *Computer Networks*, *191*(February), 108004. DOI: 10.1016/j.comnet.2021.108004

Xiao, Y., Xiao, R., Zhu, C., & Wei, S. Digital Copyright Management Strategy of Digital Library Based on Blockchain Technology. Journal of Combinatorial Mathematics and Combinatorial Computing, 119, 153–162.

Xu, S., Wang, X., Lv, W., Chang, Q., Cui, C., Deng, K., Wang, G., Dang, Q., Wei, S., Du, Y., & Lai, B. (2022). *PP-YOLOE: An evolved version of YOLO*. http://arxiv.org/abs/2203.16250

Xu, Z., Wang, X., Sun, Y., Wang, J., & Song, H. (2020). Using deep convolutional neural networks for image-based diagnosis of nutrient deficiencies in rice. *Computational Intelligence and Neuroscience*, *7307252*, 1–12. Advance online publication. DOI: 10.1155/2020/7307252 PMID: 32952543

Yağcı, M. (2022). Educational data mining: Prediction of students' academic performance using machine learning algorithms. *Smart Learning Environments*, *9*(1), 11. DOI: 10.1186/s40561-022-00192-z

Yan, F., Huang, H., Pedrycz, W., & Hirota, K. (2023). Automated breast cancer detection in mammography using ensemble classifier and feature weighting algorithms. *Expert Systems with Applications*, *227*, 120282. DOI: 10.1016/j.eswa.2023.120282

Yesuf, S. H. (2019). Breast cancer detection using machine learning techniques. *International Journal of Advanced Research in Computer Science*, *10*(5), 27–33. DOI: 10.26483/ijarcs.v10i5.6464

Yogeshwari, M., & Thailambal, G. (2023). Automatic feature extraction and detection of plant leaf disease using GLCM features and convolutional neural networks. *Materials Today: Proceedings*, *81*, 530–536. DOI: 10.1016/j.matpr.2021.03.700

Yu, C., & Wang, Y. (2022). College Student Management System Based on K-means Clustering Algorithm. *International Journal of New Developments in Education*, *4*(2), 28–33. DOI: 10.25236/IJNDE.2022.040206

Zack, T., Lehman, E., Suzgun, M., Rodriguez, J. A., Celi, L. A., Gichoya, J., Ju-rafsky, D., Szolovits, P., Bates, D. W., Abdulnour, R.-E. E., Butte, A. J., & Alsentzer, E. (2023). Coding Inequity: Assessing GPT-4's Potential for Perpetuating Racial and Gender Biases in Healthcare. Preprint. *medRxiv*. DOI: 10.1101/2023.07.13.23292577

Zaffar, A., Jawad, M., Shabbir, M., Factory, P. O., & Cantt, W. (2023). A Novel CNN-RNN Model for E-Cheating Detection Based on Video Surveillance. *UW JOURNAL OF CMPUTER SCIENCE, 5*(1). https://uwjcs.org.pk/index.php/ojs/article/view/64

Zankl, A., Seuschek, H., Irazoqui, G., & Gulmezoglu, B. (2021). "Side-channel attacks in the internet of things: threats and challenges" in Research anthology on artificial intelligenceapplications in security (IGI Global), 2058–2090.

Zhai, Q., Wang, Y., Guo, Z., Liu, M., Gao, P., & Wang, J. (2022). Rice nitrogen nutrition monitoring classification method based on the convolution neural network model: Direct detection of rice nitrogen nutritional status. *PLoS One*, *17*(11), e0273360. DOI: 10.1371/journal.pone.0273360 PMID: 36413518

Zhang, J. (2024). Research on Resource Sharing Method of Library and Document Center Under the Multimedia Background. *International Journal of Advanced Computer Science and Applications*, *15*(4). Advance online publication. DOI: 10.14569/IJACSA.2024.0150435

Zhang, Z. K., Cho, M. C. Y., Wang, C. W., Hsu, C. W., Chen, C. K., & Shieh, S. (2014). *"IoT security: ongoing challenges and research opportunities" in 2014 IEEE 7th InternationalConference on Service-Oriented Computing and Applications*. IEEE.

Zhang, Z., Gao, H., Peng, J., & Chen, Y. (2023). A high-quality rice leaf disease image data augmentation method based on a dual GAN. *IEEE Access : Practical Innovations, Open Solutions*, *11*, 21176–21191. DOI: 10.1109/ACCESS.2023.3251098

Zhao, Z. Q., Zheng, P., Xu, S. T., & Wu, X. (2019). Object Detection with Deep Learning: A Review. In *IEEE Transactions on Neural Networks and Learning Systems* (Vol. 30, Issue 11, pp. 3212–3232). Institute of Electrical and Electronics Engineers Inc. DOI: 10.1109/TNNLS.2018.2876865

Zhao, C., Shao, J., Zhang, R., Mu, W., & Zhang, J. (2023, May). Analysis of the digital transfer framework of power grid engineering based on the integration of QR code and blockchain. In *Eighth International Symposium on Advances in Electrical, Electronics, and Computer Engineering (ISAEECE 2023)* (Vol. 12704, pp. 98-103). SPIE. DOI: 10.1117/12.2680662

Zhou, Z., Li, Y., Peng, C., Wang, H., & Du, S. (2021). Image Processing: Facilitating Retinanet for Detecting Small Objects. *Journal of Physics: Conference Series*, *1815*(1), 012016. DOI: 10.1088/1742-6596/1815/1/012016

Zhu, J. K. (2002). Salt and drought stress signal transduction in plants. *Annual Review of Plant Biology*, *53*(1), 247–273. DOI: 10.1146/annurev.arplant.53.091401.143329 PMID: 12221975

Zoph, B., Cubuk, E. D., Ghiasi, G., Lin, T. Y., Shlens, J., & Le, Q. V. (2020). Learning data augmentation strategies for object detection. In Computer Vision–ECCV 2020: 16th European Conference, Glasgow, UK, August 23–28, 2020 [Springer International Publishing.]. *Proceedings*, *16*(Part XXVII), 566–583.

Zulkifli, N. (2022). Efektivitas Pelaksanaan Ujian Semester Menggunakan Computer Based Test. *Edumaspul: Jurnal Pendidikan*, *6*(1), 186–191. DOI: 10.33487/edumaspul.v6i1.3057

About The Contributors

Mosiur Rahaman earned his PhD in Computer Science and Information Engineering from Asia University in 2024. He also holds an MBA in Human Resources with IT from Osmania University (2017), an MTech in Computer Science and Engineering from JNTU Hyderabad (2014), and a BTech in Computer Science and Engineering from JNTU Hyderabad (2012). Currently, Dr. Rahaman is a Full-time Postdoctoral Researcher at the International Center for AI and Cyber Security Research and Innovations, Department of Computer Science and Information Engineering, Asia University, Taiwan. His research focuses on using advanced feature selection techniques to detect mobile communication phishing attacks, supported by the National Science and Technology Council, Taiwan. His research interests also span AI, IoT, cybersecurity, and digital transformation, and he has published numerous articles in prestigious journals. In addition to his research, Dr. Rahaman has held various teaching positions, including Adjunct Lecturer and Teaching Assistant roles at Asia University. He has also taught English at Kuang Fu Elementary and Junior High Schools in Taiwan, and served as an Assistant Professor at the Royal Institute of Technology & Science in India. Beyond these professional affliations, Dr. Rahaman has also served as Director and Technical Program Chair at the 2nd International Conference on EMI and AI: Bridging Language and Culture through Technology (2024, Taiwan) and undertaken leadership roles in conferences such as MIMSE 22 and MIMSE 23. He is the Managing Editor of the European Journal of Advanced Computing and Management and serves on the editorial boards of several other reputable journals.

Faruk Azam Shaik is a research scientist (CRCN) at the Centre National de la Recherche Scientifique (CNRS), France, affiliated with the Institut d'Électronique, de Microélectronique et de Nanotechnologie (IEMN), UMR 8520. He received his Ph.D. in Electrical Engineering and Information Systems from The University of Tokyo, Japan, in 2018, and his M.Tech. in Biomedical Engineering from the

Indian Institute of Technology Hyderabad, India, in 2015. Dr. Shaik's research is centered on microtechnology, with a particular emphasis on Bio-MEMS, sensors and actuators, microfluidics, and lab-on-a-chip systems designed to address biological and clinical challenges. He has authored and co-authored numerous peer-reviewed journal articles and conference papers, and is a co-inventor on several patents. In addition to his scientific publications, he has actively contributed to various research grant proposals and co-led a technology maturation program.

***

Shabnam is an Assistant Professor at the Centre for Distance and Online Education (CDOE), Chandigarh University, Gharuan, Punjab (Postal Code: 140413). She holds an MBA with dual specialization in Finance and Human Resources and is UGC NET-JRF qualified in Management. With over 10 years of academic experience, her research interests span Finance, Human Resource Management, Taxation, and Statistics.

Rachit Agarwal is working as an Assistant Professor at University School of Business, Chandigarh University, Mohali. He is having more than 7 years of work experience in teaching and research. His areas of interest are financial analysis, investment behavior analysis and behavioral study. He has published various papers in national and international journals including Scopus and ABDC listed journals. He has also organized many workshops and Faculty Development Programme on Research Methodology and on different statistical tools.

Mizanul Ridho Aohana is a Research Collaborator at University of Mataram, where he also earned his Bachelor's degree (2024) in Informatics Engineering. Currently, he serves as a Machine Learning Engineer at PT. Hizrian Bersaudara Sukses, focusing on the development of AI-powered solutions in real-world applications. His research interests include computer vision, deep learning, and applied machine learning, particularly in optimizing intelligent systems for industry and academia. Aohana has been involved in various collaborative research projects and has contributed to publications and AI-based innovations aimed at addressing real-world challenges. Beyond his professional work, he actively participates in academic events and discussions to promote research excellence and technological advancement. As a young researcher, he is dedicated to bridging the gap between research and industry through practical, data-driven solutions and interdisciplinary collaboration.

Santi Kumari Behera is a distinguished researcher recognized among the World's Top Scientists in 2024 by Stanford University. Her work spans across medical image processing, deep learning, and machine learning, with a focus on cancer detection and classification. Dr. Behera has made significant contributions through her research, co-authoring several papers on ovarian, cervical, breast, and brain cancer detection. Her expertise and dedication to advancing medical imaging techniques have earned her a prominent place in the scientific community.

Sunneng Sandino Berutu is a computer science academic with a strong background in both education and research. I began my academic journey by earning a Bachelor's (S1) in Computer Science from Universitas Kristen Immanuel (UKRIM) in Yogyakarta, Indonesia.

Fitri Bimantoro is an Assistant Professor and researcher specializing in computer vision. He currently teaches at Universitas Mataram, where he also earned his Bachelor's degree in Electrical Engineering in 2010. He completed his Master's degree in Information Technology from Sepuluh Nopember Institute of Technology in 2014. His research focuses on the development of intelligent visual systems and real-world applications of AI, particularly in image and video analysis. Actively engaged in both academic and applied research, Fitri Bimantoro has contributed to various projects and publications in the field. His dedication to advancing computer vision technologies is complemented by his role as an educator, where he mentors students and fosters innovation through collaborative research initiatives.

Baishnu Devi is an M.Tech student at Veer Surendra Sai University of Technology (VSSUT), Burla, specializing in image processing. Her research focuses on applying advanced techniques to solve real-world problems in this domain. She has already published three research papers, showcasing her contributions to the field and her dedication to advancing knowledge in image processing.

Agni Isador Harsapranata, He is a lecturer at Bina Sarana Informatika University in Jakarta, Indonesia, in the Informatics Engineering Division at FTI. Nusa Mandiri University is where he obtained his Master's degree in computer science. In addition to his position at UBSI Jakarta, he is an advisor to several companies that handle the field of Computer Network Infrastructure, and he is also actively involved in Research, Innovation. His work mainly focuses on machine learning, artificial neural networks, and artificial intelligence.

Amitava Kar is a distinguished faculty member in the Department of Information Technology at Gopal Narayan Singh University, located in Sasaram, Bihar (Postal Code: 821305). With a strong academic background and a passion for teaching, Dr. Kar contributes significantly to the university's mission of delivering quality education in the field of Information Technology. His expertise and dedication have made him a respected figure among students and colleagues alike.

Christian Kaunert is Professor of International Security at Dublin City University, Ireland. He is also Professor of Policing and Security, as well as Director of the International Centre for Policing and Security at the University of South Wales. In addition, he is Jean Monnet Chair, Director of the Jean Monnet Centre of Excellence and Director of the Jean Monnet Network on EU Counter-Terrorism.

Tanya Kumar is currently working at Chandigarh Group of Colleges, Landran Mohali. She has experience in teaching and research. She has completed her Ph.D. in commerce. Her areas of interest are finance and marketing. She has also presented various papers in national and international conferences. She provides supervision to students in financial analysis, blockchain technology, artificial intelligence and cybersecurity and fintech. She has guided more than 90 master's students to complete their dissertation. She has published papers in national and international journals, including Scopus, Web of Sciences and ABDC-listed journals.

Rajkumar Mandal is a respected member of the Faculty of Information Technology at Gopal Narayan Singh University in Sasaram, Bihar. Known for his expertise in the field of IT, Dr. Mandal is dedicated to advancing knowledge through teaching and research. His work contributes significantly to the development of skilled professionals equipped to meet the challenges of the ever-evolving technology landscape.

T.C. Manjunath was born on 6th February 1967. He did his BE from RV College of Engg. in 1989, ME in Robotics & Automation from LD College of Engg. in 1992 with First Rank and PhD in Systems & Control Engg. from IIT Bombay in the year 2007. He is currently working as the Dean Research R D and as Professor in the Dept. of Computer Science & Engineering with specialization in IoT, Cyber Security & Blockchain Technology in Rajarajeswari College of Engineering, Bengaluru, Karnataka. He has completed B.E. Degree from R.V. College of Engg., M.E. Govt.'s LD College of Engg., Gujarat Univ., Ahmedabad in First Rank and Ph.D. in Systems and Control Engineering from the Indian Institute of Technology Bombay IIT Bombay He has got a teaching academic, research and administrative experience of more than 35+ years 10 years as Principal, 8 years as HoD rest as faculty and has published more than 1200+ research papers in various national & international journals and conferences, with 175+ patents, 25 book-chapters, 25 awards and also published 25 text books in various field of science, engineering and technology, completed projects worth Rs. 9 Crores till date apart from guiding 10

PhD scholars. He is a member of 30 Professional Society Chapters and is a fellow of IE, IETE, AMIEE and a chartered engineer. He got the best research scholar award in 2006 in IIT Bombay amongst all the research scholars of the institute. He has got 10 awards in academics and research activities.

Selvakumar P. is currently an Associate Professor in the Department of Science and Humanities at Nehru Institute of Technology, Coimbatore, Tamil Nadu, India. Since 2023, he has served as the Department IQAC Coordinator at the institute. Prior to joining Nehru Institute of Technology, he held positions as an Assistant Professor in the Department of Science and Humanities various engineering colleges in Coimbatore, Tamil Nadu, India, spanning from 2011 to 2023. Dr.P. Selvakumar's academic and research interests lie primarily in the field of phytochemistry. He has made significant contributions to his field, with an impressive record that includes 30 journal publications, 12 patents, and 20 conference papers. Additionally, he has organized 5 seminars and conferences, further demonstrating his commitment to academic and scientific exchange. Throughout his career, Dr. P.Selvakumar has been recognized for his dedication and excellence in teaching, having been honored with several awards, including the Best Teacher Awards. He holds a prominent position in the academic community, contributing actively to research.

JM Padhi is a dedicated research scholar at Sambalpur University, specializing in rice disease identification using image processing techniques. His Ph.D. thesis focuses on leveraging advanced computational methods to detect and classify diseases in rice crops, a critical area for improving agricultural productivity. Throughout his research journey, he has authored five research papers, contributing valuable insights into the application of image processing and deep learning in agriculture. His work aims to enhance the accuracy and efficiency of disease detection in rice plants, promoting better crop management and sustainable farming practices.

Shyamalendu Paul, Assistant Professor, Department of Computer Science & Engineering, Brainware University, West Bengal, India

Hindriyanto Purnomo is a professor at Department of Information Technology, Satya Wacana Christian University, Indonesia. He received his Bachelor degree in Engineering Physics, from Gadjah Mada University, Indonesia, in 2005, Magister of Information Technology from The University of Melbourne, Australia in 2009 and Doctor of Philosophy in Industrial and System Engineering from Chung Yuan Christian University, Taiwan, in 2013. He was a visiting scholar at Sophia University, Japan, in 2022. He has received several recognitions and awards for his academic achievement. He also has published many articles in reputable journals, conferences and books. His research interests are in the field of metaheuristics, soft computing, machine learning, and deep learning.

Guru Basava Aradhya S is a seasoned academician with 17 years of experience in management education. He currently serves as the Director & Professor at Chandigarh School of Business, Mohali. He holds a Ph.D. in Management and has mentored five Ph.D. scholars. His expertise includes HR, marketing, and business analytics. A prolific researcher, he has published in SCOPUS, ABDC, and UGC CARE journals. He has organized numerous international conferences and contributed to NAAC and NBA accreditations. His industry collaborations include faculty exchange programs and consultancy projects. Passionate about academia, he actively fosters innovation, research, and student development.

Eko Sediyono is a professor at Universitas Kristen Satya Wacana in Salatiga, Indonesia, in the Division of Informatics Engineering within the FTI. The University of Indonesia is where he obtained his Master's and Doctorate degrees in computer science. Apart from his position at UKSW Salatiga, he is an assessor at BAN PT and LAM Infokom, and he is also actively involved in the Research, Innovation, and Entrepreneurship Division. His work primarily focuses on machine learning, artificial neural networks, and artificial intelligence.

Prabira Kumar Sethy is an Associate Professor in the Department of Electronics and Communication Engineering at Guru Ghasidas Vishwavidyalaya, Bilaspur. His research expertise lies in the fields of image processing, machine learning, and deep learning, with a focus on agricultural and medical imaging applications. He is the Principal Investigator of the project titled "Identification of Diseases and Mineral Deficiencies of Rice Plant based on Machine Learning Approach in the Natural Environment," funded by the Odisha State Higher Education Council under the Mukhyamantri Research and Innovation Fellowship Program (MRIP). Dr. Sethy has authored numerous research papers and is recognized as one of the World's Top 2% Scientists according to Stanford University, highlighting his significant contributions to the global scientific community.

Aditee Huparikar Shah is an Assistant Professor at Indira College of Engineering and Management in Pune. With a strong academic foundation and a passion for teaching, she is committed to nurturing future professionals in the fields of engineering and management. Her dedication to student success and academic excellence makes her a valued member of the institution.

Preeti Sharma is working as Head of Department in BBA department at Chandigarh Group of Colleges, Landran Mohali. She has more than 18 years of experience in teaching and research. She has expertise in the areas of economics and business management. She has published various research papers in conferences, edited books, case studies and journals indexed in Scopus and ABDC journal.

Bhupinder Singh is working as Professor at Sharda University, India. Also, Honorary Professor in University of South Wales UK and Santo Tomas University Tunja, Colombia. His areas of publications as Smart Healthcare, Medicines, fuzzy logics, artificial intelligence, robotics, machine learning, deep learning, federated learning, IoT, PV Glasses, metaverse and many more. He has 3 books, 139 paper publications, 163 paper presentations in international/national conferences and seminars, participated in more than 40 workshops/FDP's/QIP's, 25 courses from international universities of repute, organized more than 59 events with international and national academicians and industry people's, editor-in-chief and co-editor in journals, developed new courses. He has given talks at international universities, resource person in international conferences such as in Nanyang Technological University Singapore, Tashkent State University of Law Uzbekistan; KIMEP University Kazakhstan, All'ah meh Tabatabi University Iran, the Iranian Association of International Criminal law, Iran and Hague Center for International Law and Investment, The Netherlands, Northumbria University Newcastle UK.

Shivprakash Varimani is an experienced finance and tax professional with a strong background in tax analysis, compliance, and data analytics. Currently serving as an Assistant Professor at Chandigarh Group of Colleges, he mentors MBA students in tools such as Power BI, Tableau, SQL, and Python. With prior experience as a Credit Analyst and Tax Analyst Intern, he excels in financial assessments, tax return preparation, and compliance. Shivprakash holds a PGDM in Finance and Taxation and is certified in SAP FICO, Tally ERP 9, and Financial Modelling. Known for his analytical skills and mentoring abilities, he is passionate about teaching and finance.

I Gede Pasek Suta Wijaya is a Professor and Head of the Informatics Engineering Department at University of Mataram. He earned his Doctorate in Electrical Engineering and Computer Science from Kumamoto University in 2010, followed by a Post-Doctoral program in AI applications in 2013. In 2015, he returned to Kumamoto University for a Research Fellowship focused on computer vision. He also holds a Master's degree in Information Technology (2001) and a Bachelor's degree in Electrical Engineering (1997) from Universitas Gadjah Mada. Prof. I Gede Pasek Suta Wijaya is highly active in research, particularly in the fields of artificial intelligence and its real-world applications. He is deeply involved in collaborative projects with industry partners, aiming to bridge academic innovation with industrial needs. His work reflects a strong commitment to advancing technology and fostering impactful research in both academic and professional domains.

Index

A

B

C

D

E

F

H

I

K

L

M

N

P

R

S

T

U

Y

www.ingramcontent.com/pod-product-compliance
Lightning Source LLC
LaVergne TN
LVHW080346011026
849161LV00009B/2204